T0270987

KNOWLEDGE-BASED CLUSTERING

KNOWLEDGE-BASED CLUSTERING

From Data to Information Granules

Witold Pedrycz

Department of Electrical and Computer Engineering
University of Alberta
Edmonton, Canada

and

Systems Research Institute
Polish Academy of Sciences
Warsaw, Poland

WILEY-INTERSCIENCE

A JOHN WILEY & SONS, INC., PUBLICATION

For general information on our other products and services please contact our Customer Care Department within the U.S. at 877-762-2974, outside the U.S. at 317-572-3993 or fax 317-572-4002.

Wiley also publishes its books in a variety of electronic formats. Some content that appears in print, however, may not be available in electronic format.

Library of Congress Cataloging-in-Publication Data:

Pedrycz, Witold, 1953-
    Knowledge-based clustering : from data to information granules / Witold Pedrycz.
        p. cm.
    "A Wiley-Interscience publication."
    Includes bibliographical references and index.
    ISBN 0-471-46966-1 (cloth)
        1. Soft computing. 2. Granular computing. 3. Fuzzy systems. I. Title.
    QA76.9.S63P45 2005
    006.3–dc22

10 9 8 7 6 5 4 3 2 1

To Ewa, Adam, and Barbara

Contents

Foreword

It is always a challenging task to write a foreword to a work authored by Professor Pedrycz. The reason is that, as a rule, what he writes about goes far beyond what can be found in the existing literature. This is particularly true in the instance of *Knowledge-Based Clustering: From Data to Information Granules* or *Knowledge-Based Clustering*, for short. *Knowledge-Based Clustering* is a magnum opus which touches upon some of the most basic facets of human cognition. It does so with authority, originality, erudition, insight, and high expository skill. Profusion of examples, figures, and references make Professor Pedrycz's work a pleasure to read.

In *Knowledge-Based Clustering*, Professor Pedrycz addresses a vast array of linked subjects. Starting with an exposition of clustering and fuzzy clusters, he moves to computing with granular information, with a granule being a clump of attribute-values drawn together by indistinguishability, equivalence, similarity, proximity, or functionality. Professor Pedrycz's co-authorship of a recent text on granular computing provides him with an effective framework for linking clustering with granular computing. In granular computing, the objects of computation are granules rather than singletons. In its general form, granular computing subsumes computing with intervals, computing with rough sets, and computing with probability distributions. The linkage between granular computing and cluster analysis plays a pivotal role throughout Professor Pedrycz's work, and is an important novel feature of his approach to cluster analysis.

The chapters that are focused on granular computing serve as a foundation for the core of the book—knowledge-based clustering. In this mode of clustering, clustering is guided by the knowledge that underlies data. There is much that is new in this part of the book, especially in chapters dealing with conditioned fuzzy clustering, collaborative clustering, directional clustering, fuzzy relational clustering, and clustering of nonhomogeneous patterns. The last part of Professor Pedrycz's work is an informative exposition of applications of knowledge-based clustering to generic models. In this part, we find a range of unconventional concepts and techniques, among them hyperbox modeling, linguistic modeling, and granular mapping.

To see the importance of Pedrycz's work in a proper perspective, an observation is in order. As we move further into the age of machine intelligence and automated reasoning, a daunting problem becomes harder and harder to master. How can we cope with the explosive growth in data, information, and knowledge? How can we locate and infer from decision-relevant information that is embedded in a large database that is unstructured, imprecise, and not totally reliable?

What these issues point to is an imperative need for new ideas and new techniques in the realm of organization of data, information, and knowledge. In effect, information is organized data and knowledge is organized information.

A key concept that underlies the concept of organization is that of relatedness and, more specifically, the concepts of clustering and granulation. In this perspective, the concepts, ideas, and methods that are the objects of discussion in Professor Pedrycz's work are of direct relevance to the goal of devising organizational structures that can cope with the explosive growth in data, information, and knowledge.

There is another observation that I should like to make. There is an enormous literature dealing with cluster analysis and related subjects but, strangely enough, what cannot be found in this literature is an operational definition of the concept of a cluster. There is a brief discussion of the concept of cluster validity in Professor Pedrycz's work, but it stops short of defining the concept of a cluster.

Is this an omission or are there some problems in defining a cluster? In my view, there are two basic problems. First, the concept of a cluster is a fuzzy concept in the sense that it is a matter of degree. And second, the concept of a cluster is a second-order concept in the sense that a cluster is a set of points rather than a singleton. Examples of second-order concepts are convex set, edge, and a mountain. In fact, the concepts of a cluster and mountain have the same deep structure.

The intrinsic problem is that, in general, second-order fuzzy concepts cannot be defined within the conceptual structure of bivalent logic. This is the principal reason why an operational definition of a cluster cannot be found in the literature of cluster analysis. A question that arises is: if the concept of a cluster cannot be defined within the conceptual structure of bivalent logic, then how can it be defined? In my view, what is needed for this purpose is PNL (Precisiated Natural Language)—a language that is based on fuzzy logic—a logic in which everything is or is allowed to be a matter of degree. To define a concept through the use of PNL, with PNL serving as a definition language, two steps are needed. First, the concept is defined in a natural language; second, the natural language definition is precisiated. Unlike bivalent-logic-based definitions, PNL-based definitions are context-dependent rather than context-free. This is the price that has to be paid to achieve a closer rapport with reality.

The fact that there is no definition of a cluster in Professor Pedrycz's work or, for that matter, anywhere else in the literature, in no way detracts from its importance. *Knowledge-Based Clustering* is a major contribution that is must reading for anyone who is interested in cluster analysis and, more generally, in the conception, design, and utilization of advanced knowledge-based systems. *Knowledge-Based Clustering* is a superlative work and its author, Professor Pedrycz, and the publisher, John Wiley & Sons, Inc. deserve our loud applause and congratulations.

Lotfi A. Zadeh

Preface

Data and patterns are an integral part of the cultural fabric of our information society. The challenge we are confronted with every day is to cope with the flood of data generated by banking transactions, millions of sensors, World Wide Web log records, communication traffic of cellular calls, satellite image collection systems, and networks of intelligent home appliances, to name just a few evident examples.

Making sense of data has become a critical objective of intelligent data analysis (IDA), data mining (DM), sensor fusion, image understanding, and logic-driven system modeling. As never before, we are faced with the growing need to construct a powerful computer "eye"—a human-centric, human-interactive, and human-sensitive computer environment that helps us understand data and make sensible decisions.

Clustering is one of the well-established manifestations of such a computer eye. With its agenda of venturing into data spaces and discovering their structure—clusters of data—clustering is an ideal vehicle for exploration of vast territories of data spaces. From the early concepts of the 1930s, this field has recently undergone a rapid expansion fueled by new conceptual and computing challenges. The omnipresence of clustering today is astonishing. Even a quick and fairly unsophisticated search of the Web or a simple search of any library database returns thousands of hits revealing an impressive breadth of applications: from biomedicine to marketing, engineering, economics, biological sciences, chemistry, military, food engineering, finance, and education.

Clustering has become a synonym for a diversified suite of methodologies and algorithms that are almost exclusively data-driven and in which any optimization is predominantly, if not exclusively, data-oriented. Clustering gives rise to a variety of information granules whose use reveals the structure of data. The formalisms of granular computing help design clustering methods designed to meet user-defined objectives. In this diversified landscape of clustering, the algorithms operating within the framework of fuzzy sets have assumed an important and unique position. The reason is obvious: fuzzy sets regarded as basic information granules are human-centric. Dealing with concepts and groups (clusters) that allow for partial membership is highly appealing. Identifying data (patterns) that are of borderline character and may require special attention as potential outliers is a useful value-added feature of fuzzy clustering. Discovering the most typical patterns (with the highest membership values) in the cluster is another important feature by fuzzy sets.

In light of the recent applications and new forms of agent-based technology, Web-based pursuits, and rapidly growing dimensionality and heterogeneity of data sets, the human-centricity of clustering has become even more essential. The paradigm of *data*-centric clustering has to be augmented. The paradigm of *knowledge*-based clustering I introduce in this book is concerned with reconciling two important driving forces of clustering activities: gaining data and domain knowledge and building a coherent platform of navigation in highly dimensional and often heterogeneous data spaces. The user plays a basic role in forming an essential feedback loop in any highly interactive data analysis. Needless to say, we require a carefully selected conceptual and algorithmic layer of human-machine communication.

This book is divided into three parts. The first parts consisting of Chapters 1 to 3, provides a concise, carefully structured introduction to the subject. Three interrelated components are presented. First, I discuss the fundamentals of fuzzy clustering. Second, I review fuzzy computing, regarded as an important realization of granular computing, focused on the issues of fuzzy clustering. Third, I elaborate on the logic-based neurons and ensuing neural networks. The core of the book, Chapters 4 to 10, presents a highly diversified landscape of knowledge-based clustering. The third part of the book, consisting of Chapters 11 to 15, is devoted to generic models whose design is directly linked to the paradigm of knowledge-based clustering. First, I concentrate on hyperbox models of clusters, demonstrating how the essential structure can be captured in terms of hyperbox geometry. This is followed by studies of granular mappings and linguistic models.

Throughout the book, I adhere to the standard notation used in pattern recognition and system analysis, as well as the standard terminology used there. The terms "data" and "pattern" are used interchangeably to emphasize the unified way of treating various forms of pattern recognition, system modeling, and data analysis. The book is self-contained. While the reader can benefit from some initial familiarity with computational Intelligence (CI), this is not a must. CI helps place the material in perspective and allows the reader to fully appreciate the ideas of information granularity and information granules as building blocks of various CI architectures.

The purpose of this book is to present the main ideas in a fairly general format and not to skew the subject by limiting the discussion to selected application areas. The algorithmic aspects are also kept quite general, and no attempt is made to strive for the most efficient yet intricate implementations possible. This makes the book of interest to a broad audience. Those readers interested in clustering, fuzzy clustering, unsupervised learning, neural networks, fuzzy sets, and pattern recognition, as well as those involved in numerous tasks of data analysis, will find the book thought-provoking and intellectually stimulating. Readers involved in system modeling will view knowledge-driven clustering as an attractive vehicle of rapid prototyping of granular models.

Knowledge-based clustering has already emerged. This book outlines its fundamentals, presents the essential algorithmic developments, and discusses its application-driven aspects. No attempt has been made to cover the subject completely. However, the material selected paints a coherent picture of the most recent developments central to this rapidly evolving area.

Witold Pedrycz

1 Clustering and Fuzzy Clustering

This chapter provides a comprehensive, focused introduction to clustering, viewed as a fundamental means of exploratory data analysis, unsupervised learning, data granulation, and information compression. We discuss the underlying principles, elaborate on the basic taxonomy of numerous clustering algorithms (including such essential classes as hierarchical, objective function-based algorithms), and review the main interpretation mechanisms associated with various clustering algorithms.

1.1. INTRODUCTION

Making sense of data is an ongoing task of researchers and professionals in almost every practical endeavor. The age of information technology, characterized by a vast array of data, has enormously amplified this quest and made it even more challenging. Data collection anytime and everywhere has become the reality of our lives. Understanding the data, revealing underlying phenomena, and visualizing major tendencies are major undertakings pursued in intelligent data analysis (IDA), data mining (DM), and system modeling.

Clustering is a general methodology and a remarkably rich conceptual and algorithmic framework for data analysis and interpretation (Anderberg, 1973; Bezdek, 1981; Bezdek et al., 1999; Devijver and Kittler, 1987; Dubes, 1987; Duda et al., 2001; Fukunaga, 1990; Hoppner et al., 1999; Jain et al., 1999, 2000; Kaufmann and Rousseeuw, 1990; Babu and Murthy, 1994; Dave, 1990; Dave and Bhaswan, 1992; Kersten, 1999; Klawonn and Keller, 1998; Mali and Mitra, 2002; Webb, 2002). In this chapter, we introduce basic notions, explain the functional components essential to the formulation of clustering problems, and discuss the main classes of clustering algorithms. These algorithms are accompanied by the formalisms of granular computing, including sets, fuzzy sets, shadowed sets, and rough sets.

1.2. BASIC NOTIONS AND NOTATION

To establish a formal setting in which clustering can be carried out, we start with basic notions such as data types, distance, and similarity/resemblance.

Knowledge-Based Clustering, by Witold Pedrycz
ISBN 0-471-46966-1   Copyright © 2005 John Wiley & Sons, Inc.

1.2.1. Types of Data

The world surrounding us generates various types of data in abundance. The richness of data formats is impressive. The formal representation and organization of patterns reflect the way in which we intend to process the data. The most general taxonomy being in common use distinguishes among numeric (continuous), ordinal, and nominal variables. A numeric variable can assume any value in **R**. An ordinal variable assumes a small number of discrete states, and these states can be compared. For instance, there are four states, denoted a_1, a_2, a_3, and a_4, and we can say that a_1 and a_2 are closer (in some sense of similarity that we define in the next section) than a_1 and a_3. A nominal variable assumes a small number of states, but nothing can be said about their closeness. Regardless of this distinction, nominal and ordinal variables are represented as discrete variables. For computing purposes, we usually have several coding schemes, such as binary coding or binary coding with various options.

The variables can be organized into internal structures that reflect the specificity of the problem. If each pattern is described by a number of features, intuitively we arrange them into vectors—say, **x**, **y**, and **z**. Depending upon the character of the variables involved, the entries can be real or binary. Obviously, this can give rise to a variety of vectors, including both types of entries. Vectors and matrices are "flat" structures in the sense that all variables are at the same level as individual entries of the feature vector, and they have no structure. Hierarchical structures like trees are used to visualize the relationship between objects (patterns) we are interested in when dealing with clustering or classification.

1.2.2. Distance and Similarity

The concept of dissimilarity (or distance) or dual similarity is the essential component of any form of clustering that helps us navigate through the data space and form clusters. By computing dissimilarity, we can sense and articulate how close together two patterns are and, based on this closeness, allocate them to the same cluster. Formally, the dissimilarity $d(\mathbf{x}, \mathbf{y})$ between **x** and **y** is considered to be a two-argument function satisfying the following conditions:

$$d(\mathbf{x}, \mathbf{y}) \geq 0 \quad \text{for every } \mathbf{x} \text{ and } \mathbf{y}$$

$$d(\mathbf{x}, \mathbf{x}) = 0 \quad \text{for every } \mathbf{x} \tag{1.1}$$

$$d(\mathbf{x}, \mathbf{y}) = d(\mathbf{y}, \mathbf{x})$$

This list of requirements is intuitively appealing. We require a nonnegative character of the dissimilarity. The symmetry is also an obvious requirement. The dissimilarity attains a global minimum when dealing with two identical patterns, that is $d(\mathbf{x}, \mathbf{x}) = 0$.

Distance, (metric) is a more restrictive concept, as we require the triangular inequality to be satisfied; that is, for any pattern **x**, **y**, and **z** we have

$$d(\mathbf{x}, \mathbf{y}) + d(\mathbf{y}, \mathbf{z}) \geq d(\mathbf{x}, \mathbf{z}) \tag{1.2}$$

TABLE 1.1. Selected Distance Functions Between Patterns x and y

Distance Function	Formula and Comments
Euclidean distance	$d(\mathbf{x},\mathbf{y}) = \sqrt{\sum_{i=1}^{n}(x_i - y_i)^2}$
Hamming (city block) distance	$d(\mathbf{x},\mathbf{y}) = \sum_{i=1}^{n}|x_i - y_i|$
Tchebyschev distance	$d(\mathbf{x},\mathbf{y}) = \max_{i=1,2,\ldots,n}|x_i - y_i|$
Minkowski distance	$d(\mathbf{x},\mathbf{y}) = \sqrt[r]{\sum_{i=1}^{n}(x_i - y_i)^p}, \; p > 0$
Canberra distance	$d(\mathbf{x},\mathbf{y}) = \sum_{i=1}^{n}\frac{|x_i - y_i|}{x_i + y_i}$, x_i and y_i are positive
Angular separation	$d(\mathbf{x},\mathbf{y}) = \dfrac{\sum_{i=1}^{n}x_i y_i}{\left[\sum_{i=1}^{n}x_i^2 \sum_{i=1}^{n}y_i^2\right]^{1/2}}$

Note: this is a similarity measure that expresses the angle between the unit vectors in the direction of **x** and **y**

In the case of continuous features (variables), we have a long list of distance functions (see (Table 1.1)). Each of these functions implies a different view of the data because of their geometry. The geometry is easily illustrated when we consider only two features ($\mathbf{x} = [x_1 x_2]^T$) and compute the distance of **x** from the origin. The contours of the constant distance (Figure 1.1) show what type of geometric construct becomes a focus of the search for structure. Here we become aware that the Euclidean distance favors circular shapes of data clusters. With the distance functions come some taxonomy; the Minkowski distance comprises an infinite family of distances, including well-known and commonly used ones such as the Hamming, Tchebyschev, and Euclidean distances.

The same effect shown in Figure 1.1d can be achieved when the value of the power in the Minkowski distance is changed; see Figure 1.2.

One commonly used generalization is the Mahalanobis distance

$$d(\mathbf{x},\mathbf{y}) = \mathbf{x}^T A^{-1}\mathbf{y} \tag{1.3}$$

where A is a positive definite matrix. By choosing this matrix, we can control the geometry of potential clusters by rotating the ellipsoid (off diagonal entries of

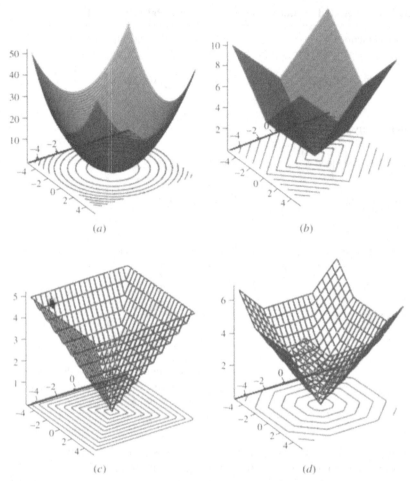

Figure 1.1. Examples of distance functions—three-dimensional and contour plots: (a) Euclidean, (b) Hamming (city block), (c) Tchebyschev, (d) "combined" type of distance max (2/3 Hamming, Tchebyschev).

A) and changing the length of its axes (the elements lying on the main diagonal of the matrix).

With binary variables, we traditionally focus on the notion of similarity rather than distance (or dissimilarity). Consider two binary vectors **x** and **y** that consist of two strings $[x_k]$, $[y_k]$ of binary data; compare them coordinatewise and do the simple counting of occurrences:

number of occurrences when x_k and y_k are both equal to 1
number of occurrences when $x_k = 0$ and $y_k = 1$
number of occurrences when $x_k = 1$ and $y_k = 0$
number of occurrences when x_k and y_k are both equal to 0

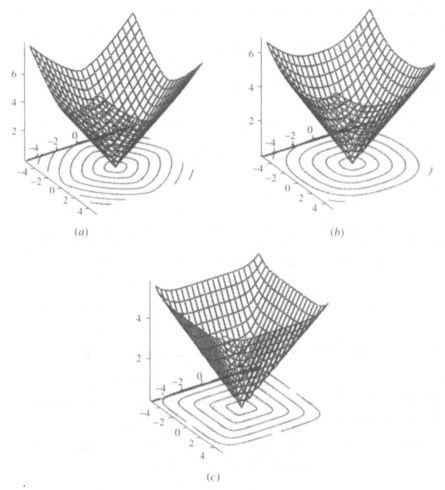

(a)   (b)

(c)

Figure 1.2. Examples of the Minkowski distance function for selected values of the power: (a) 1.5 (b) 2.5, and (c) 7.0.

These four numbers can be organized in a 2 by 2 co-occurrence matrix (contingency table) that visualizes how "close" these two strings are to each other.

	1	0
1	a	b
0	c	d

Evidently the zero nondiagonal entries of this matrix point at the ideal matching (the highest similarity). Based on these four entries, there are several commonly encountered measure of similarity of binary vectors $\mathbf{x}$ and $\mathbf{y}$. The simplest matching coefficient computes as the following ratio:

$$\frac{a+d}{a+b+c+d} \tag{1.4}$$

The Russell and Rao measure of similarity consists of the quotient

$$\frac{a}{a+b+c+d} \tag{1.5}$$

The Jacard index involves the case when both inputs assume values equal to 1:

$$\frac{a}{a+b+c} \tag{1.6}$$

The Czekanowski index is practically the same as the Jacard index, but by adding the weight factor of 2, it emphasizes the coincidence of situations where entries of **x** and **y** both assume values equal to 1:

$$\frac{2a}{2a+b+c} \tag{1.7}$$

1.3. MAIN CATEGORIES OF CLUSTERING ALGORITHMS

Clustering techniques are rich and diversified. They have been continuously developing for over a half century following a number of trends, depending upon the emerging optimization techniques, main methodology (system modeling, DM, signal processing), and application areas. At the very high end of the overall taxonomy we envision two main categories of clustering, known as hierarchical and objective function-based clustering.

1.3.1. Hierarchical Clustering

The clustering techniques in this category produce a graphic representation of data (Duda et al., 2001). The construction of graphs (as these methods reveal the structure by considering each individual pattern) is done in two ways: bottom-up and top-down. The other names used reflect the way a structure is revealed. In the bottom-up mode known as an agglomerative approach, we treat each pattern as a single-element cluster and then successively merge the closest clusters. At each pass of the algorithm, we merge the two closest clusters. The process is repeated until we get to a single data set or reach a certain predefined threshold value. The top-down approach, known as a divisive approach, works in the opposite direction: we start with the entire set treated as a single cluster and keep splitting it into smaller clusters. Considering the nature of the process, these methods are often computationally inefficient, with the possible exception of patterns with binary variables.

The results of hierarchical clustering are usually represented in the form of dendrograms (Figure 1.3). Dendrograms are visually appealing graphical constructs: they show how difficult it is to merge two clusters. The distance scale shown at the right-hand side of the graph helps us quantify the distance between the clusters. This implies a simple stopping criterion: given a certain threshold value of the distance, we stop merging the clusters once the distance between them exceeds this threshold, meaning that merging two distinct structures does not seem to be feasible.

An important issue is how to measure the distance between two clusters. Note that we have discussed how to express the distance between two patterns. Here, as each cluster may contain many patterns, computation of the distance is neither obvious nor unique. Consider two clusters, A and B, illustrated in Figure 1.4. Let us describe the distance by $d(A, B)$ and denote the number of patterns in A and B by n_1 and n_2, respectively. Intuitively, we can easily envision three typical ways of computing the distance between the two clusters.

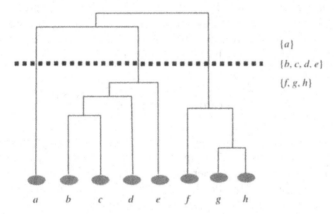

Figure 1.3. A dendrogram as a visualization of the structure of patterns; also shown are the distance values guiding the process of successive merging of the clusters.

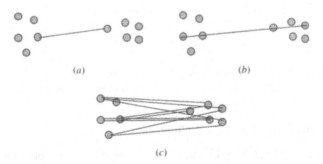

Figure 1.4. Two clusters and three main ways of computing the distance between them: (*a*) single link, (*b*) complete link, and (*c*) group average link.

Single-Link Method. The distance $d(A,B)$ is based on the minimal distance between the patterns belonging to A and B. It is computed in the form

$$d(A, B) = \min_{x \in A, y \in B} d(\mathbf{x}, \mathbf{y}) \qquad (1.8)$$

In essence, the distance supports a sort of radically "optimistic" mode of expressing vicinity between clusters where we get involved the closest patterns located in different clusters. Clustering based on this distance is one of the most commonly used methods.

Complete-Link Method. This method is at the opposite end of the spectrum, as it is based on the distance between the two farthest patterns belonging to two clusters:

$$d(A, B) = \max_{x \in A, y \in B} d(\mathbf{x}, \mathbf{y}) \qquad (1.9)$$

Group Average Link Method. In contrast to the two previous approaches, where the distance is determined on the basis of extreme values of the distance function, this method considers the average between the distances computed between all pairs of patterns, one from each cluster. We have

$$d(A, B) = \frac{1}{\mathrm{card}(A)\mathrm{card}(B)} \sum_{x \in A, y \in B} d(\mathbf{x}, \mathbf{y}) \qquad (1.10)$$

Obviously, these computations are more intensive. However, they reflect a general tendency between the distances computed for individual pairs of patterns.

Obviously, we can develop other ways of expressing the distance between A and B. For instance, the Hausdorff method of computing the distance between two sets of patterns could be an attractive alternative.

There is an interesting general expression for describing various agglomerative clustering approaches known as the Lance-Williams recurrence formula. It expresses the distance between clusters A and B and the cluster formed by merging them (C)

$$d_{A \cup B, C} = \alpha_A d_{A,C} + \alpha_B d_{B,C} + \beta d_{A,B} + \gamma |d_{A,C} - d_{B,C}| \qquad (1.11)$$

with the adjustable values of the parameters $\alpha_A (\alpha_B)$, β, and γ. This is shown in Table 1.2, where the choice of values implies a certain clustering method.

1.3.2. Objective Function-Based Clustering

The second general category of clustering is concerned with building partitions (clusters) of data sets on the basis of some performance index known also as an objective function. In essence, partitioning N patterns into c clusters (groups)

TABLE 1.2. Values of the Parameters in the Lance-Williams Recurrence Formula and the Resulting Agglomerative Clustering; n_A, n_B, and n_C Denote the Number of Patterns in the Corresponding Clusters

Clustering Method	α_A (α_B)	β	γ
Single link	$1/2$	0	$-1/2$
Complete link	$1/2$	0	$1/2$
Centroid	$\dfrac{n_A}{n_A + n_B}$	$-\dfrac{n_A n_B}{(n_A + n_B)^2}$	0
Median	$1/2$	$-1/4$	0

is a nontrivial problem. First, the number of the partitions is expressed in the following form (Webb, 2002):

$$\frac{1}{c!} \sum_{i=1}^{c} (-1)^{c-i} \binom{c}{i} i^N \qquad (1.12)$$

This number increases very quickly, making any attempt to enumerate all of the partitions unfeasible. The minimization of a certain objective function can be treated as an optimization approach leading to some suboptimal configuration of the clusters (which, in practice, is an appealing solution). The main design challenge lies in formulating an objective function that is capable of reflecting the nature of the problem so that its minimization reveals a meaningful structure in the data set. The minimum variance criterion is one of the most common options. Having N patterns in $\mathbf{R}^n$, and assuming that we are interested in forming c clusters, we compute a sum of dispersions between the patterns and a set of prototypes $\mathbf{v}_1, \mathbf{v}_2, \ldots, \mathbf{v}_c$

$$Q = \sum_{i=1}^{c} \sum_{k=1}^{N} u_{ik} \| \mathbf{x}_k - \mathbf{v}_i \|^2 \qquad (1.13)$$

with $\| \ \ \|^2$ being a certain distance between $\mathbf{x}_k$ and $\mathbf{v}_i$. The important component in the above sum is a partition matrix $U = [u_{ik}], i = 1, 2, \ldots, c, k = 1, 2, \ldots, N$ whose role is to allocate the patterns to the clusters. The entries of U are binary. Pattern k belongs to cluster i when $u_{ik} = 1$. The same pattern is excluded from the cluster when $u_{ik} = 0$. Partition matrices satisfy the following conditions:

Each cluster is nontrivial, that is, it does not include all patterns and is nonempty:

$$0 < \sum_{k=1}^{N} u_{ik} < N, \quad i = 1, 2, \ldots, c$$

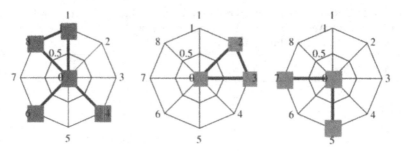

Figure 1.5. Star diagram as a graphical representation of the partition matrix for three clusters.

Each pattern belongs to a single cluster:

$$\sum_{i=1}^{c} u_{ik} = 1, \quad k = 1, 2, \ldots, N$$

The family of partition matrices (viz., binary matrices satisfying these two conditions) will be denoted by **U**. As a result of minimization of Q, we construct the partition matrix and a set of prototypes. Formally we express this in the following way, which is just an optimization problem with constraints:

$$\text{Min } Q \text{ with respect to } \mathbf{v}_1, \mathbf{v}_2, \ldots, \mathbf{v}_c \text{ and } U \in \mathbf{U} \tag{1.14}$$

Several methods are used to achieve this optimization. The most common one, named C-Means (Duda et al., 2001; Webb, 2002), is a well-established way of clustering data.

Partition matrices are an intuitively appealing form in which to illustrate the structure of the patterns. For instance, the matrix formed for $N = 8$ patterns split into $c = 3$ clusters is

$$U = \begin{bmatrix} 1 & 0 & 0 & 1 & 0 & 1 & 0 & 1 \\ 0 & 1 & 1 & 0 & 0 & 0 & 0 & 0 \\ 0 & 0 & 0 & 0 & 1 & 0 & 1 & 0 \end{bmatrix}$$

Each row describes a single cluster. Thus we have the following arrangement: the first cluster consists of patterns {1, 4, 6, 8}, the second involves patterns {2, 3}, and the third covers the remaining patterns, {5, 7}.

Graphically, the partition matrix (or, equivalently, the structure of the data set) can be shown in the form of a so-called star or radar diagram (Figure 1.5).

1.4. CLUSTERING AND CLASSIFICATION

The structure revealed through the clustering process allows us to set up a classifier. The "anchor" points of the classifier are the prototypes of the clusters. Each

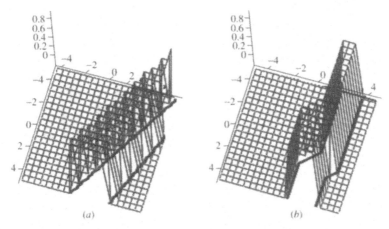

Figure 1.6. Plots of the classification regions of the first class with prototype $\mathbf{v}_1 = [1.7\ 2.5]^T$. The two other prototypes are $\mathbf{v}_2 = [0.4\ 0.3]^T$ and $\mathbf{v}_3 = [2.5\ 4.5]^T$ for two distance functions: (a) Euclidean and (b) Hamming.

cluster forms a single class ω_1, ω_2, ..., ω_c. Using the clusters, we develop a nearest neighbor classification rule by stating that $\mathbf{x}$ belongs to class ω_j if it is the closest to the prototype $\mathbf{v}_j$:

$$j = \arg\ \min_i \|\mathbf{x} - \mathbf{v}_i\|^2 \tag{1.15}$$

This classification rule generates the corresponding decision regions in the feature space. Depending on the form of the distance function, we end up with different geometry of the classification boundaries, as shown in Figure 1.6.

The classifier formed in this way is one of the simplest architectures focusing exclusively on the prototypes of the clusters.

1.5. FUZZY CLUSTERING

The binary character of partitions described so far may not always be a convincing representation of the structure of data. Consider the set of two-dimensional patterns illustrated in Figure 1.7. While we can easily detect three clusters, their character is different. The first one is quite compact, with highly concentrated patterns. The other two exhibit completely different structures. They are far less condensed, with several patterns whose allocation to a given cluster may be far less certain. In fact, we may be tempted to allocate them to two clusters with varying degrees of membership. This simple and appealing idea forms a cornerstone of fuzzy sets—collections of elements with partial membership in several categories. As illustrated in Figure 1.7, the two identified patterns could easily belong to several clusters.

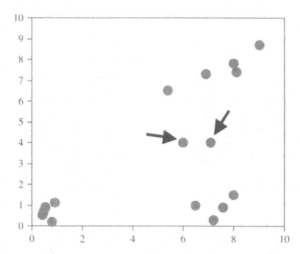

Figure 1.7. Three clusters with patterns of partial membership (belongingness) in the clusters. The patterns of borderline character are pointed to by the arrows.

These situations of partial membership occur quite often. Structures (clusters) may not be well separated for a variety of reasons. There may be noise or lack of discriminatory power of the feature space in which the patterns are represented. Some patterns could be genuine outliers. Some of them could be borderline cases and thus are difficult to classify. As a result, they may require far greater attention. A clustering algorithm that could easily provide detailed insight into the membership grades of the patterns could be a genuine asset. Let us assume that this is true and that the partition matrix now consists of grades of membership distributed in the unit interval. For the data in Figure 1.8, the partition matrix comes with the entries shown. The results are highly appealing, and they fully reflect our intuitive observations: patterns 6 and 7 have a borderline character, with membership grades in one of the clusters at the 0.5 level. The values in the partition matrix quantify the effect of partial membership.

Because we have allowed for partial membership, the clustering algorithm leading to such conceptual augmentation is regarded as a generalization of the standard FCM and is named of fuzzy C-Means or FCM, for short This generalization was introduced by Dunn (1974) and generalized by Bezdek (1981). The performance index (objective function) guiding the search through the data space assumes the form

$$Q = \sum_{i=1}^{c} \sum_{k=1}^{N} u_{ik}^{m} \|\mathbf{x}_k - \mathbf{v}_i\|^2 \qquad (1.16)$$

It seems similar to the one guiding the optimization of the Boolean (two-valued) partition matrix with only one significant exception: we consider U to be a fuzzy partition, viz., a matrix with the entries confined to the unit interval that satisfies two important requirements:

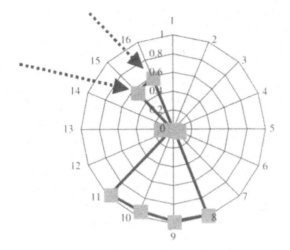

Figure 1.8. Star diagram of the fuzzy partition matrix.

- The clusters are nontrivial. For each cluster ($i = 1, 2, \ldots, c$) we end up with a nonempty construct that does not include all patterns.
- The total membership grades sum to 1, so the distribution of belongingness is equal to 1.

In the above objective function, we can rewrite the distance in a quadratic form by noting that $\|\mathbf{x}_k - \mathbf{v}_i\|^2 = (\mathbf{x}_k - \mathbf{v}_i)^T (\mathbf{x}_k - \mathbf{v}_i) = \mathbf{x}_k^T \mathbf{x}_k - 2\mathbf{x}_k^T \mathbf{v}_i + \mathbf{v}_i^T \mathbf{v}_i$. More formally, determining the structure is equivalent to the optimization task of the form given by (1.13) in the case of the binary C-Means. As before, the minimization is completed with respect to the partition matrix and the prototypes. The fuzzification factor (m), $m > 1$, helps control the shapes of the clusters and produces a balance between the membership grades close to 0 or 1 and those with intermediate values.

The derivations of the solution are completed in two steps. The first one involves the constraints accompanying the requirements imposed on the partition matrix. We incorporate the constraints with the aid of Lagrange multipliers. Then for each pattern $t = 1, 2, \ldots, N$, we formulate the augmented functional

$$V = \sum_{i=1}^{c} u_{it}^m d_{it}^2 - \lambda \left(\sum_{i=1}^{c} u_{it} - 1 \right) \qquad (1.17)$$

with λ denoting a Lagrange multiplier. Computing the derivative of V with respect to u_{st} and making it equal to 0, we obtain

$$\frac{\partial V}{\partial u_{st}} = m u_{st}^{m-1} d_{st}^2 - \lambda = 0 \qquad (1.18)$$

and

$$u_{st} = \left(\frac{\lambda}{m}\right)^{1/(m-1)} \frac{1}{(d_{st})^{\frac{2}{m-1}}} \tag{1.19}$$

Taking into account the identity constraint $\sum\limits_{j=1}^{c} u_{jt} = 1$, we have

$$\left(\frac{\lambda}{m}\right)^{1/(m-1)} \sum_{j=1}^{c} \frac{1}{(d_{jt})^{\frac{2}{m-1}}} = 1 \tag{1.20}$$

This allows us to determine the Lagrange multiplier λ:

$$\left(\frac{\lambda}{m}\right)^{1/(m-1)} = \frac{1}{\sum\limits_{j=1}^{c} \frac{1}{(d_{jt})^{\frac{2}{m-1}}}} \tag{1.21}$$

Next, we insert the above expression into (1.19), which yields

$$u_{st} = \frac{1}{\sum\limits_{j=1}^{c} \left(\frac{d_{st}}{d_{jt}}\right)^{\frac{2}{m-1}}} \tag{1.22}$$

The computations of the prototypes are straightforward, as no constraints are imposed on them. The minimum of Q computed with respect to $\mathbf{v}_s$ yields

$$\nabla_{v_s} Q = 0$$

The detailed solution depends on the distance function. In the case of the Euclidean distance, this leads to the expression

$$2 \sum_{k=1}^{N} u_{sk}^{m} (\mathbf{x}_k - \mathbf{v}_s) = 0 \tag{1.23}$$

We immediately obtain

$$\mathbf{v}_s = \frac{\sum\limits_{k=1}^{N} u_{sk}^{m} \mathbf{x}_k}{\sum\limits_{k=1}^{N} u_{sk}^{m}} \tag{1.24}$$

Other forms of the distance function, such as the Hamming or Tchebyschev distances do not lead to an immediate solution and require more optimization effort. The clustering using these options has been reported in the literature. We will discuss this extension later on.

To summarize, we can regard the FCM algorithm as an iterative process involving successive computations (updates) of the prototypes and the partition matrix. The values of the parameters are set up in advance. They consist of the following items: the number of clusters (c), the distance function $\| \cdot \|$, the fuzzification factor (m), and the termination criterion (ε).

Initialization Phase. Select values of c, m, and ε. Choose the distance function. Initialize (randomly) the partition matrix:

> *Repeat* // main iteration loop
> Compute prototypes of the clusters
> Compute the partition matrix
> *until* a given stopping criterion quantified in terms of e has been satisfied.

Let us review the design of the clustering process in more detail; in this context, it is useful to refer to the list of parameters we can choose up front the entire process. The number of clusters reflects the level of generality we are interested in setting up when dealing with the data. While this number is not known exactly, the feasible range of clusters we can establish is rather narrow and clearly reflects the domain knowledge we may have about the problem at hand and/or the objectives of the data analysis problem. For instance, when dealing with a huge customer database, it is reasonable to assume that we are interested in a few (say, five) groups of customers, which we want to characterize for further marketing of new products and services. There is no point in moving toward a very refined partition of the data into 20 or more clusters, as these could be difficult to interpret and take advantage of (perhaps we do not wish to be so specialized and launch a very narrow marketing campaign). The limit cases are trivial: $c = 1$ does not reveal any structure. The number of clusters set up to N, $c = N$, is meaningless: each pattern forms an individual cluster, and this does not make any sense. Overall, we observe that there is a strong monotonic character of the objective function treated as a function of c; when the number of clusters increases, the values of Q decrease. There may be a slight departure from this tendency, but very often it is limited and insignificant. The distance function helps focus our search for structure. As we demonstrated in Section 1.2.2 each distance function implies a certain geometry and navigates search for finding structure in data. Hopefully, the structure revealed in this way is compatible with the "internal" geometry of the data set itself. For instance, when dealing with the Euclidean distance, our favorite geometry consists of hyperballs (or hyperellipsoids in case we start weighting the features). The Hamming distance focuses clustering on the search for the structure that is compatible with diamond-like shapes of concentrations of data. The Tchebyschev distance favors a search focusing on hyperboxes. The

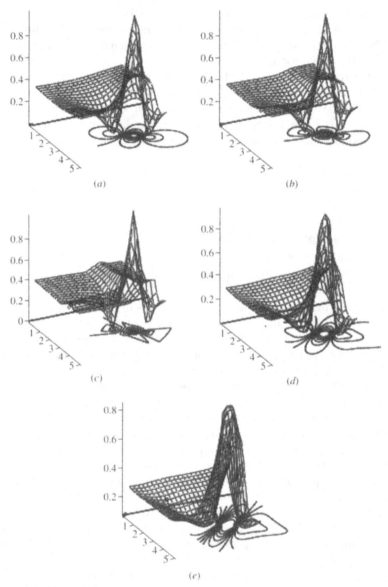

Figure 1.9. Shapes of the membership function for selected values of the fuzzification factor (m): (a) $m = 2$; (b) $m = 1.5$; (c) $m = 1.05$; (d) $m = 3$; (e) $m = 5$.

fuzzification factor influences the shape of the clusters. Typically, its value is set to 2. By changing the values of m, we can make the clusters (partition matrices) look more Boolean so that we see more membership grades close to 0 or 1. This happens when m approaches 1. On the other hand, when m increases (with values greater than 2), the resulting membership grades lead to spike-like functions. Illustrative examples are presented in Figure 1.9.

The stopping criterion quantifies the situation in which the clustering process can be stopped because it has reached a steady state and further computations are not justified. The typical approach used is to compare partition matrices produced in two successive iterations of the FCM, say $U(\text{iter} + 1)$ and $U(\text{iter})$, and, if the distance between them $\|U(\text{iter} + 1) - U(\text{iter})\|$ does not exceed a certain threshold (ε), stop the computing. The distance itself could be quantified by taking into account the biggest change in the partition matrix, that is,

$$\|U(\text{iter} + 1) - U(\text{iter})\| = \max_{i,k} |u_{ik}(\text{iter} + 1) - u_{ik}(\text{iter})| \qquad (1.25)$$

The numeric value of ε is application oriented; usually it is confined to the range of $10^{-3}-10^{-5}$.

Fuzzy partition matrices provide detailed insight into the structure of the data set that is thoroughly quantified through the membership values. Based on these values, we can easily separate the patterns that are typical of the cluster (as they have membership grades close to 1) from the data of borderline character. Obviously, the membership grades help quantify the effect of partial membership. Sometimes we are interested in cluster allocation without having any detailed information on the patterns. This is achieved by a transformation known as hardening (Bezdek, 1981). It produces a binary relation (partition matrix) based on the original partition matrix by choosing the largest membership value. More specifically, given u_{ik}, we produce Boolean entries of matrix $\tilde{u}_{ik}$ such that

$$\tilde{u}_{ik} = \begin{cases} 1 & \text{if } i = \arg\ \max_j u_{jk} \\ 0 & \text{otherwise} \end{cases} \qquad (1.26)$$

Let us briefly note the use of the clustering results in the design of the classifier. As in the K-means algorithm, we use the nearest neighbor rule, which in this case takes into account the fact that the membership grades are continuous. We then anticipate that all prototypes contribute to some extent to the determination of the membership values. Consider that, given the set of prototypes, we now have a new pattern, $\mathbf{x}$. Its membership grades to the clusters $\mathbf{u} = [u_1 u_2 \ldots u_c]^T$ take the form

$$u_i = \frac{1}{\displaystyle\sum_{j=1}^{c} \left(\frac{\|\mathbf{x} - \mathbf{v}_i\|}{\|\mathbf{x} - \mathbf{v}_j\|} \right)^{2/(m-1)}} \qquad (1.27)$$

It is easy to check that (1.27) is a solution to the optimization problem

$$\text{Min}_u \sum_{i=1}^{c} u_i^m \|\mathbf{x} - \mathbf{v}_i\|^2 \qquad (1.28)$$

with the unity constraint imposed on the membership grades, $\sum_{i=1}^{c} u_i = 1$.

1.6. CLUSTER VALIDITY

As we indicated above, the number of clusters is application-driven and user-centric. As the user is in the middle of the process of data analysis, it is beneficial to consider a varying number of clusters and analyze the results produced. Obviously, it would be helpful to have some automation in this process. This automation should come with a synthetic measure with which we can assess the quality of the discovered structure. To state the problem in a different way, what is the optimal number of clusters? Or, even better, what is the preferred number of clusters given the underlying geometry imposed on the clustering process through the use of some objective function? The characteristics of the data could be quite different from those captured by the objective function. How easily could we detect elongated clusters when the objective function favors spherical shapes? What about two elongated and crossing clusters? What about a porcupine-like distribution of patterns? To address these questions, we need a certain cluster validity measure (Dubes, 1987; Windham, 1980; Windham, 1982; Xie and Beni, 1991). Now that the problem has been identified, a number of proposals have been made. Most of them define the validity functional on the partition matrix returning a certain real number. More formally, the validity index $v(c)$ is defined as the following mapping $V: U \rightarrow \mathbf{R}$. The extreme value (either minimum or maximum) of V treated as a function of c (as we are interested in changing this parameter) points at the feasible or most likely number of clusters in the data structure. The commonly used validity functionals include the partition index and partition entropy.

Partition Index. The partition index of U, denoted by $P(U)$, produces an average of the squared values of the membership grades encountered in the partition matrix:

$$P(U) = \frac{1}{N} \sum_{i=1}^{c} \sum_{k=1}^{N} u_{ik}^2 \tag{1.29}$$

If each pattern belongs to a single cluster (hard partition), then the partition index assumes its maximal value of 1. If patterns share their membership across all clusters, with the same membership grade equal to $1/c$, this gives rise to the lowest value of $P(U)$, which in this case equals $1/c$. In other words, the index quantifies the ambiguity of the partition matrices so that we can rank them and select the one with the lowest ambiguity.

Partition Entropy. We form an entropy function defined over the partition matrix in the following manner:

$$H(U) = -\frac{1}{N} \sum_{i=1}^{c} \sum_{k=1}^{N} u_{ik} \ln(u_{ik}) \tag{1.30}$$

(we assume that for $u = 0$, $\ln = 0$)

The values of the partition entropy range from 0 to $\ln(N)$. Again, if we consider Boolean entries of the partition matrix, the entropy is equal to 0. The highest value is obtained when there is a uniform distribution of membership grades (equal to $1/c$); here we have

$$H(U) = -\frac{1}{N}\sum_{i=1}^{c}\sum_{k=1}^{N}\frac{1}{c}\ln\left(\frac{1}{c}\right) = \ln(c) \qquad (1.31)$$

If we are interested in the lowest entropy, the clusters leading to it are preferred over other structures discovered in the collection of patterns.

While the partition index and partition entropy exhibit interesting properties that are useful in quantifying the ambiguity of partition matrices and identifying those with the lowest values as the most preferred, their use in revealing the most plausible number of clusters may not be obvious. Even though they have been used for this purpose, both indexes tend to be quite monotonic, without a well-delineated minimum (which could have been used as a sound indicator of the plausible data structure). The literature provides other cluster validity criteria [proportion exponent (Windham, 1980, 1982), separation index, fuzzy hypervolume, average partition density], but to some extent, all of them are affected by monotonicity. Furthermore when applied together, they may lead to conflicting findings concerning the most preferred number of clusters. These indexes can be useful, but we should keep in mind their limited role and treat the findings implied by them as only useful guidelines. In any case, we should be concerned about the range of the plausible number of clusters rather than a single value.

1.7. EXTENSIONS OF OBJECTIVE FUNCTION-BASED FUZZY CLUSTERING

The objective function given by (1.16) is generic in terms of its composition and the distance function being used. There are several interesting extensions of this objective function that are helpful in exploring a broad range of geometry of the clusters.

1.7.1. Augmented Geometry of Fuzzy Clusters: Fuzzy C Varieties

The search for structure (clusters) is always biased by the selection of the distance function. Changing the distance function imposes different "computing lenses" of the algorithm and causes the clustering technique to favor some geometry of the clusters. There are many possible variations of the distance function. Instead of expressing the distance between two patterns (one of them being a prototype), we may imagine that the prototypes are more complex geometric constructs. This is the rationale behind constructs such as fuzzy c-lines, fuzzy c-ellipsoids, and so on. The concept of fuzzy c-varieties studied by Bezdek et al. (1981) addresses the geometry of clusters. So far, these are just points. Here we consider a variety of geometric constructs. Each cluster represents an r-dimensional variety, where $r \in \{0, 1, 2, \ldots, n - 1\}$, and is described by a prototype $\mathbf{v}_i$ and the collection of orthogonal unit vectors $\mathbf{e}_{i1}, \mathbf{e}_{i2}, \ldots, \mathbf{e}_{ir}$ and in essence becomes an affine subspace of $\mathbf{R}^n$:

$$\{\mathbf{y} \in \mathbf{R}^n | \mathbf{y} = \mathbf{v}_i + \sum_{j=1}^{r} \mathbf{t}_j \mathbf{e}_{i1}, \mathbf{t} \in \mathbf{R}^r\} \qquad (1.32)$$

In the case where $r = 0$, we end up with the generic version of the FCM confined to a single prototype. If $r = 1$, this variety represents a straight line. The value of r set to 2 returns a plane. In general, if $r = p - 1$, we end up with a hyperplane. The distance between any pattern $\mathbf{x}$ and the ith c-variety is computed in the form

$$d(\mathbf{x}; \mathbf{v}_i, \mathbf{e}_i) = \|\mathbf{x} - \mathbf{v}_i\|^2 - \sum_{j=1}^{r} ((\mathbf{x} - \mathbf{v}_i)^T \mathbf{e}_{ij})^2 \qquad (1.33)$$

These distances for $n = 2$ (here the c-variety reduces to a straight line) are illustrated in Figure 1.10.

The geometry of fuzzy clusters becomes of paramount importance when we use clustering for image processing, that is, processing of geometric objects such as circles, ovals, boxes, elongated edges of machine parts, and the like. While in this domain we are generally confined to two-dimensional spaces (digital images), it is apparent that the diversity of geometric shapes poses a serious challenge. If we are interested in searching for a specific geometric form, then this search should be driven by the geometry in question. For example, if we are interested in circular shapes, the distance function of the form $d(\mathbf{x}; \mathbf{v}, r) = |\|\mathbf{x} - \mathbf{v}\| - r|$ would reflect our interest in looking for these type of clusters (see Figure 1.11). Its minimal values are distributed on the circle we are interested in revealing in the data.

The number of clustering studies in this area is extensive. The research monograph by Hoppner et al. (1999) forms a comprehensive compendium.

1.7.2. Possibilistic Clustering

Possibilistic clustering arose as a challenge to the probabilistic-like character required for the membership grades in clusters. The limitation of the form does not allow us to distinguish between a situation in which a pattern is shared between clusters and one in which it is simply atypical and we would like to see this effect quantified in some way. The unity constraint does not allow this. Possibilistic clustering, as advocated by Krishnampuram and Keller (1993, 1996), drops the requirement that the sum of membership grades must equal 1. This is reflected in the form of the augmented objective function:

$$Q = \sum_{i=1}^{c} \sum_{k=1}^{N} u_{ik}^m \|\mathbf{x}_k - \mathbf{v}_i\|^2 + \sum_{i=1}^{c} \beta_i \sum_{k=1}^{N} (1 - u_{ik})^m \qquad (1.34)$$

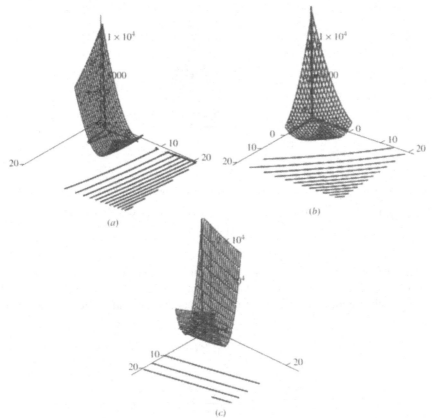

Figure 1.10. Three-dimensional and contour plots of the distance for the c-variety for selected values of $\mathbf{e}$: (a) $\mathbf{e} = [1.0\ \ 5.0]^T$; (b) $\mathbf{e} = [3.0\ \ 3.0]^T$; (c) $\mathbf{e} = [7.0\ \ 0.5]^T$. In all cases $\mathbf{v} = [1\ \ 2]^T$.

Note that in addition to the standard term, the second sum expresses our desire to have the membership grades sum to 1 (but this is not the constraint articulated in the standard FCM algorithm). The positive weight factor β_i (which depends on a specific cluster) helps us set up a suitable balance between the structure-seeking term and the departures from the unity requirement of the membership grades. From the standpoint of the second term, we prefer the membership grades to be closer to 1; this may have led to very high values of the first sum. This weighted sum is critical when we want to articulate a sound balance. Because of the character of the membership constraint, we refer to the resulting algorithm as a possibilistic fuzzy C-Means (P-FCM). The derivations of the partition matrix lead to the following expression:

$$u_{ik} = \frac{1}{1 + \left(\dfrac{\|\mathbf{x}_k - \mathbf{v}_i\|^2}{\beta_i}\right)^{\frac{1}{m-1}}} \qquad (1.35)$$

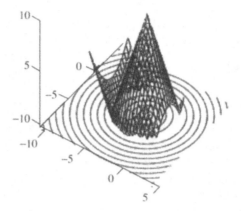

Figure 1.11. Expressing the distance between $\mathbf{x}$ and a circle with center $\mathbf{v}\,(=[1\ 1]^T)$ and radius $r(=5)$; $\|.\ \|$ is specified as the Euclidean distance.

The formula for the prototypes is the same as that used the FCM method. The choice of an optimal value of β_k impacts the results; a sound selection of these weight factors was proposed by Krishnampuram and Keller (1996) to be in the form

$$\beta_i = \frac{\sum\limits_{k=1}^{N} u_{ik}^m \|\mathbf{x}_k - \mathbf{v}_i\|}{\sum\limits_{k=1}^{N} u_{ik}^m} \tag{1.36}$$

1.7.3. Noise Clustering

The method used to cope with noisy data is another important application-driven issue in searching for structure in experimental data. The technique of Ohashi (1984) and Dave (1991) is to introduce a special cluster, the noise cluster, whose role is to "localize" the noise and place it in a single auxiliary cluster. By assigning patterns to this noise cluster, we declare them to be outliers in the data set. The objective function that helps capture this effect is represented in the form

$$Q = \sum\sum u_{ik}^m \|\mathbf{x}_k - \mathbf{v}_i\|^2 + \sum_{k=1}^{N} \delta^2 \left(1 - \sum_{i=1}^{c} u_{ik}\right)^m \tag{1.37}$$

The weight coefficient δ^2 reflects the distance between all data and the noise cluster. Note that we end up with $c + 1$ clusters, with the extra cluster serving as the noise cluster. The difference in the second term of the objective function expresses the degree of membership of each pattern in the noise cluster. The sum over the first c is less than or equal to 1. The derivation of the partition matrix produces the following expression:

$$u_{ik} = \frac{1}{\sum_{j=1}^{c} \left(\frac{\|\mathbf{x}_k - \mathbf{v}_i\|}{\|\mathbf{x}_k - \mathbf{v}_j\|} \right)^{\frac{2}{m-1}} + \sum_{j=1}^{c} \left(\frac{\|\mathbf{x}_k - \mathbf{v}_i\|}{\delta} \right)^{\frac{2}{m-1}}} \qquad (1.38)$$

1.8. SELF-ORGANIZING MAPS AND FUZZY OBJECTIVE FUNCTION-BASED CLUSTERING

Objective function-based clustering forms one of the main optimization paradigms of data discovery. To put this in a broader perspective, we elaborate on the alternative arising in neurocomputing, namely, self-organizing maps. This helps us contrast the underlying optimization mechanisms and look at various formats of results generated by different methods.

The term self-organizing map (SOM) was coined by Kohonen (1982, 1989, 1995 Kohonen et al, 1996). As usually emphasized in the literature, SOMs are regarded as regular neural structures (neural networks) composed of a grid of artificial neurons that attempt to visualize highly dimensional data in a low-dimensional structure, usually in the form of a two- or three-dimensional map. To make such visualization meaningful, this low-dimensional representation of the originally high-dimensional data has to preserve the *topological* properties of the data set. This means that two data points (patterns) that are close to each other in the original feature space should retain this similarity (or closeness) in their representation (mapping) in the reduced, low-dimensional space in which they are visualized. And, reciprocally, two distant patterns in the original feature space should retain their distant location in the low-dimensional space. Put it differently, we can state that the SOM acts as a *computer eye* that helps us gain insight into the structure of the data set and observe relationships occurring between the patterns originally located in a highly dimensional space. In this way, we can confine ourselves to the two-dimensional map that apparently reveals all essential relationships between the data, as well as dependencies between the software measures themselves. In spite of the existing variations, the generic SOM architecture (as well as the learning algorithm) remains basically the same. Below we summarize the essence of the underlying self-organization algorithm that achieves a certain form of unsupervised learning (Kohonen et al., 1996; Vesanto and Alhoniemi, 2000).

Before proceeding with the detailed computations, we introduce the necessary notation. Here n software measures are organized in a vector $\mathbf{X}$ of real numbers situated in the n-dimensional space of real numbers, $\mathbf{R}^n$. The SOM is a collection of linear neurons organized in the form of a regular two-dimensional grid (array) (Figure 1.12).

In general, the grid may consist of p rows and r columns; commonly we confine ourselves to the square array of $p \times p$ elements (neurons). Each neuron

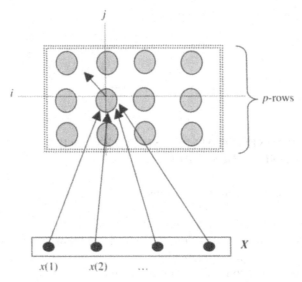

Figure 1.12. A basic topology of the SOM constructed as a grid of identical processing units (neurons).

is equipped with modifiable connections $\mathbf{w}(i,j)$ that form an n-dimensional vector of connections $\mathbf{w}(i, j) = [w_1(i, j)w_2(i, j)\ldots w_n(i, j)]$. It completes computing of the distance function $d(.\,,\,.)$ between its connections and the corresponding input $\mathbf{x}$

$$y(i, j) = d(\mathbf{w}(i, j), \mathbf{x}) \tag{1.39}$$

where the pair (i,j) denotes a certain (i,j) position of the neuron in the array. $\mathbf{x}$ is an input to all neurons. The distance can be chosen from the list of alternatives presented earlier. The same input $\mathbf{x}$ affects all neurons. The neuron with the shortest distance between the input and the connections becomes activated to the highest extent and is called the winning neuron. Let us denote its coordinates by $(i0, j0)$. More precisely, we have

$$(i0, j0) = \arg \min_{(i,j)} d(\mathbf{w}(i, j), \mathbf{x}) \tag{1.40}$$

The winning neuron matches (responds to) $\mathbf{x}$. Because it is the winner of this competition, we reward the neuron and allow it to modify the connections so that they are even closer to the input data. The update mechanism is governed by the expression

$$\mathbf{w\_new}(i0, j0) = \mathbf{w}(i0, j0) + \alpha(\mathbf{x} - \mathbf{w}(i0, j0)) \tag{1.41}$$

where α denotes a learning rate, $\alpha > 0$. The higher the learning rate, the more intensive the updates of the connections. In addition to the changes of the connections of the winning node (neuron), we allow this neuron to affect its neighbors (viz., the neurons located at similar coordinates of the map). The way in which this influence is quantified is expressed via a neighbor function $\Phi(i, j, i0, j0)$. In general, this function satisfies two intuitively appealing conditions: (a) it attains a maximum equal to 1 for the winning node, $i = i0$, $j = j0$, and (b) when the node is apart from the winning node, the value of the function gets lower (in other words, the updates are less vigorous). Evidently, there are also nodes where the neighbor function equals 0. Considering the above, we rewrite (1.41) in the following form:

$$\mathbf{w\_new}(i, j) = \mathbf{w}(i0, j0) + \alpha\Phi(i, j, i0, j0)(\mathbf{x} - \mathbf{w}(i, j)) \qquad (1.42)$$

The typical neighbor function comes in the form

$$\Phi(i, j, i0, j0) = \exp(-\beta((i - i0)^2 + (j - j0)^2)) \qquad (1.43)$$

with the parameter β usually assuming small positive values.

The above update expression (1.42) applies to all the nodes (i,j) of the map. As we iterate (update) the connections, the neighbor function shrinks: at the beginning of updates, we start with a large region of updates, and when the learning settles down, we start reducing the size of the neighborhood. For instance, its size may decrease linearly.

The number of iterations is specified in advance or the learning terminates once there are no significant changes in the connections of the neurons.

SOM and FCM are complementary, and so are their advantages and shortcomings. FCM requires the number of groups (clusters) to be defined in advance. It is guided by a certain performance index (objective function), and the solution comes in the clear form of a certain partition matrix. In contrast, SOM is more user-oriented. No specific number of clusters (group) needs to be specified in advance.

1.9. CONCLUSIONS

We have reviewed the paradigm of clustering, and fuzzy clustering in particular, and have discussed its role in revealing structure in data sets. Fuzzy clustering leads to information granulation in terms of fuzzy sets or fuzzy relations. Membership grades are important indicators of the typicality of patterns or their borderline character. We discussed various categories of fuzzy clustering and extensions of the underlying objective functions. The cluster indexes are useful in identifying the most relevant number of clusters. Each objective function implies a search for structure in data driven by some superimposed geometry, and we have shown that different distance functions emphasize a certain geometry we intend to find in the data set. Stated differently, clustering predisposes the search toward a certain geometry that is favored when the clusters are built. For instance, the Euclidean distance focuses the search on spherical shapes of the clusters. The Mahalanobis distance expands this search by allowing hyperellipsoidal shapes in the data set. Having said that, we should interpret the notion of unsupervised learning (which is often synonymous with clustering) in a suitable manner. There is no direct supervision (as encountered, for example, in classifier design), but there is still a mechanism of implicit supervision in the form of the geometric bias of search accompanying the accepted distance function.

REFERENCES

M.R. Anderberg, *Cluster Analysis for Applications*, Academic Press, New York, 1973.

G.P. Babu, M.N. Murthy, Clustering with evolutionary strategies, *Pattern Recognition*, 27, 1994, 321–329.

J.C. Bezdek, *Pattern Recognition with Fuzzy Objective Function Algorithms*, Plenum Press, New York, 1981.

J.C. Bezdek, C. Coray, R. Guderson, J. Watson, Detection and characterization of cluster substructure, *SIAM Journal of Applied Mathematics*, 40, 1981, 339–372.

J.C. Bezdek, J. Keller, R. Krishnampuram, N.R. Pal, *Fuzzy Models and Algorithms for Pattern Recognition and Image Processing*, Kluwer Academic Publishers, Dordercht, 1999.

R.N. Dave, Fuzzy shell clustering and application to circle detection in digital images, *Int. J. General Systems*, 16, 1990, 343–355.

R.N. Dave, Characterization and detection of noise in clustering, *Pattern Recognition Letters*, 12, 1991, 657–664.

R.N. Dave, K. Bhaswan, Adaptive c-shells clustering and detection of ellipses, *IEEE Trans. on Neural Networks*, 3, 1992, 643–662.

P.A. Devijver, J. Kittler (eds.), *Pattern Recognition Theory and Applications*, Springer-Verlag, Berlin, 1987.

R. Dubes, How many clusters are the best?—an experiment, *Pattern Recognition*, 20, 6, 1987, 645–663.

R.O. Duda, P.E. Hart, D.G. Stork, *Pattern Classification*, 2nd edition, John Wiley, New York, 2001.

J.C. Dunn, A fuzzy relative of the ISODATA process and its use in detecting compact well-separated clusters, *J. of Cybernetics*, 3, 3, 1974, 32–57.

H. Frigui, R. Krishnapuram, A comparison of fuzzy shell clustering methods for the detection of ellipses, *IEEE Trans. on Fuzzy Systems*, 4, 1996, 193–199.

K. Fukunaga, *Introduction to Statistical Pattern Recognition*, 2nd edition, Academic Press, London, 1990.

F. Hoppner, F. Klawonn, R. Kruse, T. Runkler, *Fuzzy Cluster Analysis*, John Wiley, Chichester, England, 1999.

A.K. Jain, R.P.W. Duin, J. Mao, Statistical pattern recognition: a review, *IEEE Trans. on Pattern Analysis and Machine Intelligence*, 22, 1, 2000, 4–37.

A.K. Jain, M.N. Murthy, P.J. Flynn, Data clustering: a review, *ACM Comput. Survey*, 31, 3, 1999, 264–323.

L. Kaufmann, P.J. Rousseeuw, *Finding Groups in Data: An Introduction to Cluster Analysis*, John Wiley, New York, 1990.

P.R. Kersten, Fuzzy order statistics and their applications to fuzzy clustering, *IEEE Trans. on Fuzzy Systems*, 7, 7, 1999, 708–712.

F. Klawonn, A. Keller, Fuzzy clustering with evolutionary algorithms, *Int. J. of Intelligent Systems*, 13, 1998, 975–991.

T. Kohonen, Self-organized formation of topologically correct feature maps, *Biological Cybernetics*, 43, 1982, 59–69.

T. Kohonen, *Self-organization and Associative Memory*, Springer-Verlag, Berlin, 1989.

T. Kohonen, *Self-organizing Maps*, Springer-Verlag, Berlin, 1995.

T. Kohonen, S. Kaski, K. Lagus, T. Honkela, Very large two-level SOM for the browsing of newsgroups, in: *Proceedings of ICANN96*, Lecture Notes in Computer Science, 1112, Springer-Verlag, Berlin, 1996, 269–274.

R. Krishnapuram, J. Keller, A possibilistic approach to clustering, *IEEE Trans. on Fuzzy Systems*, 1, 1993, 98–110.

R. Krishnapuram, J. Keller, The possibilistic C-Means algorithm: insights and recommendations, *IEEE Trans. on Fuzzy Systems*, 4, 1996, 385–393.

K. Mali, S. Mitra, Clustering of symbolic data and its validation, in: N.R. Pal and M. Sugeno (eds.), *Advances in Soft Computing—AFSS 2002*, Springer-Verlag, Heidelberg, 2002, 339–344.

Y. Ohashi, Fuzzy Clustering and robust estimation, Proceedings of 9th Meeting SAS User Group Int, Hollywood Beach, Florida, 1984.

J. Vesanto, A. Alhoniemi, Clustering of the self-organizing map, *IEEE Trans. on Neural Networks*, 11, 2000, 586–600.

A. Webb, *Statistical Pattern Recognition*, 2nd edition, John Wiley, Hoboken, NJ, 2002.

M.P. Windham, Cluster validity for fuzzy clustering algorithms, *Fuzzy Sets and Systems*, 3, 1980, 1–9.

M.P. Windham, Cluster validity for the fuzzy C-Means clustering algorithms, *IEEE Trans. on Pattern Analysis and Machine Intelligence*, 11, 1982, 357–363.

X.L. Xie, G. Beni, A validity measure for fuzzy clustering, *IEEE Trans. on Pattern Analysis and Machine Intelligence*, 13, 1991, 841–847.

2 Computing with Granular Information: Fuzzy Sets and Fuzzy Relations

This chapter provides an introduction to the fundamentals of fuzzy sets, regarded as essential information granules supporting human-centric computing. We discuss the key concepts and show how they translate into detailed algorithmic constructs. This discussion provides only limited coverage of the fundamentals of fuzzy sets. The reader may refer to one of the many comprehensive treatises on this subject (see, e.g., Pedrycz and Gomide, 1998; Zadeh and Kacprzyk, 1999; Zimmermann, 2001). We start with a general framework of granular computing that helps cast fuzzy sets in the setting of information granules and enhances understanding of the differences between them and other forms of knowledge representation.

2.1. A PARADIGM OF GRANULAR COMPUTING: INFORMATION GRANULES AND THEIR PROCESSING

Information granules tend to permeate human endeavors (Bargiela and Pedrycz, 2003; Pedrycz, 2001; Zadeh, 1979, 1996, 1997, 1999; Zadeh and Kacprzyk, 1999). Regardless of the task, we usually cast it into a certain conceptual framework of basic entities we consider relevant in the given formulation. This is the framework in which we formulate generic concepts at some level of abstraction, carry out processing, and communicate the results. Consider image processing: in spite of continuous progress in this area, human beings assume a dominant position when it comes to understanding and interpreting images. We do not focus on individual pixels and process them afterward, but instead group them together into semantically meaningful constructs—objects we deal with in everyday life. They involve regions that consist of pixels or groups of pixels drawn together because of their proximity in the image, similarity of texture, color, and so on. This remarkable and unchallenged ability is based on our effortless ability to construct information granules and manipulate them. As another example, consider a collection of time series. We can describe them in a semi-qualitative manner by pointing at specific regions of such signals. Specialists can

Knowledge-Based Clustering, by Witold Pedrycz
ISBN 0-471-46966-1   Copyright © 2005 John Wiley & Sons, Inc.

effortlessly interpret electrocardiographic signals. They distinguish some segments of such signals and interpret their combinations. Again, the individual samples of the signal are not the focus of the analysis and signal interpretation. We granulate all phenomena (regardless of whether they are originally discrete or analog). Time is a variable that is subject to granulation. We use seconds, minutes, days, months, and years. Depending on the specific problem and the user, the size of the information granules (time intervals) can vary dramatically. For high-level management, a time interval of quarters of a year or a few years could be a meaningful basis on which to develops a model. For the designer of high-speed digital systems, the temporal information granules consist of nanoseconds, microseconds, and, rarely, microseconds. Even such common and simple examples are convincing enough to lead us to believe that (a) information granules are key components of knowledge representation and processing; (b) the level of granularity of information granules (their size, to be more descriptive) is crucial to the problem's description; and (c) there is no universal level of granularity of information; the size of granules is problem-oriented and user-dependent.

What has been said so far touches on the qualitative aspect of the problem. The challenge is to develop a computing framework within which all these representation and processing endeavors can be formally achieved. The common platform emerging within this context is called granular computing. In essence, it is an emerging paradigm of information processing. While we have already noted a number of important conceptual and computational constructs developed in the domains of system modeling, machine learning, image processing, pattern recognition, and data compression in which various abstractions (and ensuing information granules) came into existence, granular computing is innovative and intellectually proactive in several fundamental ways:

- It identifies the essential commonalities among surprisingly diversified problems and technologies, which can be cast into a unified framework we usually refer to as a granular world. This is an operational processing entity that interacts with the external world (which could be another granular or numeric world) by collecting necessary granular information and returning the outcomes of the granular computing.

- The unified framework of granular processing, allows us to understand more clearly the role of the interaction among various formalisms and visualize the way in which they communicate.

- It brings together the existing formalisms of set theory (interval analysis), fuzzy sets, rough sets, and so on under the same roof by clearly demonstrating that in spite of their visibly distinct underpinnings (and ensuing processing), they also have fundamental commonalities. In this sense, granular computing establishes a stimulating environment of synergy among the individual approaches.

- By focusing on the commonalities of the existing formal approaches, granular computing helps build heterogeneous and multifaceted models of processing of information granules by clearly recognizing the orthogonal nature of some of the existing and well-established frameworks (e.g., probability theory, with its probability density functions, and fuzzy sets with their membership functions).

- Granular computing fully acknowledges a notion of variable granularity ranging from detailed numeric entities to abstract and general information granules. It considers the compatibility of such information granules and the ensuing communication mechanisms of the granular worlds.

To cast granular computing in historic perspective, we should note that there have been several fundamental mechanisms of granulation ranging from interval analysis (Hansen, 1975; Jaulin et al., 2001; Moore, 1966; Warmus, 1956) to fuzzy sets (Zadeh, 1965), uncertain variables (Bubnicki, 2002, 2004), and rough sets (Pal and Skowron, 1999; Pawlak, 1991; Polkowski and Skowron, 1998; Skowron, 1989) (see Figure 2.1).

In spite of the main emphasis of this chapter on fuzzy sets, it is helpful to contrast these to key technologies such as sets and probability.

Sets and Interval Analysis. The two-valued world of sets and interval analysis (Moore, 1966) ultimately focuses on a collection of intervals in the line of reals: $[a,b]$, $[c,d]$, and so on. Conceptually, sets are rooted in a two-valued logic with their fundamental predicate of membership ($\in$). There is an important isomorphism between the structure of two-valued logic endowed with truth values (false-true) and set theory, with sets described by characteristic functions. Interval analysis is a cornerstone of so-called reliable computing that is ultimately associated with digital computing in which the accuracy of any variable

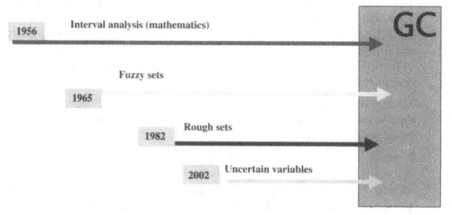

Figure 2.1. Some milestones contributing to the development of granular computing (GC); note that their emergence and growth were independent of each other.

comes is finite (implied by the fixed number of bits used to represent numbers). This limited accuracy gives rise to a certain pattern of propagation of computing error. For instance, the addition of two intervals $[a,b]$ and $[c,d]$ leads to a broader interval of the form $[a + c, b + d]$. Here the accumulation of uncertainty (or decreased granularity of the result) depends upon the specific algebraic operation completed for given intervals. Interestingly, intervals formed uniformly in a certain space achieve an analog-to-digital conversion; the greater the number of bits, the finer the intervals and the higher their number. The well-known fundamental relationship states that with n bits we can build a collection of 2^n intervals of width $(b - a)/2^n$ for the original range of $[a,b]$. Traditionally, given a universe of discourse **X**, a family of intervals (sets) defined therein is denoted by $P(\mathbf{X})$.

Probability and Probabilistic Information Granules. The theory of probability concerns constructs of probability functions or probability density functions (pdf's) that can be treated as information granules. We determine the frequency of occurrence of some event and on this basis combine our observations in the format of the corresponding pdf. In this sense, the pdf captures a general view of the given phenomenon. The granularity of such information granules is conveniently captured through its standard deviation.

2.2. FUZZY SETS AS HUMAN-CENTRIC INFORMATION GRANULES

The striking difference between sets (intervals) and fuzzy sets (Zadeh, 1965) is that in fuzzy sets we admit a concept of partial membership so that we can discriminate between elements that are typical of the concept and those of borderline character. Information granules such as *high* speed, *warm* weather, and *fast* car are examples of information granules falling into this category. We cannot specify a single well-defined element that forms a solid border between full belongingness and full exclusion. Fuzzy sets, with their soft transition boundaries, are an ideal vehicle to capture the notion of partial membership. Formally, fuzzy set A defined in **X** is defined by its membership function

$$A : \mathbf{X} \to [0, 1] \tag{2.1}$$

where $A(x)$ denotes a degree of membership of x in A. A family of fuzzy sets defined in **X** is denoted $F(\mathbf{X})$. As the semantics of A is far richer than that encountered in sets, fuzzy sets have several important characteristics. These are summarized in Table 2.1.

Fuzzy sets are provided in the form of membership functions in either a continuous or discrete format. In the first case, we have an analytical expression for the membership function:

$$A(x) = \begin{cases} \exp(-0.2x) & \text{if } x \geq 0 \\ 0 \text{ otherwise} \end{cases} \tag{2.2}$$

TABLE 2.1. Main Descriptors Used in the Characterization of Fuzzy Set A

Notion and Definition	Description
α-cut $A_\alpha = \{x \mid A(x) \geq \alpha\}$	Set induced by some threshold consisting of elements belonging to A to an extent not lower than α. By choosing a certain threshold, we convert A into the corresponding representative set. α-cuts provide important links between fuzzy sets and sets.
Height of A, $\mathrm{hgt}(A) = \sup_x A(x)$	Supremum of the membership grades; A is normal if $\mathrm{hgt}(A) = 1$. The core of A is formed by all elements of the universe for which $A(x)$ equals 1.
Support of A, $\mathrm{supp}(A) = \{x \mid A(x) > 0\}$	Set induced by all elements of A belonging to it with nonzero membership grades.
Cardinality of A, $\mathrm{card}(A) = \int_X A(x)\, dx$ (assuming that the integral does exist)	Counts the number of elements belonging to A; characterizes the granularity of A. Higher $\mathrm{card}(A)$ implies higher granularity (specificity) or, equivalently, lower generality.

In the second case, the membership function is defined in discrete elements of the universe of discourse and could be expressed by some formula (say, $A(x_i) = i/100$, $i = 1, 2, \ldots, 100$) or in a tabular format:

x_i	1	2	3	4	5	6
$A(x_i)$	0.6	1.0	0.0	0.5	0.3	0.9

Fuzzy sets defined in the line of real numbers ($\mathbf{R}$) whose membership functions have several intuitively appealing properties such as (a) unimodality, (b) continuity, and (c) convexity are referred to as fuzzy numbers. They generalize the concept of a single numeric quantity by providing an envelope of possible values it can assume.

The calculus of fuzzy numbers generalizes the idea of operations on intervals and follows the so-called extension principle. Given two fuzzy numbers A and B, their function $C = f(A, B)$ returns a fuzzy number with the membership function

$$C(z) = \sup_{x, y: z = f(x, y)}[\min(A(x), B(y))] \tag{2.3}$$

For the sake of completeness, Table 2.2 presents the results of algebraic operations (addition, multiplication, and division) on two intervals, $A = [a, b]$ and $B = [c, d]$.

2.3. OPERATIONS ON FUZZY SETS

Fuzzy sets defined in the same space are combined logically through logic operators of intersection, union, and complement. These operations are completed via

TABLE 2.2. Algebraic Operations on Numeric Intervals A and B

Operation	Result
Addition $[a, b] + [c, d]$	$[a + c, b + d]$
Multiplication $[a, b]^*[c, d]$	$[\min(ac, ad, bc, bd), \max(ac, ad, bc, bd)]$
Division $[a, b]/[c, d]$	$[\min(a/d, a/c, b/c, b/d), \max(a/d, a/c, b/c, b/d)]$

t- and s-norms commonly used as models of the logic operators *and* and *or*, respectively. This gives rise to the expression

$$A \cap B : \quad (A \cap B)(x) = A(x) \ t \ B(x)$$
$$A \cup B : \quad (A \cup B)(x) = A(x) \ s \ B(x) \qquad (2.4)$$
$$\overline{A} : \quad \overline{A}(x) = 1 - A(x)$$

The negation operation, denoted by an overbar, is usually performed by subtracting the membership function from 1, that is, $\overline{A}(x) = 1 - A(x)$. Let us recall that a t-norm is a two-argument function

$$t: [0, 1]^2 \to [0, 1]$$

that satisfies the following conditions (Butnariu and Klement, 1983):

$xty = ytx$	commutativity
$xt \ (ytz) = (xty) \ tz$	associativity
if $x \le y$ and $w \le z$, then $xtw \le ytz$	monotonicity
$0 \ tx = 0, \ 1 \ tx = x$	boundary conditions

Several examples of commonly encountered t-norms are presented in Table 2.3.

The most commonly used t-norms are the minimum (min), product, and Lukasiewicz *and* operator. Given any t-norm, we can generate a so-called dual s-norm through the expression

$$asb = 1 - (1 - a) \ t \ (1 - b) \qquad (2.5)$$

$a, b \in [0, 1]$, which is nothing but the Morgan law (recall that $\overline{A \cup B} = \overline{A} \cap \overline{B}$). In other words, we do not need a separate table for s-norms, as these can be easily generated. Again, the list of s-norms in common use involves the operation of the maximum (max), the probabilistic sum, and the Lukasiewicz *or* operator.

2.4. FUZZY RELATIONS

Fuzzy sets are defined in a given space. Fuzzy relations are defined in Cartesian products of some spaces and represent composite concepts. For instance, the notion "*high* price and *fast* car" can be represented as a fuzzy relation R defined in the Cartesian product of price and speed. Note that R can be formally treated

TABLE 2.3. Examples of Selected t-Norms

t-Norm	Comments
$x t y = \min(x, y)$	The first model of the *and* operation used in fuzzy sets (as proposed by Zadeh in his seminal 1965 paper). Note that this model is noninteractive, that is, the result depends on only a single argument. More specifically, $\min(x, x + \varepsilon) = x$, regardless of the value of ε.
$x t y = \max(0, (1 + p)(x + y - 1) - p x y), \; p \geq -1$	For $p = 0$, this yields the Lukasiewicz *and* operator
$x t y = x y$	The product operator is commonly encountered in applications; the operator is interactive.
$x t y = 1 - \min(1, \sqrt[p]{(1 - x)^p + (1 - y)^p}), \; p > 0$	The parametric flexibility is assured by the choice of values of p.
$x t y = \dfrac{x y}{p + (1 - p)(x + y - x y)}, \; p \geq 0$	As above, this family of t-norms is indexed by the auxiliary parameter, whose value can be adjusted.
$x t y = \dfrac{x y}{\max(x, y, p)}, \; p \in [0, 1]$	As above.
$x t y = \begin{cases} x & \text{if } y = 1 \\ y & \text{if } x = 1 \\ 0 & \text{otherwise} \end{cases}$	Drastic product—exhibits a "drastic" behavior, that is, it returns a nonzero argument if one of the arguments is equal to 1; otherwise, it returns 0.

as a two-dimensional fuzzy set, $R: \mathbf{X} \times \mathbf{Y} \to [0, 1]$, with $\mathbf{X}$ and $\mathbf{Y}$ being the corresponding spaces of price and speed.

Fuzzy partition matrices generated by fuzzy clustering provide a discrete characterization of fuzzy relations. Given c clusters, the partition matrix consists of c fuzzy relations $A_1, A_2, \ldots, A_c$ whose membership grades consist of individual rows of the matrix. In other words, U can be written down in the form

$$U = \begin{bmatrix} A_1 \\ A_2 \\ \\ A_c \end{bmatrix} \qquad (2.6)$$

The partition matrix results from granulation of the original numeric data.

2.5. COMPARISON OF TWO FUZZY SETS

Given two fuzzy entities (fuzzy sets, relations, numbers, etc.) defined in the same universe of discourse **X**, a fundamental question arises as to their similarity or proximity. How similar (or different) are two fuzzy sets? To address this matter, we should note that there is no unique way of producing a single numeric characterization of such matching. In fact, this problem is the same as that of expressing similarity between two real functions (and membership functions are just special cases) and coming up with a single number.

Several fundamental approaches have been introduced in the literature and are encountered in various applications.

Possibility Measure. The possibility measure, denoted by Poss(A,X), describes a level of overlap between two fuzzy sets or fuzzy relations and is expressed in the form

$$\text{Poss}(A, X) = \sup_{x \in \mathbf{X}}[A(x)tX(x)] \tag{2.7}$$

The plot visualizing the computations of the possibility is shown in Figure 2.2a. Computationally, we note that the possibility measure is concerned with the determination of the intersection between A and X, $A(x)tX(x)$ that is followed by the optimistic assessment of this intersection. It is done by picking up the highest values among the intersection grades of A and X that are taken over all elements of the universe of discourse **X**.

Necessity Measure. The necessity measure expresses a pessimistic degree of inclusion of A in X and is computed as follows:

$$\text{Nec }(A, X) = \inf_{x \in \mathbf{X}}[(1 - A(x))sX(x)] \tag{2.8}$$

The computational details are presented in Figure 2.2b. In contrast to the possibility measure, the necessity measure is asymmetric (which is obvious, as we are concerned with the inclusion predicate).

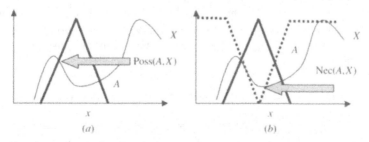

Figure 2.2. Computations of possibility (a) and necessity (b) measures; t-norm: minimum, s-norm: maximum. The dotted line in (b) shows a complement of A, $1 - A(x)$.

By its nature (as a measure of overlap), the possibility measure is symmetric, $\text{Poss}(A, X) = \text{Poss}(X, A)$. The necessity measure, expressing the extent of inclusion, is not symmetric. Because only a single number is returned as the result of this operation, this may not be enough to secure highly discriminative abilities of the possibility and necessity measures. In other words, we may have two quite different fuzzy sets, X_1 and X_2, with the possibility measure assuming the same value, $\text{Poss}(A, X_1) = \text{Poss}(A, X_2)$.

Compatibility Measure. Rather than returning a single number, this measure (Zadeh, 1979) produces a function defined in [0,1] that is computed as follows:

$$\text{Comp}(X, A)(u) = \sup_{x:u=A(x)} X(x), \quad u \in [0, 1] \tag{2.9}$$

Note that the compatibility measure is not symmetric, so $\text{Comp}(A,X)$ is different from $\text{Comp}(X,A)$. As we return here a function (rather than a single number, as in the two previous measures), the result is more informative and has higher discriminative abilities (that is, we can quantify the difference in a more refined manner even when the possibility or necessity measure may not be able to produce any discrimination abilities). If $A = X$, then $\text{Comp}(A,X)$ returns an identity function, $\text{Comp}(A, X)(u) = u, u \in [0,1]$.

Various Distance Functions. These matching instruments view membership functions as real functions and compute the distance between A and X. in general, the Minkowski distance $d(A, X)$ is taken as an integral (we assume that it makes sense):

$$d(A, X) = \sqrt[p]{\int_X |A(x) - X(x)|^p \, dx}, \quad p \geq 1 \tag{2.10}$$

By changing the value of the parameter (p) used in the above definition, we sweep through the entire collection of distance functions including the Hamming ($p = 1$), Euclidean ($p = 2$), and Tchebyschev ($p = \infty$) functions. Note that these are the same distance functions we will be discussing in conjunction with objective function–based clustering.

A Hausdorff distance, another example of distance, is used to compare two information granules (Diamond and Kloeden, 1994). It was originally proposed for intervals or sets. Let A and B be two sets defined in the same space **X**. Then the Hausdorff distance is expressed in the form

$$d(A, B) =, \max\{\sup_{a \in A} \inf_{b \in B} \|a - b\|, \sup_{a \in A} \inf_{b \in B} \|a - b\|\} \tag{2.11}$$

where $\| \, \|$ is the distance between two elements.

For illustrative purposes, consider two sets, $A = \{1, 2, 3, 4\}$ and $B = \{6, 7, 8\}$. It is convenient to compute distances, say Hamming ones, between the pairs of elements belonging to A and B and organize them into a matrix where the rows are indexed by values of a and the columns by values of b:

$$
\begin{array}{c}
1 \\ 2 \\ 3 \\ 4
\end{array}
\left[
\begin{array}{ccc}
5 & 6 & 7 \\
4 & 5 & 6 \\
3 & 4 & 5 \\
2 & 3 & 4
\end{array}
\right]
$$
$$\quad\quad 6\ \ 7\ \ 8$$

the operations inf and sup are replaced here by min and max, respectively. Then the first component in the Hausdorff distance ($\sup_a \inf_b$) is computed by finding the minima in each row of the matrix above and then taking their maximal value. In other words, $\max_a \min_b \|a - b\| = \max(5, 4, 3, 2) = 5$. The second expression, $\max_b \min_a \|a - b\|$, is realized by taking minima over each column and returning their maximal value, that is, $\max_b \min_a \|a - b\| = \max(2, 3, 4) = 4$. Finally, the maximum operation is taken over these two results: $\max(5, 4) = 5$.

In fuzzy sets, the generalization is straightforward: we consider an α-cut of A and B, determine the corresponding Hausdorff distance for this cut, denote it by $d_\alpha(A, B)$, and integrate these specific distances over all possible α-cuts:

$$
D(A, B) = \int_0^1 d_\alpha(A, B)\, d\alpha \tag{2.12}
$$

2.6. GENERALIZATIONS OF FUZZY SETS

Fuzzy sets are constructs with membership grades in the unit interval. There are several interesting generalizations (Mendel and John, 2002; Pal and Skowron, 1999) whose use is justified by specific applications. Two of them are of particular interest.

Second-Order Fuzzy Sets. These are fuzzy sets whose membership grades are fuzzy sets defined in [0,1]. Thus we depart from the individual numeric membership grades and acknowledge that the degrees of membership themselves could be stated in an approximate way. This implies the model of fuzzy sets as membership valuation. In particular, we can admit some ranges of membership, thus arriving at so-called interval-valued fuzzy sets. The generalization along this line is of particular interest when dealing with situations where the granularity in quantification of membership cannot be ignored and has to be incorporated in further processing. An example of second-order fuzzy sets is shown in Figure 2.3; there we illustrate both an interval-valued fuzzy set and a second-order fuzzy set, with fuzzy sets regarded as their memberships.

Fuzzy Sets of Type 2. This is another conceptual extension of fuzzy sets where we define a certain fuzzy set over a universe of several reference fuzzy sets.

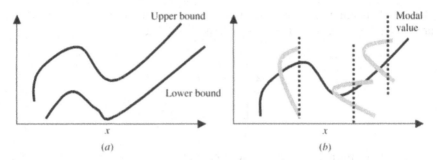

Figure 2.3. Examples of second-order fuzzy sets: (*a*) interval-valued and (*b*) membership functions as grades of belongingness.

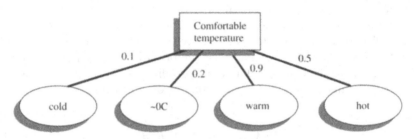

Figure 2.4. Fuzzy set of type 2 of *comfortable* temperature; note the collection of referential fuzzy sets over which the type-2 fuzzy set is formed.

For instance, a term of *comfortable* temperature can be defined in a collection of referential terms such as *cold* temperature, *around* zero, *warm, hot*, and so on. Fuzzy sets of type-2 are more abstract than fuzzy sets of type-1 (that is, standard fuzzy sets). We portray fuzzy set of type 2 of comfortable temperature in graphical form in Figure 2.4.

Some further generalizations and extensions become possible as well; for example, we can consider fuzzy sets of type-2 of order 2.

2.7. SHADOWED SETS

Fuzzy sets help describe and quantify concepts of continuous boundaries. By introducing a certain α-cut, we convert a fuzzy set into a set. By choosing a threshold level (α) that is high enough, we admit elements whose membership grades are sought meaningful (as viewed from the standpoint of the imposed threshold). This may create the impression that any fuzzy set could be made equivalent to some set. This view is highly deceptive. In essence, by building any α-cut, we elevate some membership grades to 1 (full membership) and eliminate others with lower membership grades (total exclusion). Surprisingly, no account is taken of the distribution of elements with partial membership, so this effect cannot be quantified in the resulting construct. The idea of shadowed sets (Pedrycz, 1998; Pedrycz, 1999; Pedrycz and Vukovich, 2000) is aimed at alleviating this problem by constructing regions of complete ignorance about membership grades. In essence, a shadowed set $A^{\sim}$ induced by given fuzzy set A defined in **X** is an interval-valued set in **X** that maps elements of **X** into 0,1 and the unit interval [0,1]. Formally, $A^{\sim}$ is a mapping

$$A^{\sim} : \mathbf{X} \rightarrow \{0, 1, [0, 1]\} \tag{2.13}$$

Given $A^{\sim}(x)$, the two numeric values, 0 and 1, take on a standard interpretation: 0 denotes complete exclusion from $A^{\sim}$, and 1 stands for complete inclusion in A. $A^{\sim}(x)$ equal to [0,1] represents complete ignorance—nothing is known about membership of x in $A^{\sim}$: we *neither* confirm its belongingness to $A^{\sim}$ *nor* commit to its exclusion. In this sense, such an x is the most questionable point and should be treated as such (e.g., triggering action to analyze this element in more detail, exclude it from further analysis). The name shadowed set is a descriptive reflection of a set that comes with "shadows" positioned around the edges of the characteristic function (see Figure 2.5).

Shadowed sets are isomorphic with a three-valued logic. Operations on shadowed sets are the same as those proposed in this logic. The underlying principle is to retain the vagueness of the arguments (shadows of the shadowed sets being used in the aggregation). The following tables capture the description of the operators:

Union

$$
\begin{array}{c}
0 \\
1 \\
[0, 1]
\end{array}
\left[
\begin{array}{ccc}
0 & 1 & [0, 1] \\
1 & 1 & 1 \\
[0, 1] & 1 & [0, 1]
\end{array}
\right]
$$
$$
\begin{array}{ccc}
0 & 1 & [0, 1]
\end{array}
$$

Intersection

$$
\begin{array}{c}
0 \\
1 \\
[0, 1]
\end{array}
\left[
\begin{array}{ccc}
0 & 0 & 0 \\
0 & 1 & [0, 1] \\
0 & [0, 1] & [0, 1]
\end{array}
\right]
$$
$$
\begin{array}{ccc}
0 & 1 & [0, 1]
\end{array}
$$

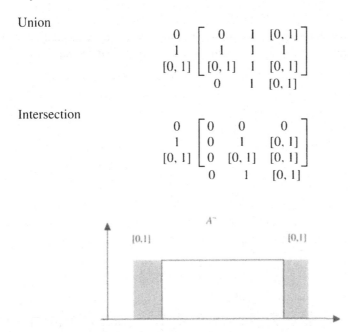

Figure 2.5. An example of a shadowed set $A^{\sim}$; observe the "shadows" produced at the edges of the characteristic function.

Complement

$$\begin{array}{c} 0 \\ 1 \\ [0, 1] \end{array} \left[\begin{array}{cc} 1 \\ 0 \\ [0, 1] \end{array} \right]$$

From the design point of view, shadowed sets are induced by fuzzy sets, and in this setting their role is to help interpret results given in the form of fuzzy sets and eventually reduce the computational overhead. Since shadowed sets do not concentrate on detailed membership grades and process only 0, 1, and 1/2 (considering that the last numeric value is used to code the shadow), all ensuing processing is very simple and computationally appealing.

The development of shadowed sets starts from a given fuzzy set. The transformation criterion governing this transformation is straightforward: maintain a balance of uncertainty in the sense that while reducing low membership grades to 0 and bringing high membership grades to 1, maintain the overall balance of membership grade changes. The changes of membership grades to 0 and 1 are compensated for by the construction of the shadow that "absorbs" the previous elimination of partial membership at low and high ranges of membership. This design principle for a unimodal fuzzy set is illustrated in Figure 2.6. This transformation is guided by the value of the threshold β; specifically, we are concerned with the two individual thresholds, that is, β and $1 - \beta$.

The retention of balance translates into the following dependency:

$$\Omega_1 + \Omega_2 = \Omega_3$$

where the corresponding regions are illustrated in Figure 2.6. Note that we are dealing with the increasing and decreasing portions of the membership functions

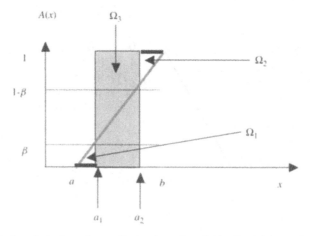

Figure 2.6. Induced shadowed set: elimination of regions of partial membership is balanced by the formation of the shadows "absorbing" the reduction in the region of partial membership grades.

separately. The integral form of the above relationship is in the form

$$\int_{a}^{a_1} A(x)\,dx + \int_{a_2}^{b} (1 - A(x))\,dx = \int_{a_1}^{a_2} dx \qquad (2.14)$$

See again Figure 2.6. A certain threshold value of β, $\beta \in [0,1/2)$, that satisfies this expression is treated as a solution to the problem. Based on this, we form a shadow of the shadowed set. In the commonly encountered membership functions, the optimal value of β can be determined in an analytical manner. For the triangular membership function, we consider that each segment (viz., the increasing and decreasing portions of the membership function) is treated separately and concentrate on the linearly increasing portion of the membership function governed by the expression $(x - a)/(b - a)$. Simple calculations reveal that the cutoff points a_1 and a_2 are equal to $a + \beta(b - a)$ and $a + (1 - \beta)(b - a)$. Subsequently the resulting optimal value of β is equal to $\dfrac{2^{3/2} - 2}{2} = 0.4142$. Similarly, when dealing with a square root type of membership function of the form $A(x) = \sqrt{\dfrac{x - a}{b - a}}$ in $x \in [a, b]$ and 0 outside this interval, we get $a_1 = a + \beta^2(b - a)$ and $a_2 = a + (1 - \beta)^2(b - a)$. The only root satisfying the requirements imposed on the threshold values is equal to 0.405.

In discrete membership grades, the computations follow the sum-based version of the criterion where the individual components of the expression assume the form of finite sums. To analyze this in more detail, let us consider a partition matrix $U = [u_{ik}], i = 1, 2, \ldots, c, k = 1, 2, \ldots, N$. Then the segments of membership grades are given in the form

$$\Omega_1 = \sum_{i,k:u_{ik} \leq \beta} u_{ik} \qquad \Omega_2 = \text{card}\{u_{ik} | \beta < u_{ik} < 1 - \beta\} \qquad \Omega_3 = \sum_{i,k:u_{ik} \geq 1-\beta} (1 - u_{ik})$$

$$(2.15)$$

There is no analytical solution to this problem. It is then advisable to form a performance index V such that

$$V(\beta) = |\Omega_1 + \Omega_3 - \Omega_2| \qquad (2.16)$$

whose values are minimized with respect to β, that is, $\beta_{\text{opt}} = \arg \text{Min}_\beta\, V(\beta)$.

The shadowed sets are instrumental in fuzzy cluster analysis, especially the results produced there. Consider the data set shown in Figure 2.7. The standard FCM run for $c = 2$ clusters returned the partition matrix whose membership grades are then transformed into the shadowed set.

The prototypes are equal to $v_1 = [5.51\ 2.48]^T$, $v_2 = [1.05\ 4.71]^T$ and reflect the structure of the data set. The membership function of one cluster (fuzzy relation) is visualized in Figure 2.8.

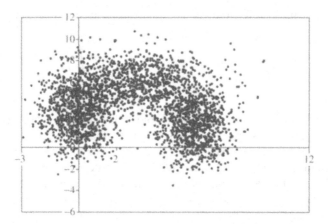

Figure 2.7. Two-dimensional synthetic data set.

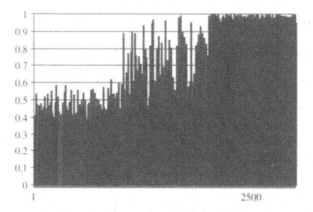

Figure 2.8. Membership grades of one of the clusters developed by the clustering algorithm.

The optimization of the threshold level (β) inducing the shadowed set is completed through a simple enumeration; following the graph in Figure 2.9, we obtain $\beta = 0.4322$. This, in turn, highlights several patterns to be treated as potential candidates for further thorough analysis.

When we complete clustering with $c = 12$ (which is substantially higher than in the first case), the results become quite different (see Figure 2.10). First, the optimal value of β becomes equal to 0.3636; refer to Figure 2.11. The overlap between the clusters has also been visibly reduced.

Interestingly, with the optimized threshold β, the standard classification rule

Accept **x** to class i if membership grade $u_i(\mathbf{x})$ is maximal ($i = \arg \max_{j=1,2,\ldots,c} u_j(x)$)

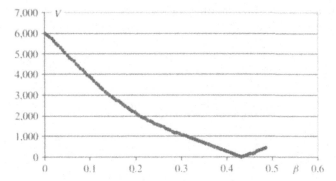

Figure 2.9. V viewed as a function of β.

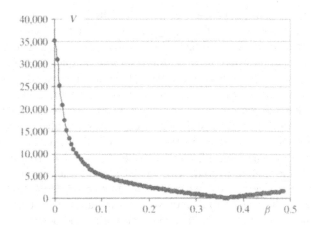

Figure 2.10. V viewed as a function of β for $c = 12$ clusters.

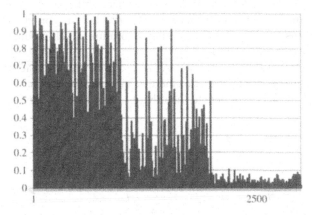

Figure 2.11. Distribution of membership grades of one of the clusters developed by the clustering algorithm.

can be now augmented by admitting the threshold level, eliminating the pattern being classified if it falls within the region of the shadow:

Accept $\mathbf{x}$ to class i if membership grade $u_i(\mathbf{x})$ is maximal ($i = \arg \max_{j=1,2,\ldots,c} u_j(x)$) and $u_i(\mathbf{x}) > 1 - \beta$.

The addition provision takes the threshold b into account:

If $u_i(\mathbf{x}) \in [\beta, 1 - \beta]$, then consider $\mathbf{x}$ to be a boundary character with respect to the ith class.

2.8. ROUGH SETS

The fundamental concept represented and quantified by rough sets concerns a description of a given concept in the language of a certain collection (vocabulary) of rather generic terms. Depending upon this collection relative to the concept, there can be situations where it cannot be fully and uniquely described with the aid of the given vocabulary (Pawlak, 1991). This may give rise to an approximate or, better, rough description of the given concept. To illustrate this effect, let us consider a concept of temperature in range $[a, b]$ that we intend to describe with the vocabulary of uniformly distributed intervals, as visualized in Figure 2.12. Apparently the concept (shown in the solid thick line) to bc described fully "covers" (includes) one interval, that is, I_3. There are also some intervals that have at least some limited overlap with it, that is, I_2, I_3, and I_4. In other words, we say that the concept, when characterized by the predefined vocabulary of intervals, does not lend itself to a unique description and the best characterization we can produce consists of some bounds. The tighter the bounds, the better the description.

Rough sets are directly related to the families of clusters that could be formed through fuzzy clusters. As schematically visualized in Figure 2.13, the first clustering algorithm operates on data set $\mathbf{X}$ and returns its partition into $c[1]$ clusters. The same occurs for the second clustering algorithm, which produces a different partition of $\mathbf{X}$ into $c[2]$ clusters.

Denote these clusters by $A_1, A_2, \ldots, A_{c[1]}$. Some other clustering algorithm (for the sake of argument, let us assume that this one operates on the same data) produces $c[2]$ clusters. We denote them by $B_1, B_2, \ldots, B_{c[2]}$, respectively. Each

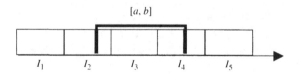

$[a, b]$

$I_1 \qquad I_2 \qquad I_3 \qquad I_4 \qquad I_5$

Figure 2.12. Concept (set) $[a. b]$ represented in the language of uniformly distributed intervals; note the upper and lower bounds forming its representation.

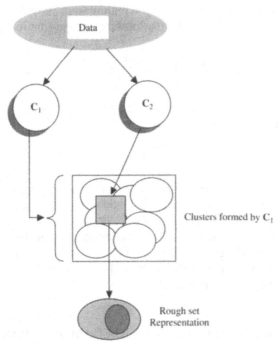

Figure 2.13. Emergence of rough sets through the description problem: clusters developed by clustering algorithm C_2 are represented (described) in the setting of information granules constructed by C_1.

cluster B_i can be characterized in the language of information granules designed by the first clustering algorithm. The result of this description is a rough set. More specifically, we identify those A_j's that are fully "covered" by B_i's (that is, they are fully included there):

$$B_{i^*} = \{A_j | A_j \subset B_i\} \qquad (2.17)$$

Similarly, we list all A_j's such that their intersection with B_i is nonempty:

$$B_i^* = \{A_j | A_j \cap B_i \neq \emptyset\} \qquad (2.18)$$

For the given family $\{A_i\}$ the resulting rough set is fully characterized by its upper and lower bounds: $\langle B_{i^*}, B_i^* \rangle$.

As an illustrative example, let us consider that the partition matrix of the first clustering algorithm is given in the form ($c[1] = 5$)

$$U[1] = \begin{bmatrix} 100101100001000 \\ 110000000000000 \\ 000100100000000 \\ 000000010100000 \\ 000000000000111 \end{bmatrix}$$

The partition matrix, being the result of optimization provided by the second clustering with $c[2] = 2$, yields the partition matrix with the following entries:

$$U[2] = \begin{bmatrix} 000101100110000 \\ 110100110011111 \end{bmatrix}$$

Then B_1 expressed in the language of the first partition matrix comes with the lower and upper bounds where

$$B_{1*} = \{A_1, A_3\} \quad B_1^* = \{A_1, A_3, A_4\} \qquad (2.19)$$

In other words, we have arrived at the rough set representation of B_1. From the application standpoint, note that the granularity of the information granules formed by the first clustering (that is, directly implied by the number of clusters) implies the character of the lower and upper boundaries of the rough set representation. The larger the granules (the smaller the number of clusters), the closer the bounds of the resulting rough set. It is also apparent that the choice of the value of $c[1]$ is application-oriented.

2.9. GRANULAR COMPUTING AND DISTRIBUTED PROCESSING

Information granulation and the resulting information granules form a conceptual framework within which all processing activities take place (Pedrycz, 2001). As already noted, the selection of information granules, their size, and their formal representation framework are problem-dependent issues. The flexibility we are provided in this manner is essential to the efficient handling of the problem. The size of the information granules helps us capture the level of required specificity of the solution to the problem. The decision about the design of the formal setting is essential and is based on the available domain knowledge, the format of the experimental numeric or granular data, and the expected outcomes of the modeling activities (that could result in control, classification, or decision-making algorithms). As various types of computing processing could be going on at the same time with ongoing communication, we can envision a distributed model in which each process can be treated as an individual agent (Figure 2.14).

Each agent operates in its environment of granular computing and communicates with others. From the high-end architectural standpoint, we can envision two-layer architectures with a core supporting all computing, with a communication layer surrounding it. Computing processes occur within the framework set up via some formalism of granular computing. For instance, there could be an agent operating on fuzzy sets and fuzzy relations. For the other one the pertinent computing framework results in interval calculations. The purpose of the communication layer is to facilitate acceptance of data and support the exchange of results produced by other agents. As the heterogeneity of information granulation is profound, the functional demands on this layer are immense. First, it is necessary to ensure

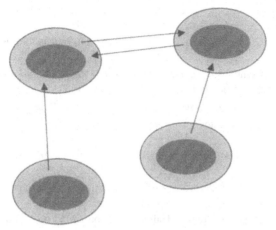

Figure 2.14. Granular computing in the realization of autonomous agents; their computing and communication layers are distinguished.

communication between the same formal frameworks whose information granules
exhibit very different levels of granularity (specificity). In other words, we should
be able to handle specificity (abstraction) incompatibilities. Second, there should
be mechanisms that support the exchange and translation of information granules
formed within different formal settings (intervals, fuzzy sets, rough sets, etc.). This
critical matter of communication is discussed in Bargiela and Pedrycz (2003).

2.10. CONCLUSIONS

We have covered the fundamentals of fuzzy sets that are regarded as a solid operating framework for fuzzy clustering. Information granules subsume fuzzy sets
as one of their conceptual architectures. Fuzzy sets have several obvious advantages. Their nature is user-centric. Modeling linguistic concepts and articulating
them in the format of membership function is highly appealing. The diversity
of membership functions available for the quantification of information granules,
along with a vast array of realizations of logic connectives supported in terms
of t- and s-norms, is an important operational facet of fuzzy computing. The
associated machinery of shadowed sets (which, as discussed here, are constructs
induced by fuzzy sets) serves as a computationally efficient processing vehicle.
We have also discussed several interesting generalizations of fuzzy sets that are of
particular relevance in the context of fuzzy clustering. The essence of computing
with rough sets has to be highlighted as well.

REFERENCES

A. Bargiela, W. Pedrycz, *Granular Computing: An Introduction*, Kluwer Academic Publishers, Dordrecht, 2003.

Z. Bubnicki, *Uncertain Logics, Variables and Systems*, Lecture Notes in Control and Information Sciences, no. 276, Springer-Verlag, Berlin, 2002.

Z. Bubnicki, *Analysis and Decision Making in Uncertain Systems*, Springer-Verlag, London, 2004.

D. Butnariu, E.P. Klement, *Triangular Norm Based Measures and Games with Fuzzy Coalitions*, Kluwer Academic Publishers, Dordrecht, 1983.

P. Diamond, P. Kloeden, *Metric Spaces of Fuzzy Sets*, World Scientific, Singapore, 1994.

E. Hansen, A generalized interval arithmetic, *Lecture Notes in Computer Science*, 29, 1975, 7–18.

L. Jaulin, M. Kieffer, O. Didrit, E. Walter, *Applied Interval Analysis*, Springer-Verlag, London, 2001.

J.M. Mendel, R.I.B. John, Type-2 fuzzy sets made simple, *IEEE Trans. on Fuzzy Systems*, 10, 2002, 117–127.

R. Moore, *Interval Analysis*, Prentice Hall, Englewood Cliffs, NJ, 1966.

S.K. Pal, A. Skowron (eds.), *Rough Fuzzy Hybridization: A New Trend in Decision-Making*, Springer-Verlag, Singapore, 1999.

Z. Pawlak, *Rough Sets: Theoretical Aspects of Reasoning About Data*, Kluwer Academic Publishers, Dordrecht, 1991.

W. Pedrycz, Shadowed sets: representing and processing fuzzy sets, *IEEE Trans. on Systems, Man, and Cybernetics, Part B*, 28, 1998, 103–109.

W. Pedrycz, Shadowed sets: bridging fuzzy and rough sets, in: S.K. Pal, A. Skowron (eds.), *Rough Fuzzy Hybridization: A New Trend in Decision-Making*, Springer-Verlag, Singapore, 1999, 179–199.

W. Pedrycz (ed.), *Granular Computing: An Emerging Paradigm*, Physica-Verlag, Heidelberg, 2001.

W. Pedrycz, F. Gomide, *An Introduction to Fuzzy Sets: Analysis and Design*, MIT Press, Cambridge, MA, 1998.

W. Pedrycz, G. Vukovich, Investigating a relevance of fuzzy mappings, *IEEE Trans. on Systems, Man, and Cybernetics*, 30, 2000, 249–262.

L. Polkowski, A. Skowron (eds.), *Rough Sets in Knowledge Discovery*, Physica-Verlag, Heidelberg, 1998.

A. Skowron, Rough decision problems in information systems, *Bulletin de l'Academie Polonaise des Sciences (Tech)*, 37, 1989, 59–66.

M. Warmus, Calculus of approximations, *Bulletin de l'Academie Polonaise des Sciences*, 4, 5, 1956, 253–259.

L.A. Zadeh, Fuzzy sets, *Information and Control*, 8, 1965, 338–353.

L.A. Zadeh, Fuzzy sets and information granularity, in: M.M. Gupta, R.K. Ragade, R.R. Yager (eds.), *Advances in Fuzzy Set Theory and Applications*, North Holland, Amsterdam, 1979, 3–18.

L.A. Zadeh, Fuzzy logic = computing with words, *IEEE Trans. on Fuzzy Systems*, 4, 1996, 103–111.

L.A. Zadeh, Toward a theory of fuzzy information granulation and its centrality in human reasoning and fuzzy logic, *Fuzzy Sets and Systems*, 90, 1997, 111–117.

L.A. Zadeh, From computing with numbers to computing with words—from manipulation of measurements to manipulation of perceptions, *IEEE Trans. on Circuits and Systems*, 45, 1999, 105–119.

L.A. Zadeh, J. Kacprzyk (eds.), *Computing with Words in Information/Intelligent Systems*, Physica-Verlag, Heidelberg, 1999.

H.J. Zimmermann, *Fuzzy Set Theory and Its Applications*, 4th edition, Kluwer Academic Publishers, Boston, 2001.

3   Logic-Oriented Neurocomputing

Neurocomputing is one of the pillars of computational intelligence. It brings a host of learning abilities to intelligent systems that can be realized in both supervised and unsupervised modes. Our objective is to develop a logic-based mode of neurocomputing in which learning goes hand in hand with the transparency of the resulting structure. The logic facet of the introduced types of neurons is essential in delivering significant interpretation abilities. Once developed, the network can be interpreted in the form of logic expressions.

3.1. INTRODUCTION

Neural networks (Golden, 1996; Jang et al., 1997; Kosko, 1991; Mitra and Pal, 1994, 1995; Pal and Mitra, 1999) are regarded as synonymous with nonlinear and highly plastic (adaptive) systems equipped with significant learning. The universal approximation theorem accompanying neural networks is highly appealing, at least at the theoretical end. By stating that any continuous function can be approximated to any desired accuracy by a certain neural network, assuming that we are given enough neurons organized in a certain multilayer topology, the basic processing unit (neuron) achieves a certain level of nonlinear processing. The plethora of learning paradigms is impressive. We have many fundamental schemes of supervised learning, including perceptron learning and backpropagation. In unsupervised learning mode, we often refer to SOMs (Kohonen maps) as a typical neural architecture. Learning itself may pertain to the optimization of the parameters of the network or it can deal with the structural optimization of the network where its topology (configuration) becomes affected. Parametric learning involves various gradient-based techniques. Structural optimization (for which we cannot compute any gradient) requires other optimization tools. Here we usually confine ourselves to evolutionary optimization (Michalewicz, 1996) with this category, including genetic algorithms, genetic programming, evolutionary programming, and alike.

Owing to the highly distributed character of processing achieved by neural networks and a lack of underlying semantics of processing carried out at the level of the individual neuron, we end up with a "black-box" character of computing. In essence, once the network has been designed (trained), we

Knowledge-Based Clustering, by Witold Pedrycz
ISBN 0-471-46966-1   Copyright © 2005 John Wiley & Sons, Inc.

do not have any mechanism with which to directly examine the character of the produced mapping and investigate it vis-à-vis the data at hand. This may hamper its future usage because of the lack of comprehension of the structure achieved through the optimization (learning) process. The black-box character of the network does not increase our confidence in the generalization abilities of the network.

This lack of interpretability raises an important question concerning the future development of the networks (Casillas et al., 2003; Dickerson and Lan, 1995). Undoubtedly it would be highly desirable to design *transparent* neural networks. There are evident benefits behind them. First, we can easily interpret the result of learning and produce the corresponding highly compact description of data. Second, the learning of such networks could be greatly facilitated. In solving any problem, we usually have some prior domain knowledge. We can take advantage of it by "downloading" such knowledge onto the given structure of the network. This could set up a highly promising starting point for further weight adjustments carried out through some well-known scheme of supervised or unsupervised learning. In order to take advantage of this preliminary knowledge, we have to be in a position to do this download in an efficient manner. This, however, requires the transparency of the network itself so that we know how to affect its structure or set up initial values of the connections. In this context, it is worth stressing that when using the standard learning schemes, we usually assume random values of the connections and start from this configuration (which might result in slow, inefficient learning). The transparency of the network in this case becomes a genuine asset.

The category of fuzzy neurons (or fuzzy logic neurons) discussed in this chapter addresses these burning issues of transparency of neural networks. We build a network with the aid of conceptually simple and logically appealing nodes (neurons) that complete generic *and* and *or* logic operations. By equipping the neurons with a set of connections, we furnish them with the badly required plasticity; the values of the connections can be easily adjusted by standard gradient-based learning schemes. Likewise, the resulting network could be transformed into a collection of conditional logic statements (rules), thus resulting in a rule-based system.

We start by introducing the main categories of the fuzzy neurons, elaborate on their main properties, move on to the architectures of networks composed of such neurons, and then discuss various facets of network interpretation.

3.2. MAIN CATEGORIES OF FUZZY NEURONS

The logic aspect of neurocomputing we intend to achieve requires a clearly delineated logic structure of the processing elements. We discuss several types of aggregative and referential neurons. Each of them comes with a clearly defined semantics of its underlying logic expression and has the parametric flexibility necessary to facilitate substantial learning.

3.2.1. Aggregative Neurons

Formally, these neurons realize a logic mapping from $[0, 1]^n$ to $[0, 1]$. Two main classes of processing units exist in this category (Hirota and Pedrycz, 1994, 1999; Pedrycz, 1991a, 1991b; Pedrycz and Rocha, 1993; Pedrycz et al., 1995).

OR Neurons. OR neurons achieve an *and* logic aggregation of inputs $\mathbf{x} = [x_1, x_2, \ldots, x_n]$ with the corresponding connections (weights) $\mathbf{w} = [w_1, w_2, \ldots, w_n]$ and then summarize the partial results in an *or*-wise manner (hence the name of the neuron). The concise notation underlines this flow of computing, $y = \mathrm{OR}(\mathbf{x}; \mathbf{w})$, while the realization of the logic operations gives rise to the expression (referred to as an *s-t* combination)

$$y = \overset{n}{\underset{i=1}{\mathrm{S}}} (x_i t w_i) \tag{3.1}$$

Bearing in mind the interpretation of the logic connectives (*t*- and *s*-norms), the OR neuron achieves the following logic expression, viewed as an underlying logic description of the processing of the input signals:

$$(x_1 \text{ and } w_1) \text{ or } (x_2 \text{ and } w_2) \text{ or } \ldots \text{ or } (x_n \text{ and } w_n) \tag{3.2}$$

Apparently the inputs are logically "weighted" by the values of the connections before producing the final result. In other words, we can treat y as a truth value of the above statement where the truth values of the inputs are affected by the corresponding weights. Notably, lower values of w_i discount the impact of the corresponding inputs; higher values (especially those being positioned close to 1) do not affect the original truth values of the inputs resulting in the logic formula. In limit, if all connections w_i, $i = 12, \ldots, n$ are set to 1, then the neuron produces a plain *or*-combination of the inputs, $y = x_1$ *or* x_2 *or* $\ldots$ *or* x_n. The values of the connections set to 0 eliminate the corresponding inputs. Computationally, the OR neuron exhibits nonlinear characteristics (that is, inherently implied by the use of the *t*- and *s*-norms that are evidently nonlinear mappings). The plots of the characteristics of the OR neuron shown in Figure 3.1 visualize this effect (note that the characteristics are affected by the use of some norms). The connections of the neuron contribute to its adaptive character; the changes in their values form the crux of the parametric learning.

AND Neuron. AND neurons, denoted by $y = \mathrm{AND}\ (\mathbf{x}; \mathbf{w})$, with $\mathbf{x}$ and $\mathbf{w}$ being defined as in the case of the OR neuron, are governed by the expression

$$y = \overset{n}{\underset{i=1}{\mathrm{T}}} (x_i s w_i) \tag{3.3}$$

Here the *or* and *and* connectives are used in the reverse order: first, the inputs are combined with the use of the *s*-norm and the partial results are aggregated *and*-wise. Higher values of the connections reduce the impact of the corresponding

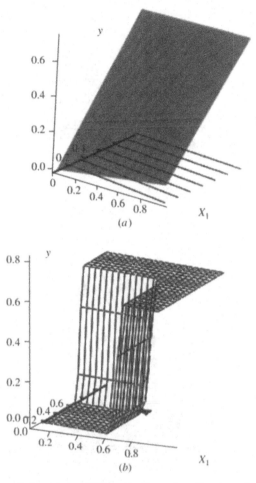

Figure 3.1. Characteristics of the OR neuron for selected pairs of t- and s-norms. In all cases, the corresponding connections are set to 0.1 and 0.7 to visualize their effect on the input-output characteristics of the neuron: (a) product and probabilistic sum; (b) Lukasiewicz *and* and *or* connectives.

inputs. In limit, $w_i = 1$ eliminates the relevance of x_i. With all w_i set to 0, the output of the AND neuron is just an *and* aggregation of the inputs

$$y = x_1 \text{ and } x_2 \text{ and } \dots \text{ and } x_n \tag{3.4}$$

The characteristics of the AND neuron are shown in Figure 3.2; note the influence of the connections and the specific realization of the triangular norms on the mapping completed by the neuron.

Let us conclude that the neurons are highly nonlinear processing units depending upon the specific realizations of the logic connectives. They also come with

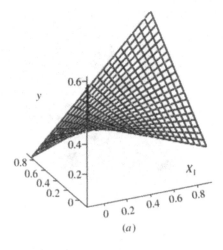

(a)

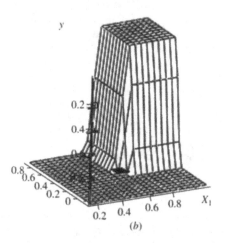

(b)

Figure 3.2. Characteristics of AND neurons for selected pairs of t- and s-norms. In all cases, the connections are set to 0.1 and 0.7 to visualize their effect on the characteristics of the neuron: (a) product and probabilistic sum; (b) Lukasiewicz logic connectives.

potential plasticity whose usage becomes critical when learning the networks involving these neurons.

At this point, it is worthwhile to contrast these two categories of logic neurons with standard neurons we encounter in neurocomputing. The typical construct comes in the form of a weighted sum followed by a nonlinear (usually monotonically increasing) function:

$$y = g\left(\sum_{i=1}^{n} w_i x_i + \tau\right)$$

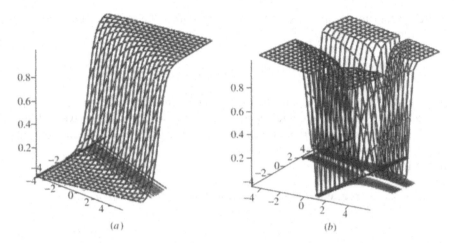

Figure 3.3. Characteristics of the neurons: (*a*) additive, with $\tau = 0.2$, $w_1 = 1.0$, $w_2 = 2.0$, and (*b*) multiplicative, where $w_1 = 0.5$, $w_2 = 2.0$, $t_1 = 1.0$, and $t_2 = 0.7$. In both cases the nonlinear function is a sigmoid function, $g(u) = 1/(1 + \exp(-u))$.

where **w** is a vector of connections, τ is a constant term (bias), and g denotes the nonlinear mapping. The other, less commonly encountered neuron is a so-called π-neuron. There can be variations in the parametric details of this construct; we can envision the following expression

$$y = g \left(\prod |x_i - t_i|^{w_i} \right)$$

where $\mathbf{t} = [t_1\ t_2 \ldots t_n]$ denotes a vector of translations and **w** (> 0) collects all connections. The plots of the two types of the neurons are presented in Figure 3.3. As before, the nonlinear function is denoted by g. While some superficial and quite loose analogy between these processing units and logic neurons could be derived, one must realize that these neurons do not demonstrate any underlying logic fabric.

3.2.2. Referential (Reference) Neurons

The essence of referential computing deals with processing logic predicates. The two-argument (or generally multivariable) predicates such as *similar, included in, and dominates* (Pedrycz and Rocha, 1993) are essential components of any logic description of a system. In general, the truth value of the predicate is the degree of satisfaction of the expression $P(x, a)$, where a is a certain reference value (reference point). Depending upon the meaning of the predicate (P), the expression $P(x, a)$ means "x is similar to a," "x is included in a," "x dominates a," and so on. In case of many variables, the compound predicate comes in the form $P(x_1, x_2, \ldots, x_n, a_1, a_2, \ldots, a_n)$ or, more concisely, $P(\mathbf{x}; \mathbf{a})$, where **x** and

a are vectors in the n-dimensional unit hypercube. We envision the following realization of $P(\mathbf{x}; \mathbf{a})$:

$$P(\mathbf{x}; \mathbf{a}) = P(x_1, a_1) \; and \; P(x_2, a_2) \; and \; \ldots \; and \; P(x_n, a_n) \qquad (3.5)$$

meaning that the satisfaction of the multivariable predicate relies on the satisfaction achieved for each variable separately. As the variables may have different levels of relevance to the overall satisfaction of the predicates, we represent this effect by some weights (connections) $w_1, w_2, \ldots, w_n$ so that (3.5) is rewritten in the form

$$P(\mathbf{x}; \mathbf{a}, \mathbf{w}) = [P(x_1, a_1) \; or \; w_1] \; and \; [P(x_2, a_2) \; or \; w_2]$$

$$and \; \ldots \; and \; [P(x_n, a_n) \; or \; w_n] \qquad (3.6)$$

Taking another look at the above expression and using a notation $z_i = P(x_i, a_i)$, it converts to a certain AND neuron $y = \text{AND}(\mathbf{z}; \mathbf{w})$, with the vector of inputs $\mathbf{z}$ resulting from the computations done for the logic predicate. Then the general notation to be used is $\text{REF}(\mathbf{x}; \mathbf{w}, \mathbf{a})$, and using the explicit notation we have

$$y = \mathop{\text{T}}_{i=1}^{n} (\text{REF}(x_i, a_i) s w_i) \qquad (3.7)$$

In essence, as visualized in Figure 3.4, we may conclude that the reference neuron is realized in a two-stage construct where we first determine the truth values of the predicate (with a treated as a reference point) and then treat these results as the inputs to the AND neuron.

So far, we have used the general term predicate computing, not confining ourselves to any specific nature of the predicate. Among a number of possibilities,

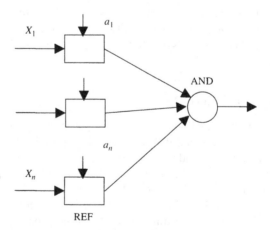

Figure 3.4. A schematic view of computing achieved by a reference neuron a involving two processing phases (referential computing and aggregation).

we discuss three predicates that tend to play an important role: inclusion, dominance, and match (similarity) predicates. As their names indicate, these predicates
return truth values of satisfaction of the relationships of inclusion, dominance,
and similarity of a certain argument x with respect to the given reference a. The
essence of all these calculations is the determination of the truth values. This
is done in a carefully developed logic framework so that the operations retain
their semantics and interpretability. What makes our discussion coherent is the
fact that the proposed operations originate from triangular norms. The inclusion
operation, denoted by $\subset$, is modeled by an implication $\rightarrow$ that is induced by a
certain left continuous t-norm (Pedrycz and Gomide, 1998):

$$a \rightarrow b = \sup\{c \in [0, 1] | atc \leq b\}, a, b \in [0, 1] \qquad (3.8)$$

For instance, for the product the inclusion takes the form $a \rightarrow b = \min(1,$
$b/a)$. The intuitive form of this predicate is self-evident: the statement "x is
included in a" and modeled as $\text{INCL}(x, a) = x \rightarrow a$ comes with the truth value
equal to 1 if x is less than or equal to a (which means that x is included in a) and
produces lower truth values once x starts exceeding the truth values of a. Higher
values of x (those above the reference point a) start generating lower truth values
of the predicate. The dominance predicate acts in a dual manner. It returns 1 once
x dominates a (so that its values exceed a) and values below 1 for x lower than
the given threshold. The formal model can be stated as $\text{DOM}(x, a) = a \rightarrow x$.
With regard to the reference neuron, the notation is equivalent to the one used
in the previous case, that is, $\text{DOM}(\mathbf{x}; \mathbf{w}, \mathbf{a})$ with the same meaning of $\mathbf{a}$ and $\mathbf{w}$.

The similarity (match) operation is an aggregate of these two, $\text{SIM}(x, a) =$
$\text{INCL}(x, a)t \, \text{DOM}(x,a)$, which is appealing from the intuitive standpoint: we say
that x is similar to a if x is included in a and x dominates a. Notably, if $x = a$,
the predicate returns 1; if x moves away from a, the truth value of the predicate
is reduced. The resulting similarity neuron is denoted by $\text{SIM}(\mathbf{x}; \mathbf{w}, \mathbf{a})$ and is
stated as

$$y = \mathop{\text{T}}_{i=1}^{n} (\text{SIM}(x_i, a_i)sw_i) \qquad (3.9)$$

The reference operations form an interesting generalization of the threshold
operations. Consider that we are viewing x as a signal of time whose behavior
needs to be monitored with respect to some bounds (α and β). If the signal does
not exceed some threshold α, then the acceptance signal should go off. Likewise,
we require another acceptance mechanism indicating a situation where the signal
does not go below another threshold value, β. In fuzzy predicates, the level
of acceptance assumes values in the unit interval rather than being a Boolean
variable. The strength of acceptance reflects how much the signal adheres to the
assumed thresholds. An example illustrating this behavior is shown in Figure 3.5.
Here the values of α and β are set to 0.6 and 0.5, respectively.

The plots of the referential neurons with two input variables are shown in
Figures 3.6 and 3.7. Here we have included two realizations of the t-norms to
illustrate their effect on the nonlinear characteristics of the processing units.

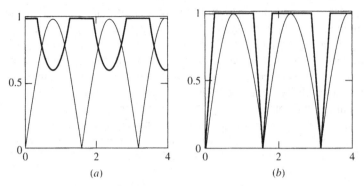

Figure 3.5. Temporal signal $x(t)$ and its acceptance signals (levels of the signals—thick lines) formed with respect to its lower and upper thresholds, (a) and (b). The complements of the acceptance are then treated as warning signals.

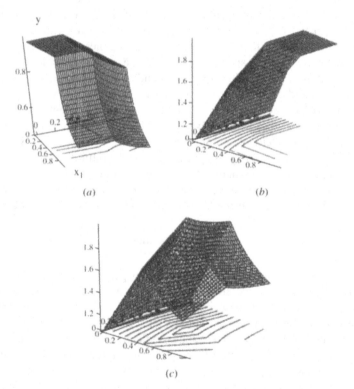

Figure 3.6. Characteristics of the reference neurons for the product (t-norm) and probabilistic sum (s-norm). In all cases the connections are set to 0.1 and 0.7 to visualize the effect of the weights on the relationships produced by the neuron. The point of reference is set to (0.5, 0.5): inclusion neuron (a), dominance neuron (b), similarity neuron (c).

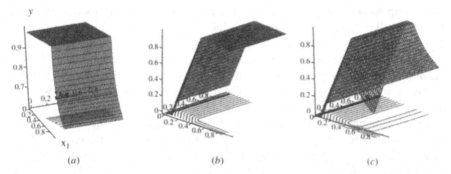

(a) (b) (c)

Figure 3.7. Characteristics of the reference neurons for the Lukasiewicz t-norm and s-norm (that is, $atb = \max(0, a + b - 1)$ and $asb = \min(1, a + b)$). In all cases the connections are set to 0.1 and 0.7 to visualize the effect of the weights. The point of reference is set to $(0.5, 0.5)$: inclusion neuron (a), dominance neuron (b), similarity neuron (c).

It is worth noting that by moving the reference point to the origin and **1**-vertex of the unit hypercube (with all its coordinates being set up to 1), the referential neuron starts resembling the aggregative neuron. In particular:

For $\mathbf{a} = \mathbf{1} = [1\ 1\ 1\dots 1]$ the inclusion neuron reduces to the AND neuron.

For $\mathbf{a} = \mathbf{0} = [0\ 0\ 0\dots 0]$ the dominance neuron reduces to the AND neuron.

One can draw a loose analogy between some types of referential neurons and the two categories of processing units encountered in neurocomputing. The analogy is based upon the *local* versus *global* character of processing realized therein. Perceptrons involve global processing. Radial basis functions involve local processing focused on receptive fields. In the same vein, the inclusion and dominance neurons are concerned with global processing, while the similarity neuron deals with, local processing.

3.3. ARCHITECTURES OF LOGIC NETWORKS

The logic neurons (aggregative and referential) can serve as building blocks of more comprehensive and functionally appealing architectures. The diversity of the topologies one can construct with the aid of these neurons is surprisingly high. This diversity is important from the application point of view, as we can fully reflect the nature of the problem in a flexible manner. It is essential to capture the problem in a logic format and then set up the logic skeleton (by forming and refining it parametrically through a thorough optimization of the connections). Throughout the development process, we can monitor the optimization of the network as well as interpret its meaning (an issue that will be discussed later on).

The typical logic network that is at the center of logic processing originates from the two-valued logic and consists of the famous Shannon theorem of decomposition of Boolean functions. Let us recall that any Boolean function $\{0, 1\}^n \rightarrow \{0, 1\}$ can be represented as a logic sum of its corresponding minterms or a logic product of maxterms. By a minterm of n logic variables $x_1, x_2, \ldots, x_n$ we mean a logic product involving all these variables in either direct or complemented form. Having n variables, we have 2^n minterms starting from the one involving all complemented variables and ending at the logic product with all direct variables. Likewise, by a maxterm we mean a logic sum of all variables or their complements. Now, in terms of the decomposition theorem, we note that the first representation scheme involves a two-layer network where the first layer consists of AND gates whose outputs are combined in a single OR gate. The converse topology occurs for the second decomposition mode: there is a single layer of OR gates followed by a single AND gate aggregating *or*-wise all partial results.

The proposed network (referred to here as a logic processor) generalizes this concept as shown in Figure 3.8. The OR-AND mode has the two types of aggregative neurons swapped.

The logic neurons generalize digital gates. The design of the network (viz., any fuzzy function) is realized through learning. If we confine ourselves to $\{0, 1\}$ values, the network's learning becomes an alternative to a standard digital design, especially a minimization of logic functions. The logic processor translates into a compound logic statement (we skip the connections of the neurons to underline the underlying logic content of the statement)

if (input$_1$ *and* ... *and* input$_j$) or (input$_d$ and ... and input$_f$), then output

The logic processor's topology (and underlying interpretation) is standard. Two logic processors can vary in terms of the number of AND neurons and their connections, but the format of the resulting logic expression is quite uniform (as a sum of generalized minterms).

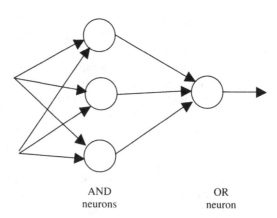

AND   OR

neurons   neuron

Figure 3.8. A topology of the logic processor in its AND-OR mode.

3.4. INTERPRETATION ASPECTS OF THE NETWORKS

While each neuron in the network comes with a well-defined semantics and can be easily interpreted, the result of interpretation could lead to a lengthy description in multivariable systems. To facilitate the process of interpretation and reduce the detailed logic expression to its essential substructure with the most meaningful topology, we prune the weakest (unnecessary) connections. Following is a discussion of the detailed thresholding expressions supporting such pruning activities.

OR Neurons. For OR neurons $y = \text{OR}(\mathbf{x}; \mathbf{w})$, we note that low values of the connection make the contribution to the corresponding input x_i quite limited. If wi tends to 0, then given of the properties of t-norms, the expression $x_i \ t \ w_i$ approaches 0 as well: $\lim (x_i \ t \ w_i) = 0$. The inputs associated with connections of low values can be dropped, and the neurons become pruned in this way. The most essential connections and input variables are retained. In this way, we end up with a compact description of the underlying logic. Two ways of pruning are envisioned:

(a) Retention of the most significant connections. For a given threshold level λ in [0, 1], we retain the connection w_i if its value is equal to or exceeds λ. Otherwise, it is dropped (or effectively replaced by 0).

(b) Binarization of the most significant connections. Here we not only retain the most significant connections (following the previous rule) but convert them into 1s.

AND Neurons. Here the pruning procedure is complementary to the one developed for OR neurons. The connections of high values are less relevant and, as such, could be eliminated. We have two ways of pruning:

(a) Retention of the most significant connections. For a given threshold level λ in [0, 1], we retain the connection w_i if its value is equal to or less than λ. Otherwise, it is dropped (or effectively replaced by 1).

(b) Binarization of the most significant connections. Here we not only retain the most significant connections (following the previous rule) but convert them into 0s.

Referential Neurons. $y = \text{REF}(\mathbf{x}; \mathbf{w}, \mathbf{a})$. Here we use the following pruning mechanism: connection w_i is binarized, producing a connection $\tilde{w_i}$ assuming Boolean values

$$\tilde{w_i} = \begin{cases} 1 & \text{if } w_i > \lambda \\ 0 & \text{if } w_i \leq \lambda \end{cases} \tag{3.10}$$

where λ denotes a certain threshold level. Because that we are concerned with the AND neurons, the connections higher than the assumed threshold are practically eliminated from the computing. Apparently we have $(w_i^{\sim} s x_i) t A \approx 1 t = A$, where A denotes the result of computing realized by the neuron for the rest of its inputs.

In referential neurons, their reference point a_i requires different treatment, depending upon the specific referential operation. For the inclusion operation, $\text{INCL}(x, a_i)$ we can use the threshold operation to come in the form

$$\text{INCL}^{\sim}(x, a_i) = \begin{cases} \text{INCL}(x, a_i) & \text{if } a_i \leq \mu \\ 1 - x & \text{if } a_i > \mu \end{cases} \tag{3.11}$$

with μ being some fixed threshold value. In other words, we consider that $\text{INCL}(x, a_i)$ is approximated by the complement of x (where this approximation is implied by the interpretational feasibility rather than being dictated by any formal optimization problem), $\text{INCL}(x, a_i) \approx 1 - x$. For the dominance neuron we have the expression for the respective binary version of DOM, $\text{DOM}^{\sim}$

$$\text{DOM}^{\sim}(x, a_i) = \begin{cases} \text{DOM}(x, a_i) & \text{if } a_i \leq \mu \\ x & \text{if } a_i > \mu \end{cases} \tag{3.12}$$

The connection set up to 1 is deemed essential. If we accept a single threshold level of 0.5, apply this consistently to all the connections of the network, and set the threshold of 0.1 for the inclusion neuron, the statement

$$y = [x_1 \text{ included in } 0.6] \text{ } or \text{ } 0.2 \text{ } and \text{ } [x_2 \text{ included in } 0.9] \text{ } or \text{ } 0.7$$

translates into a concise (yet approximate) version assuming the form of the following logic expression:

$$y = x_1 \text{ included in } 0.6$$

The choice of the threshold value could be the subject of a separate optimization phase, but we can also admit some arbitrarily values, especially if we focus on the interpretation issues.

3.5. GRANULAR INTERFACES OF LOGIC PROCESSING

The logic processing discussed in this chapter operates on information granules, viz., the model operates on logic values in [0, 1] and returns truth values as their outputs. Information granules serve as a communication layer between the external world and the logic processing module operating on truth values. As is well known in fuzzy modeling, there are two categories of communication mechanisms, commonly referred to as fuzzifiers (granular coders) and defuzzifiers (granular decoders; Figure 3.9). The role of the granular coder is to convert a numeric input coming from the external environment into the internal format of membership grades of the fuzzy sets (or relations) defined for input variable(s). The decoder takes the results produced at the level of information

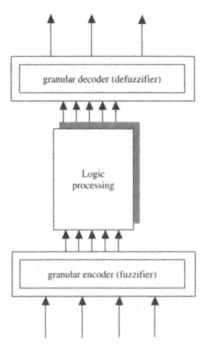

Figure 3.9. A general layered structure of logic processing; the use of granular encoders and decoders is essential in the development of communication mechanisms with the modeling environment.

granules coming through the module of logic processing and transforms them into a numeric value that is compatible with the numeric environment.

There are several ways of building a granular encoder. The most common one deals with the calculations of the given input numeric vector $\mathbf{x}$ with respect to the prototypes of the clusters formed over the input space. In this sense, the scheme of logic processing is closely related to fuzzy clustering (this will be revealed in detail in the following chapters).

The purpose of the granular decoder is to return a numeric value given some truth values as the consequence of logic processing and associated with information granules defined in the output space. For the finite number of information granules $B_1, B_2, \ldots, B_c$, each of them coming with its prototypes $\mathbf{v}_1, \mathbf{v}_2, \ldots, \mathbf{v}_c$ and the results of logic processing coming in the form of truth values μ_1, $\mu_2, \ldots, \mu_c$, we produce a numeric result $\mathbf{y}^\sim$ through some weighted average of the prototypes

$$\mathbf{y}^\sim = \frac{\displaystyle\sum_{i=1}^{c} \mu_i \mathbf{v}_i}{\displaystyle\sum_{i=1}^{c} \mu_i}$$

3.6. CONCLUSIONS

Neurocomputing combined with the underlying logic fabric builds a unique architecture of fuzzy neurocomputing. We have shown that fuzzy neurons with clearly defined semantics give rise to transparent models (Casillas et al., 2003; Dickerson and Lan, 1995; Setnes et al., 1998) whose interpretation results in a certain logic characterization of experimental data. The parametric flexibility accompanying the connections of the neurons supports all necessary learning abilities. The introduced topologies of the networks lead to the logic-based approximation of data. The unique aggregation of learning and transparency is of paramount importance to user-centric models.

REFERENCES

J. Casillas et al. (eds.), *Interpretability Issues in Fuzzy Modeling*, Springer-Verlag, Berlin, 2003.

J.A. Dickerson, M.S. Lan, Fuzzy rule extraction from numerical data for function approximation, *IEEE Trans on Systems, Man, and Cybernetics, Part B*, 26, 1995, 119–129.

R.M. Golden, *Mathematical Methods for Neural Network Analysis and Design*, MIT Press, Cambridge, MA, 1996.

K. Hirota, W. Pedrycz, OR/AND neuron in modeling fuzzy set connectives, *IEEE Trans. on Fuzzy Systems*, 2, 1994, 151–161.

K. Hirota, W. Pedrycz, Fuzzy relational compression, *IEEE Trans. on Systems, Man, and Cybernetics, Part B*, 29, 1999, 407–415.

J.S.R. Jang, C.T. Sun, E. Mizutani, *Neuro-Fuzzy and Soft Computing*, Prentice Hall, Upper Saddle River, NJ, 1997.

B. Kosko, *Neural Networks and Fuzzy Systems*, Prentice Hall, Englewood Cliffs, NJ, 1991.

Z. Michalewicz, *Genetic Algorithms + Data Structures = Evolution Programs*, 3rd edition, Springer-Verlag, Heidelberg, 1996.

S. Mitra, S.K. Pal, Logical operation based fuzzy MLP for classification and rule generation, *Neural Networks*, 7, 1994, 353–373.

S. Mitra, S.K. Pal, Fuzzy multiplayer perceptron, inferencing and rule generation, *IEEE Trans. on Neural Networks*, 6, 1995, 51–63.

S.K. Pal, S. Mitra, *Neuro-Fuzzy Pattern Recognition*, John Wiley, New York, 1999.

W. Pedrycz, Processing in relational structures: fuzzy relational equations, *Fuzzy Sets and Systems*, 40, 1991a, 77–106.

W. Pedrycz, Neurocomputations in relational systems, *IEEE Trans. on Pattern Analysis and Machine Intelligence*, 13, 1991b, 289–297.

W. Pedrycz, Fuzzy neural networks and neurocomputations, *Fuzzy Sets and Systems*, 56, 1993, 1–28.

W. Pedrycz, F. Gomide, *An Introduction to Fuzzy Sets: Analysis and Design*, MIT Press, Boston, 1998.

W. Pedrycz, P. Lam, A.F. Rocha, Distributed fuzzy modelling, *IEEE Trans. on Systems, Man, and Cybernetics, Part B*, 5, 1995, 769–780.

W. Pedrycz, A. Rocha, Knowledge-based neural networks, *IEEE Trans. on Fuzzy Systems*, 1, 1993, 254–266.

M. Setnes, R. Babuska, H. Vebruggen, Rule-based modeling: precision and transparency, *IEEE Trans on Systems, Man, and Cybernetics, Part C*, 28, 1998, 165–169.

4 Conditional Fuzzy Clustering

So far, the idea of clustering regarded as an algorithmic framework for building information granules has not been directed by external hints (points of view or navigation guidelines). Conditional clustering is an interesting way to develop clusters by focusing on a certain portion of the original data. This gives rise to the modularization of clustering and helps carry out more focused data analysis. The provided point of focus puts the ensuing clustering in a certain context; we also refer to the proposed scheme as context-based clustering. We develop an algorithmic setting, elaborate on its application aspects (including those arising in the setting of granular models), and study the numeric efficiency of the proposed approach by looking at the decomposition (modularization) aspect of the problem of conditional clustering.

4.1. INTRODUCTION

Let us imagine a situation where, in addition to a multivariable data set X, we have been advised that our search for structure must be confined to (narrowed down to) a certain context. The idea looks attractive: not only must we focus on a subset of X where the discovery of structure looks more promising, but we can make computing more manageable, especially when dealing with large data sets. We can regard the advice (context) as a certain vehicle to decompose the problem. Given several contexts, we concentrate on the corresponding subsets of X. More descriptively, a scenario of context-based clustering can be treated as a conditional clustering task. Consider a large data set of customers of a chain store. We are interested in learning about the structure of customers (X) doing regular shopping. Many interesting questions could be posed: Is this a homogeneous population? Could we distinguish between some groups of customers falling within this category? How different are these groups? Here the characterization of a regular shopper serves as a context we would like to focus on in our cluster analysis. The task can be reformulated in the following way:

Determine structure in X under condition (context) of regular shoppers.

Similarly, we could envision a task of data clustering when the context concerns customers characterized by significant spending patterns:

Knowledge-Based Clustering, by Witold Pedrycz
ISBN 0-471-46966-1   Copyright © 2005 John Wiley & Sons, Inc.

Determine structure in **X** under condition (context) of customers characterized by significant spending.

We note that the quest for conditional clustering can be formulated in a uniform manner as follows:

Determine structure in **X** under condition (context) D.

In each case, the clustering algorithm focuses only on a subset (eventually a fuzzy subset) of **X** implied (or conditioned) by the context D. In contrast, the clustering we have discussed so far can be referred to as condition-free clustering:

Determine structure in **X**.

Quite often the context D can be treated as a fuzzy set defined in the space over which the condition has been expressed. We can also take another look at the conditional clustering treating such tasks within the domain of software agents whose activities are focused on the exploration of data structure in the given context. In fact, we are concerned with a number of software agents, each equipped with its own specific context. Thus conditional clustering can be viewed as a generic clustering task endowed with some context, with the crux of the problem illustrated in Figure 4.1. Here we portray several clustering tasks implied by different contexts (F, G, H) focusing attention on pertinent subsets of **X**.

In the sequel, each conditional clustering gives rise to a certain number of clusters (say, c, c', c'', etc.), so in essence we develop a net of information granules (fuzzy relations); see Figure 4.2. It is important to realize that by introducing contexts we imposed some directionality on our data analysis by starting from the given context and carrying out the clustering implied by it. The directionality becomes important in fuzzy modeling and will be explored intensively further on.

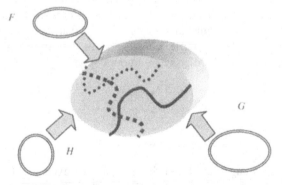

Figure 4.1. The concept of conditional (context-based) fuzzy clustering; note several individual clustering tasks running on the same data set **X** and induced by contexts F, G, and H.

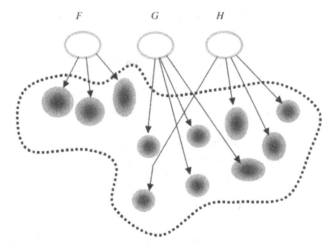

Figure 4.2. Conditional clustering in the development of a web of information granules; we construct families of clusters induced by given contexts.

In the remainder of this chapter, we present a formal formulation of context-based clustering, develop a detailed algorithm, illustrate its performance, and work on some extensions.

4.2. PROBLEM STATEMENT: CONTEXT FUZZY SETS AND OBJECTIVE FUNCTION

Formulation of the clustering problem must incorporate the context constraints. As we have learned through the illustrative examples, the context is viewed as a fuzzy set or fuzzy relation defined in the context space (viz., the space of conditional variables). The data set $\mathbf{X}$ is endowed with context F, meaning that with each $\mathbf{x}_k$, $k = 1, 2, \ldots, N$, we are provided with the value of the context (membership grade) associated with it. We denote its membership grade by f_k. In other words, we form pairs of objects $(\mathbf{x}_k, f_k)$ that are then involved in the clustering process. As an example, consider $\mathbf{x}_k$ as descriptors of customers. We record each transaction where z_k is an amount spent. This gives rise to the data in the following format:

$$
\begin{matrix}
\mathbf{x}_1 & z_1 \\
\mathbf{x}_2 & z_2 \\
\ldots & \\
\mathbf{x}_N & z_N
\end{matrix}
\tag{4.1}
$$

Over the amount being spent by each customer we express a context fuzzy set of "significant spending." Denote it by F; see also Figure 4.3.

Then each z_k can be transformed through F by computing its membership degree, $f_k = F(z_k)$. The original data set (4.1) is now augmented by the context

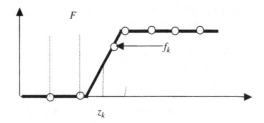

Figure 4.3. Construction of a context fuzzy set (F) with constraints f_k by transforming z_k's through the given context fuzzy set F (significant spending).

(condition) and is expressed as pairs of data:

$$
\begin{aligned}
&\mathbf{x}_1 \ f_1 \\
&\mathbf{x}_2 \ f_2 \\
&\ldots \\
&\mathbf{x}_N \ f_N
\end{aligned}
\tag{4.2}
$$

The data set in this format becomes a starting point of the clustering procedure. The formal formulation of the problem has to be expressed in such a way that we directly take into account the context values f_k. The objective function (Q) assumes the standard form

$$
Q = \sum_{k=1}^{N} \sum_{i=1}^{c} u_{ik}^{m} \| \mathbf{x}_k - \mathbf{v}_i \|^2
\tag{4.3}
$$

with the prototypes $\mathbf{v}_i$ and the partition matrix U being the components to be optimized. The constraints on the membership grades (partition matrix) are modified and incorporate the values of the context f_k. We introduce the following requirement:

$$
\sum_{i=1}^{c} u_{ik} = f_k
\tag{4.4}
$$

By doing this, we request that the membership grades of the kth pattern sum to the context constraint f_k. Note that the context set up identically to 1, $f_k = 1$ for all k, returns the original formulation of the standard FCM. The partition matrices implied by context F can be denoted by $U(F)$, with the argument being used there to emphasize the dependency upon the context constraint.

As mentioned, the context F can assume various forms. It could be a fuzzy set or a fuzzy relation (which happens with a composite requirement such as "significant spending *and* good credit record"), which is not relevant to the formulation of the optimization problem—we still perceive this constraint through the membership grades of F. The optimization problem of conditional clustering can be expressed as follows (Pedrycz, 1996; Pedrycz and Sosnowski, 2000):

$$\min_{U \in U(F), \mathbf{v}_1, \mathbf{v}_2, \dots, \mathbf{v}_c} \sum_{k=1}^{N} \sum_{i=1}^{c} u_{ik}^m \|\mathbf{x}_k - \mathbf{v}_i\|^2 \qquad (4.5)$$

where $U(F)$ denotes a family of partition matrices induced by condition F, meaning that it satisfies the normalization condition expressed by (4.4).

On the subject of the formation of fuzzy sets or fuzzy relations, it is worth stressing that these could be in two different forms:

- Expert-oriented constructs. We consider that the domain knowledge consists of linguistic terms formed by the user who intends to focus data analysis within a certain scope. The contexts are these interesting focal points. They could be dynamically adjusted, depending on the previous findings about the structure in the data set. In particular, one can move on to more specific (detailed) contexts if the structure detected so far is not informative enough. The contexts can be made more general (abstract) if the structure already revealed concerns very few patterns and cannot be made meaningful enough.
- The result of auxiliary clustering completed separately for the context variable(s). The main advantage of this approach is that these contexts are based on experimental evidence available as part of the data set.

Context-based clustering has several features that are implied by the character of the contexts. More specific contexts imply more focused, refined clustering. Given contexts A and A' such that $A' \subset A$, context-based clustering guided by A' tends to reveal more structural details than clustering guided by A. Likewise, we get a straightforward boundary condition: $D = \mathbf{X}$ leads to context-free clustering. Composite contexts are built by forming Cartesian products of their contributing components, say, $D = A \times B \times C$. The membership grade of the vector of outputs $\mathbf{z} = [z_1 \; z_2 \; z_3]^T$ is expressed as $D(\mathbf{z})$. When A, B, and C are specified in an explicit manner, we obtain $D(\mathbf{z}) = A(z_1) \; t \; B(z_2) \; t \; C(z_3)$, with a certain t-norm used to model the type of aggregation of the coordinates.

4.3. THE OPTIMIZATION PROBLEM

The optimization algorithm involves two phases: computations of the partition matrix and the prototypes. The latter is straightforward: we end up with the same formula as in the standard FCM. This is not surprising. We compute the prototypes based on the objective function, which does not depend upon the context constraints. The calculations of the partition matrix are concerned with the use of the context constraints. Following a standard technique of Lagrange multipliers, we form the expression

$$V = \sum_{i=1}^{c} u_{ik}^m \|\mathbf{x}_k - \mathbf{v}_i\|^2 - \lambda \left(\sum_{i=1}^{c} u_{ik} - f_k \right) \qquad (4.6)$$

which holds for each pattern; $k = 1, 2, \ldots, N$. The minimization of V is completed with respect to the elements of the partition matrix. The constraint f_k enters the picture and gives rise to the following expression:

$$u_{ik} = \frac{f_k}{\displaystyle\sum_{j=1}^{c} \left(\frac{\|\mathbf{x}_k - \mathbf{v}_i\|}{\|\mathbf{x}_k - \mathbf{v}_j\|} \right)^{2/(m-1)}} \tag{4.7}$$

As noted, the calculations of the prototypes are the same as those in context-free FCM. Interestingly, the contexts do not show up explicitly in the formula. Obviously they still impact the prototypes, and this happens through the values of the partition matrix.

Let us move on to some illustrative examples.

Example 1. We start with a simple two-dimensional set of 10 patterns with the given context variable, which assumes values for the patterns given in Table 4.1.

The plot of the patterns is included in Figure 4.4.

TABLE 4.1. Two-Dimensional Patterns to Be Clustered Along with Their Context Values (f_k)

x_{k1}	0.6	1.5	1.6	2.2	3.5	5.1	5.0	6.1	7.8	7.3
x_{k2}	0.9	2.0	1.8	1.5	3.3	5.4	4.7	5.4	9.6	8.9
f_k	0.8	0.7	1.0	1.0	0.5	0.2	1.0	1.0	0.9	0.6

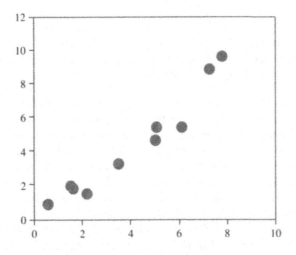

Figure 4.4. Two-dimensional patterns.

We consider $c = 2$ clusters and carry out the series of iterations starting from a random configuration of the partition matrix. The obtained partition matrix is

$$\begin{bmatrix} 0.72 & 0.08 \\ 0.70 & 0.00 \\ 1.00 & 0.00 \\ 1.00 & 0.00 \\ 0.42 & 0.08 \\ 0.05 & 0.15 \\ 1.00 & 0.00 \\ 0.14 & 0.86 \\ 0.04 & 0.86 \\ 0.00 & 0.60 \end{bmatrix}$$

The prototypes of the clusters are equal to $\mathbf{v}_1 = [\,1.71 \quad 1.68\,]$ and $\mathbf{v}_2 = [\,6.98 \quad 7.72\,]$.

It is instructive to compare these results with the structure produced by 'standard' (viz., context-free) clustering, where f_k are set to 1 for all patterns. Again, considering $c = 2$ and starting with the same random configuration of the initial values of the partition matrix, we produce the following results:

$$\begin{bmatrix} 0.96 & 0.04 \\ 1.00 & 0.00 \\ 1.00 & 0.00 \\ 1.00 & 0.00 \\ 0.82 & 0.18 \\ 0.15 & 0.85 \\ 0.27 & 0.73 \\ 0.07 & 0.93 \\ 0.08 & 0.92 \\ 0.05 & 0.95 \end{bmatrix}$$

with the prototypes being equal to $\mathbf{v}_1 = [\,1.86 \quad 1.89\,]$ and $\mathbf{v}_2 = [\,6.38 \quad 7.00\,]$. In the most visible way, we end up with different prototypes that have been shifted because of the existing context constraints.

Example 2. We consider a Boston housing data set that concerns a number of houses characterized by several features. As each house comes with its own price, it is of interest to treat the price as a conditional variable that sets up a context (framework) for the search of the structure in the data. In particular, it would be of interest to learn about the characteristics of the houses, depending on their price. We set up three contexts for the price variable by considering linguistic terms of *low, medium,* and *high* price. All of them are described by some membership functions and shown in Figure 4.5.

For each context we consider three clusters. Again, this number is arbitrary fashion and can be revisited in the course of the experiments. The fuzzification factor is set up to 2 and the clustering is run for 20 iterations.

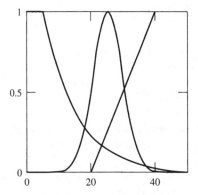

Figure 4.5. Membership functions of three contexts of conditional clustering: *small*, *medium*, and *high*.

The corresponding prototypes computed for the contexts are as follows:

Price is *low*:

v_1 = [15.53   0.57   18.17   0.02   0.68   6.04   90.64   2.00   21.70   640.86
          19.94   155.63   22.61]

v_2 = [1.79   6.14   9.69   0.02   0.53   5.99   71.12   4.19   5.34   342.22
          19.05   373.33   13.85]

v_3 = [17.11   0.43   18.26   0.01   0.68   5.72   94.60   1.79   21.95   645.65
          20.01   351.93   23.49]

Price is *medium*:

v_1 = [0.75   6.59   8.25   0.07   0.50   6.38   57.71   4.01   5.95   318.04
          18.09   384.31   8.98]

v_2 = [0.26   41.76   5.04   0.03   0.44   6.57   32.65   6.39   4.95   312.12
          17.27   386.63   6.67]

v_3 = [1.68   5.33   10.96   0.12   0.54   6.34   70.01   3.41   8.49   380.26
          18.05   374.80   10.11]

Price is *high*:

v_1 = [1.06   9.06   7.35   0.14   0.52   7.33   69.55   3.29   6.85   316.61
          16.84   385.49   5.58]

v_2 = [0.23   65.18   3.47   0.06   0.43   7.25   31.67   6.02   4.18   302.23
          15.93   390.11   4.43]

$$\mathbf{v}_3 = [\,1.27 \quad 9.50 \quad 8.21 \quad 0.17 \quad 0.53 \quad 7.31 \quad 71.82 \quad 3.15 \quad 7.31 \quad 331.20$$
$$16.68 \quad 384.26 \quad 5.63\,]$$

The membership grades organized in successive rows of the partition matrix for the last context are illustrated in Figures 4.6 to 4.8. This gives us a certain view of the revealed structure. We can learn which patterns form the corresponding clusters.

As the feature space is multidimensional, it is more illustrative to underline some variables that seem to be quite descriptive of the characterization of the real estate.

The web of the information granules visualizing the contexts and the associated prototypes generated by the corresponding contexts is shown in Figure 4.9. It shows that with the increasing price of housing the quality of the environment goes up: the crime rate decreases, the nitric oxide concentration decreases, and the student-teacher ratio changes significantly.

The clustering can be repeated for different number of clusters for each context, which helps us learn about the induced structure of the parts of the data set involved in the clustering process. The plot in Figure 4.10 shows how the objective function varies with respect to the number of clusters c. Interestingly, while we note that its value decreases, there seem to be "critical" values of c, and searching for more detailed structure (more induced clusters) does not reveal too much.

Example 3. Here we consider synthetic two-dimensional patterns whose inputs (x_1, x_2) are shown in Figure 4.11, while the output variable is shown in Figure 4.12.

The context fuzzy sets have trapezoidal (T) membership functions, and we consider four of them (the parameters of these membership functions denote its lower bound, region of modal values—two intermediate numbers on the list and the upper bound):

$$T_1(x; 8, 9, 12, 15) \qquad T_2(x; 12, 15, 23, 29)$$

$$T_3(x; 23, 29, 39, 32) \qquad T_4(x; 39, 42, 50, 55)$$

The prototypes obtained for different number of clusters ($c = 2$, 3, and 4) are equal to:

Context T_1:

$c = 2$: $\mathbf{v}_1 = [\,152.26 \quad 13.57\,]^T$ $\mathbf{v}_2 = [\,199.65 \quad 12.75\,]^T$

$c = 3$: $\mathbf{v}_1 = [\,181.83 \quad 11.52\,]^T$ $\mathbf{v}_2 = [\,205.80 \quad 14.41\,]^T$ $\mathbf{v}_3 = [\,148.23 \quad 13.77\,]^T$

$c = 4$: $\mathbf{v}_1 = [\,167.91 \quad 12.10\,]^T$ $\mathbf{v}_2 = [\,205.31 \quad 14.64\,]^T$ $\mathbf{v}_3 = [\,147.07 \quad 14.04\,]^T$
$\mathbf{v}_4 = [\,208.69 \quad 10.92\,]^T$

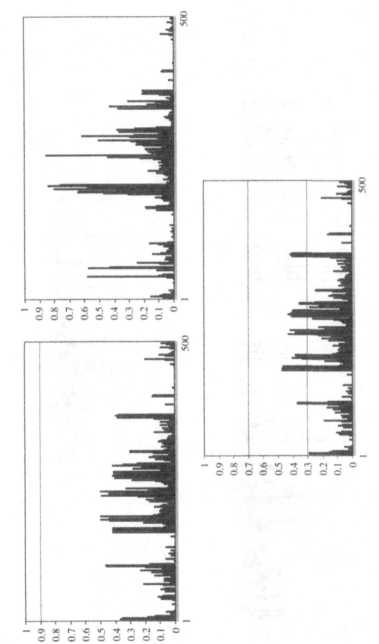

Figure 4.6. Membership grades of clusters induced by the *high* price of real estate.

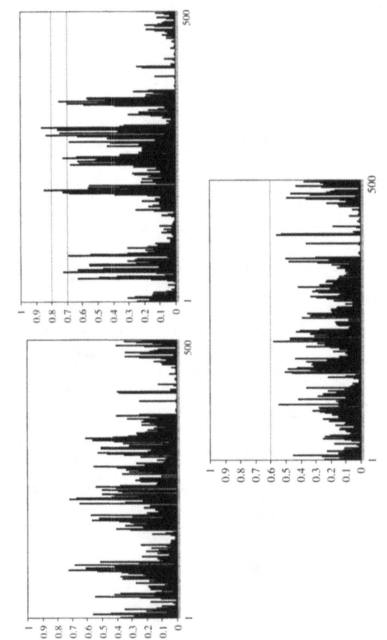

Figure 4.7. Membership grades of clusters induced by the *medium* price of real estate.

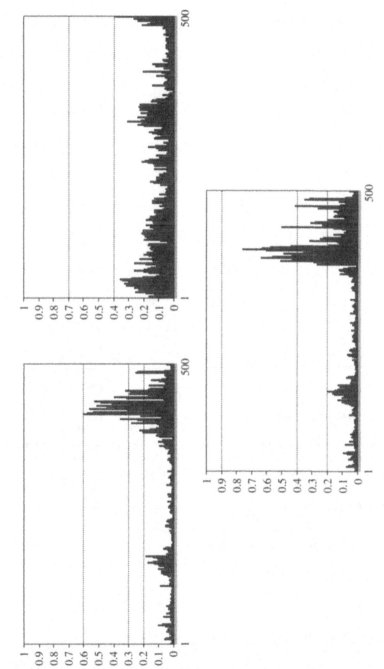

Figure 4.8. Membership grades of clusters induced by the *low* price of real estate.

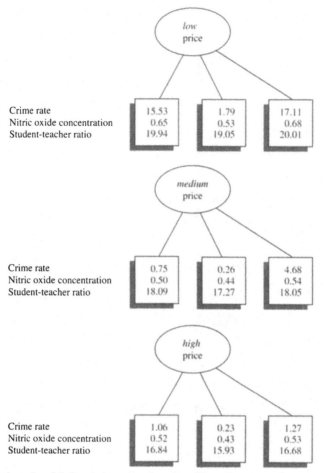

Figure 4.9. A web of information granules describing Boston real estate with prices quantified as *low, medium,* and *large* (characterization in terms of the indicators of quality of life: crime rate, nitric oxide concentration, and student-teacher ratio).

Context T_2:

$c = 2$: $\mathbf{v}_1 = [\,148.30 \quad 12.67\,]^T$ $\mathbf{v}_2 = [\,96.06 \quad 16.73\,]^T$
$c = 3$: $\mathbf{v}_1 = [\,103.28 \quad 15.14\,]^T$ $\mathbf{v}_2 = [\,91.21 \quad 18.38\,]^T$ $\mathbf{v}_3 = [\,155.88 \quad 12.23\,]^T$
$c = 4$: $\mathbf{v}_1 = [\,140.10 \quad 13.41\,]^T$ $\mathbf{v}_2 = [\,90.34 \quad 18.68\,]^T$ $\mathbf{v}_3 = [\,176.02 \quad 10.92\,]^T$
        $\mathbf{v}_4 = [\,99.21 \quad 15.51\,]^T$

Context T_3:

$c = 2$: $\mathbf{v}_1 = [\,79.10 \quad 14.84\,]^T$ $\mathbf{v}_2 = [\,68.53 \quad 18.56\,]^T$
$c = 3$: $\mathbf{v}_1 = [\,81.21 \quad 14.27\,]^T$ $\mathbf{v}_2 = [\,67.12 \quad 19.38\,]^T$ $\mathbf{v}_3 = [\,72.01 \quad 16.47\,]^T$
$c = 4$: $\mathbf{v}_1 = [\,80.34 \quad 15.34\,]^T$ $\mathbf{v}_2 = [\,66.66 \quad 19.58\,]^T$ $\mathbf{v}_3 = [\,81.36 \quad 13.88\,]^T$
        $\mathbf{v}_4 = [\,69.02 \quad 16.97\,]^T$

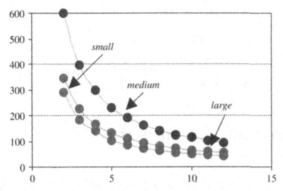

Figure 4.10. The values of objective functions treated as a function of c for the three contexts defined in the price of real estate.

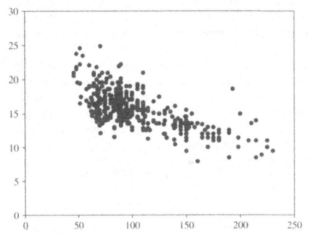

Figure 4.11. Two-dimensional synthetic data.

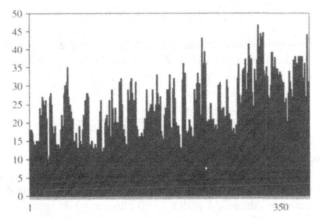

Figure 4.12. The output data over which context fuzzy sets are defined.

Context T_4:

$$c = 2: \quad \mathbf{v}_1 = [\, 68.98 \quad 15.38\,]^T \quad \mathbf{v}_2 = [\, 49.56 \quad 22.72\,]^T$$
$$c = 3: \quad \mathbf{v}_1 = [\, 70.55 \quad 14.17\,]^T \quad \mathbf{v}_2 = [\, 49.07 \quad 23.04\,]^T \quad \mathbf{v}_3 = [\, 64.49 \quad 18.34\,]^T$$
$$c = 4: \quad \mathbf{v}_1 = [\, 65.07 \quad 18.15\,]^T \quad \mathbf{v}_2 = [\, 48.05 \quad 21.60\,]^T \quad \mathbf{v}_3 = [\, 70.54 \quad 14.15\,]^T$$
$$\quad \mathbf{v}_4 = [\, 50.07 \quad 24.16\,]^T$$

4.4. COMPUTATIONAL CONSIDERATIONS OF CONDITIONAL CLUSTERING

The context introduced into the clustering leads to problem decomposition. Rather than dealing with the complete data set all at once, we decompose it by using some context and focus on its corresponding segment. This segmentation, or decomposition of data before clustering, helps reduce the computational effort (Hirota and Pedrycz, 1999). For the sake of argument, let us assume that the data set is split into P subsets, each of which is partitioned into c/P clusters (we assume that this ratio is an integer number). In this way, P subsets taken together have the same number of clusters as the overall data set. The original data set is now dealt with in the reduced form consisting of N/P patterns (again, we assume that this ratio returns an integer number). As most computing is concerned with the determination of the partition matrix $[u_{ik}]$, we use this in calculations of the computational effort. For the entire data set with N patterns and c clusters, the determination of the partition matrix requires $c^*\,(c+1)^*N$ computing of the distance function. The cost of computing a single entry of the partition matrix is assumed to be μ. In total, we obtain

$$T_0 = c^*(c+1)^*N^*\mu \tag{4.8}$$

For a single subset of patterns, their partition matrix requires the computing effort equal to

$$T_i = c'^*(c'+1)^*N'^*\mu \tag{4.9}$$

$i = 1, 2, \ldots, P$. Here $c' = c/P$ and $N' = N/P$. As we are concerned with P subsets, the overall effort leads to $T = P^*T_i$. To quantify how much improvement has been gained, we compute the ratio $\kappa = T/T_0$:

$$\kappa = \frac{\frac{N^*c}{P}\left(\frac{c}{P}+1\right)^*\mu}{c^*(c+1)^*N^*\mu} = \frac{1}{P}*\frac{\left(\frac{c}{P}+1\right)}{c+1} \tag{4.10}$$

The value of this relationship depends upon the number of clusters (c) and the number of splits of the data set (P). To get a sense of the reduction of the computing effort, let us plug in some numbers. Take $c = 8$ and $P = 4$; this yields $\kappa = 0.083$. If we are concerned with $c = 50$ clusters while splitting the data set into $P = 5$ subsets, this combination produces $\kappa = 0.043$ The split to $P = 10$ sets

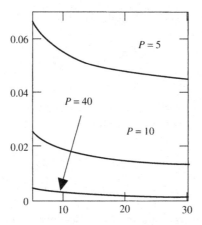

Figure 4.13. Plots of κ as a function of c for selected values of P.

while we are still looking for the same total number of clusters yields $\kappa = 0.012$. The detailed graphs of the above relationship are shown in Figure 4.13.

It is interesting to note that the decomposition of the data set into smaller segments is highly beneficial from the computational standpoint. One should be aware that the results of such decomposition tend to become more "brittle," as each subset of data within a certain context becomes processed independently of the others. Obviously, some interaction is still retained because of the overlap between the context fuzzy sets, but this tends to be confined to the relatively small segments of the output space.

4.5. GENERALIZATIONS OF THE ALGORITHM THROUGH THE AGGREGATION OPERATOR

The context-driven nature of this class of fuzzy clustering can be generalized in several different ways. The first thing to note is that the constraint formed by the context that is $\sum\limits_{i=1}^{c} u_{ik} = f_k$ is nothing but a certain equation involving a collection of the corresponding membership grades. Let us rewrite the above expression in the following form:

$$\frac{1}{c}\sum_{i=1}^{c} u_{ik} = \frac{1}{c}f_k \qquad (4.11)$$

Here we can immediately recognize that the mean value of the corresponding row of the partition matrix is equal to the average context value:

$$\overline{u}_k = \overline{f}_k \qquad (4.12)$$

where the overbar denotes a mean value. Put it differently,

$$c\left(\frac{1}{c}\sum_{i=1}^{c}u_{ik}\right) = f_k \qquad (4.13)$$

This observation led to generalization of the aggregation of the membership grades to the weighted generalized mean, where we have (Leski, 2003)

$$\text{Ave}(u_{1k}, u_{2k}, \ldots, u_{ck}; \alpha, \beta_1, \beta_2, \ldots, \beta_c) = \left(\sum_{i=1}^{c}\beta_i(u_{ik})^\alpha\right)^{\frac{1}{\alpha}} \qquad (4.14)$$

The weighted generalized mean, as shown above (Dyckhoff and Pedrycz, 1984), comes equipped with the weights $\beta_i \geq 0$ that sum to 1 and exponent $\alpha \in \mathbf{R}\backslash\{0\}$. This aggregation operator can be treated as a general logic connective of membership values. It encompasses a number of well-known cases, such as the weighted arithmetic mean ($\alpha = 1$), weighted geometric mean ($\alpha \to 0$), and weighted harmonic mean ($\alpha = -1$ and $u_{ik} \neq 0$).

Having this aggregation criterion, the derivations of detailed context-based clustering (Leski, 2003) lead to the following expressions for the entries of the partition matrix:

$$u_{ik} = \frac{\dfrac{f_k}{c}}{\left[\displaystyle\sum_{j=1}^{c}\beta_j\left(\frac{\|\mathbf{x}_k - \mathbf{v}_i\|}{\|\mathbf{x}_k - \mathbf{v}_j\|}\right)^{\frac{2\alpha}{1-\alpha}}\right]^{\frac{1}{\alpha}}} \qquad (4.15)$$

with the general flow of optimization remaining the same as that presented in the earlier discussion.

Some other types of logic operators, such as ordered weighted average (OWA), are also worth pursuing (Karayiannis, 2000; Pedrycz, 1997).

4.6. FUZZY CLUSTERING WITH SPATIAL CONSTRAINTS

So far, the generic objective functions used to guide the clustering process view individual patterns as unrelated entities. This means that we treat them as completely independent, and cluster allocation of one pattern has no impact on the allocation of any other patterns. In some situations, we may envision a natural spatial or temporal neighborhood. For instance, if there are two closely located pixels or regions of an image, we may expect them to be potentially linked together in terms of their cluster membership. In other words, their cluster allocation could be potentially similar; otherwise, we may encounter a great surprise. Analogously, if we are concerned with a time series and its differences, the samples of such signals taken in successive discrete time moments could potentially have similar membership grades or at least their cluster assignment should not be radically different.

The formulation of the clustering problem of this nature requires some modifications to the objective function so that the information about spatial, temporal, or some other form of closeness (neighborhood) is incorporated into the function. The augmented objective function comes in the following form:

$$Q = \sum_{i=1}^{c}\sum_{k=1}^{N} u_{ik}^2 d_{ik}^2 + \beta \sum_{i=1}^{c}\sum_{k=1}^{N}\sum_{l=1}^{N}(u_{ik}-u_{il})^2 \psi_{kl} d_{ik}^2 \tag{4.16}$$

The second term of this function weighted by some nonnegative weight factor β deals with the differences between patterns that have been identified by the neighborhood (vicinity) relationship. It identifies the pairs of patterns that are deemed similar because of their neighborhood Ω. In the simplest case, we can assume that the neighborhood function has a binary character such that

$$\psi_{kl} = \begin{cases} 1 & \text{if } \mathbf{x}_k \text{ and } \mathbf{x}_l \text{ belong to } \Omega \\ 0 & \text{otherwise} \end{cases} = \begin{cases} 1 & \text{if } |k-l| \le \varepsilon \\ 0 & \text{otherwise} \end{cases} \tag{4.17}$$

so the difference $u_{ik}-u_{il}$ becomes relevant if the pair k and l gives rise to the value equal to 1. In other words, the difference $|k-l|$ should no exceed some threshold level ε. Obviously there are many other candidates for the neighborhood function that may reflect the specificity of the problem. In general, this function should have finite support. The reader may refer to other available alternatives when formulating the objective function with some spatial constraints (Liew et al., 2000, Liew and Yan, 2001, 2003; Pham, 2001, 2002).

The optimization is completed in the usual manner and is divided into two tasks: calculation of the partition matrix and the prototypes. Starting with the optimization of the partition matrix, we use the technique of Lagrange multipliers and consider each pattern separately. For the tth pattern we have the constraint-free objective function

$$V = \sum_{i=1}^{c} u_{it}^2 d_{it}^2 + \beta \sum_{i=1}^{c}\left\{\sum_{\substack{l=1\\l\neq t}}^{N}(u_{it}-u_{il})^2 \psi_{tl} d_{it}^2 + \sum_{\substack{k=1\\k\neq t}}^{N}(u_{ik}-u_{it})^2 \psi_{kt} d_{ik}^2\right\}$$

$$- \lambda\left(\sum_{i=1}^{c} u_{it} - 1\right) \tag{4.18}$$

Taking its partial derivative with respect to u_{st} and zeroing it, we obtain

$$\frac{\partial V}{\partial u_{st}} = 2u_{st}d_{st}^2 + 2\beta \left\{ \sum_{\substack{l=1 \\ l \neq t}}^{N} (u_{st} - u_{sl})\psi_{tl}d_{st}^2 - \sum_{\substack{k=1 \\ k \neq t}}^{N} (u_{sk} - u_{st})\psi_{kt}d_{sk}^2 \right\} - \lambda = 0$$

(4.19)

Then, after regrouping, we produce the following expression:

$$u_{st} \left[2d_{st}^2 + 2\beta \sum_{\substack{l=1 \\ l \neq t}}^{N} \psi_{tl}d_{st}^2 - 2\beta \sum_{\substack{k=1 \\ k \neq t}}^{N} \psi_{kt}d_{sk}^2 \right]$$

$$- 2\beta \sum_{\substack{l=1 \\ l \neq t}}^{N} u_{sl}\psi_{tl}d_{st}^2 - 2\beta \sum_{\substack{k=1 \\ k \neq t}}^{N} u_{sk}\psi_{kt}d_{sk}^2 - \lambda = 0 \qquad (4.20)$$

With some additional notation

$$A_{st} = 2 \left[d_{st}^2 + \beta \sum_{\substack{l=1 \\ l \neq t}}^{N} \psi_{tl}d_{st}^2 - \beta \sum_{\substack{k=1 \\ k \neq t}}^{N} \psi_{kt}d_{sk}^2 \right]$$

$$B_{st} = 2\beta \left[\sum_{\substack{l=1 \\ l \neq t}}^{N} u_{sl}\psi_{tl}d_{st}^2 + \sum_{\substack{k=1 \\ k \neq t}}^{N} u_{sk}\psi_{kt}d_{sk}^2 \right] \qquad (4.21)$$

Then we rewrite the expression above in the format

$$u_{st}A_{st} - B_{st} - \lambda = 0$$

and then

$$u_{st} = \frac{\lambda + B_{st}}{A_{st}}$$

Given the constraint imposed on the partition matrix, one has $\sum_{i=1}^{c} u_{it} = 1$, and from this

$$\sum_{i=1}^{c} \frac{\lambda + B_{it}}{A_{it}} = 1$$

Hence

$$\lambda = \frac{1 - \sum_{i=1}^{c} \dfrac{B_{it}}{A_{it}}}{\sum_{i=1}^{c} \dfrac{1}{A_{it}}} \qquad (4.22)$$

Finally, the partition matrix is in the form

$$u_{st} = \frac{1 + \sum_{i=1}^{c} \dfrac{B_{st} - B_{it}}{A_{it}}}{\sum_{i=1}^{c} \dfrac{A_{st}}{A_{it}}} \qquad (4.23)$$

Confining ourselves to the Euclidian distance, we compute the prototypes. As they do not involve any constraints, the computations follow the zero gradient of the objective function taken with respect to the prototypes $\mathbf{v}_s$, $\nabla_{\mathbf{v}_s} Q = \mathbf{0}$. To make the calculations more explicit, we rewrite the objective function as follows:

$$Q = \sum_{i=1}^{c} \sum_{k=1}^{N} u_{ik}^2 \sum_{j=1}^{n} (x_{kj} - v_{ij})^2 + \beta \sum_{i=1}^{c} \sum_{k=1}^{N} \sum_{l=1}^{N} (u_{ik} - u_{il})^2 \psi_{kl} \sum_{j=1}^{n} (x_{kj} - v_{ij})^2 \qquad (4.24)$$

Then, taking the derivative $\dfrac{\partial Q}{\partial v_{st}}$, $s = 1, 2, \ldots, c$; $t = 1, 2, \ldots, N$ to be equal to 0, we obtain the expression for the corresponding coordinates of the prototypes:

$$-2 \sum_{k=1}^{N} u_{sk}^2 (x_{kt} - v_{st}) - 2\beta \sum_{k=1}^{N} \sum_{l=1}^{N} (u_{sk} - u_{il})^2 \psi_{kl} (x_{kt} - v_{st}) = 0 \qquad (4.25)$$

Regrouping the terms produces

$$v_{st} \left[\sum u_{sk}^2 + \beta \sum_{k=1}^{N} \sum_{l=1}^{N} (u_{sk} - u_{sl})^2 \psi_{kl} \right]$$
$$= \sum u_{sk}^2 x_{kt} + \beta \sum_{k=1}^{N} \sum_{l=1}^{N} (u_{sk} - u_{sl})^2 \psi_{kl} x_{kt} \qquad (4.26)$$

which leads to the following expression:

$$\mathbf{v}_s = \frac{\sum u_{sk}^2 \mathbf{x}_k + \beta \sum_{k=1}^{N} \sum_{l=1}^{N} (u_{sk} - u_{sll})^2 \psi_{kl} \mathbf{x}_k}{\sum u_{sk}^2 + \beta \sum_{k=1}^{N} \sum_{l=1}^{N} (u_{sk} - u_{sl})^2 \psi_{kl}} \qquad (4.27)$$

4.7. CONCLUSIONS

Conditional clustering is a form of directed clustering where a direction or focus is achieved through some context fuzzy set. This results in knowledge-based navigation of clustering, as the origin of the context is usually a direct manifestation of domain knowledge available in the problem. Obviously, the context fuzzy sets or relations could be sought as a result of some auxiliary clustering completed in advance on an external data set.

REFERENCES

H. Dyckhoff, W. Pedrycz, Generalized means as a model of compensative connectives, *Fuzzy Sets and Systems*, 14, 1984, 143–154.

K. Hirota, W. Pedrycz, Fuzzy sets in data mining, *Proc. of the IEEE*, 87, 1999, 1575–1600.

N.B. Karayiannis, Soft learning vector quantization and clustering algorithms based on ordered weighted aggregation operators, *IEEE Trans. on Neural Networks*, 11, 2000, 1093–1105.

J.M. Leski, Generalized weighted conditional fuzzy clustering, *IEEE Trans. on Fuzzy Systems*, 11, 6, 2003, 709–715.

A.W.C. Liew, S.H. Leung, W.H. Lau, Fuzzy image clustering incorporating spatial continuity, *IEE Proc.—Vision, Image and Signal Processing*, 147, 2, 2000, 185–192.

A.W.C. Liew, H. Yan, Adaptive spatially constrained fuzzy clustering for image segmentation, *Proc. 10th IEEE Int. Conference on Fuzzy Systems*, 2–5 Dec. 2001, vol. 3, 801–804.

A.W.C. Liew, H. Yan, An adaptive spatial fuzzy clustering algorithm for 3D MR image segmentation, *IEEE Trans. on Medical Imaging*, 22, 9, 2003, 1063–1075.

W. Pedrycz, Conditional fuzzy C-Means, *Pattern Recognition Letters*, 17, 1996, 625–632.

W. Pedrycz, OWA-based computing: learning algorithms, in: R.R. Yager, J. Kacprzyk (eds.), *The Ordered Weighted Averaging Operators: Theory, Methodology and Applications*, Kluwer Academic Publishers, Boston 1997, 309–320.

W. Pedrycz, Z.A. Sosnowski, Designing decision trees with the use of fuzzy granulation, *IEEE Trans. on Systems, Man, and Cybernetics*, 30, 2, 2000, 151–159.

D.L. Pham, Spatial models for fuzzy clustering, *Computer Vision and Image Understanding*, 84, 2, 2001, 285–297.

D.L. Pham, Fuzzy clustering with spatial constraints, *Proc. Int. Conf. on Image Processing*, 22–25 Sept. 2002, vol. 2, II-65–II-68.

5 Clustering with Partial Supervision

Commonly, fuzzy clustering is used to discover structure in data on the basis of unlabeled patterns. This structure determination is guided (or predisposed) by an introductory objective function whose value is minimized through an allocation of patterns to different clusters. We have shown that different distance functions used in the objective function help focus the search for structures with specific geometry of the clusters (spheres, hyberboxes, etc.). In this chapter, we develop a scheme of partial supervision in which the subset of patterns has been labeled, and we use it for guiding the clustering achieved for the remaining part of the data set.

5.1. INTRODUCTION

Partially supervised fuzzy clustering is concerned with a subset of patterns whose labels have been provided. A mixture of labeled and unlabeled patterns may be encountered in many practical situations.

Consider a large data set of handwritten characters (say, digits extracted from various postal codes). We want to build a classifier for these characters. The structure of the character sets (groups of digits) revealed through clustering becomes helpful in the design of the classifier. The characters are not labeled; hence, the unsupervised mode of learning is an obvious alternative. Suppose that we are now provided with some knowledge-based hints, that is, a small subset of labeled digits. These characters could be labeled by an expert (those that are difficult to decipher might have been a reasonable choice). Such labeled characters can play an important role in enhancing the clustering process. More descriptively, they serve as "anchor" points when launching clustering: we expect that the structure discovered in the data will conform to the class membership of the reference (labeled) patterns. Practically, with hundreds of thousand of handwritten characters, only a small fraction of them could be labeled. This class assignment comes with an extra cost, and it is always worth while to analyze how much labeling is useful and helpful in the ensuing clustering.

Another scenario in which partial supervision could play an important role originates at the conceptual end. Consider that the patterns have been labeled,

Knowledge-Based Clustering, by Witold Pedrycz
ISBN 0-471-46966-1 Copyright © 2005 John Wiley & Sons, Inc.

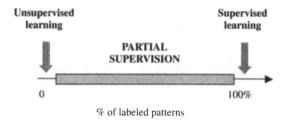

Figure 5.1. A schematic visualization of (infinite) possibilities of partial supervision quantified on the basis of the fraction of labeled patterns.

so we can conclude that they imply the use of full supervision and call for the standard mechanisms of supervised learning used in classifier design. The labeling may have been unreliable, and therefore our confidence in the already assigned labels could be questioned. Then we resort to the clustering mode and accept only a small fraction of patterns that we deem to be labeled reliably. The design scenarios similar to those presented above occur often. We need to remember that labeling is a time-consuming process and comes with an extra cost. Clustering may be far more effective. There is a spectrum of learning between "pure" models of unsupervised and supervised learning, as schematically visualized in Figure 5.1. Here the criterion discriminating among various cases is a mixture of labeled and unlabeled patterns. In the two extreme situations, 100% of patterns fall into one of the two modes.

There are more refined situations where we are provided with different formats of supervision: proximity-based guidance, entropy-oriented hints, and the like. These issues are discussed in Chapter 6. As the use of partially labeled data seems to be common, here we discuss this matter in great detail.

5.2. PROBLEM FORMULATION

Partial supervision involves a subset of labeled patterns, which come with their class membership. These knowledge-based hints have to be included in the objective function and indicate that some patterns have been labeled. In optimization, we expect that the structure to be discovered conforms to the membership grades already provided for these selected patterns. More descriptively, we can treat the labeled patterns as forming a grid of anchor points with which we attempt to discover the entire structure of the data set. Stated differently, the labeled data should help us use our discovery process. An objective function becomes an essential component in the clustering process. Bearing this in mind, we consider an additive form of the objective function that has a plausible interpretation (Pedrycz, 1985; Pedrycz and Waletzky, 1997a, 1997b):

$$Q = \sum_{i=1}^{c} \sum_{k=1}^{N} u_{ik}^2 d_{ik}^2 + \alpha \sum_{i=1}^{c} \sum_{k=1}^{N} (u_{ik} - f_{ik} b_k)^2 d_{ik}^2 \qquad (5.1)$$

The first term is used to discover the structure of data and is the same as in the standard FCM. We have used the standard notation. The second term (weighted by the scaling factor α) addresses the effect of partial supervision. It requires careful attention because of the way in which it has been introduced into the objective function and the role it plays during its optimization. There are two essential data structures containing information about the initial labeling process:

The vector of labels, denoted by $\mathbf{b} = [b_1 \ b_2 \ldots b_N]^T$. Each pattern $\mathbf{x}_k$ comes with a Boolean indicator: we make b_k equal to 1 if the pattern has been already labeled and equal to 0 otherwise.

The partition matrix $F = [f_{ik}]$, $i = 1, 2, \ldots, c; k = 1, 2, \ldots N$, which contains membership grades assigned to the selected patterns (already identified by the nonzero values of $\mathbf{b}$). If $b_k = 1$, then the corresponding column shows the provided membership grades. If $b_k = 0$, then the entries of the corresponding kth column of F do not matter; technically; we could set them equal to 0.

The nonnegative weight factor (α) helps set up a suitable balance between the supervised and unsupervised modes of learning. Apparently, when $\alpha = 0$, we end up with the standard FCM. If there are no labeled patterns ($\mathbf{b} = \mathbf{0}$), then the objective function is

$$Q = (1 + \alpha) \sum_{i=1}^{c} \sum_{k=1}^{N} u_{ik}^2 d_{ik}^2 \tag{5.2}$$

and becomes merely a scaled version of the standard objective function encountered in the FCM optimization process. If the values of α increase significantly, we start discounting any structural aspect of optimization (where properly developed clusters tend to minimize) and rely primarily on the information in the labels of the patterns. Subsequently, any departure from the values in F would lead to a significant increase in the values of the objective function.

One could consider a slightly modified version of the objective function

$$Q = \sum_{i=1}^{c} \sum_{k=1}^{N} u_{ik}^2 d_{ik}^2 + \alpha \sum_{i=1}^{c} \sum_{k=1}^{N} (u_{ik} - f_{ik})^2 b_k d_{ik}^2 \tag{5.3}$$

where the labeling vector b shows up in a slightly different format. This function captures the essence of partial supervision.

For slight variations on the issue of partial supervision, the reader is referred to the work by Bensaid et al. (1996), Abonyi and Szeifert (2003), Coppi and D'Urso (2003), Kersten (1996), Liu and Huang (2003), and Timm et al. (2002).

5.3. DESIGN OF THE CLUSTERS

As usual, the optimization of the objective function Q is completed with respect to the partition matrix and prototypes of the clusters. The first step is a constraint-based minimization, which involves Langrage multipliers to accommodate the constraints of the membership grades. The augmented objective function takes the form

$$V = \sum_{i=1}^{c} u_{ik}^2 d_{ik}^2 + \alpha \sum_{i=1}^{c} (u_{ik} - f_{ik}b_k)^2 d_{ik} - \lambda \left(\sum_{i=1}^{c} u_{ik} - 1 \right) \qquad (5.4)$$

To compute the gradient of V with respect to the partition matrix U, we note that making the value of the fuzzification factor equal to 2 is helpful; by doing that, we avoid solving a high-order polynomial equation with respect to the entries of the partition matrix.

The resulting entries of the partition matrix U take the form

$$u_{ik} = \frac{1}{1+\alpha} \left[\frac{1 + \alpha \left(1 - b_k \sum_{i=1}^{c} f_{ik} \right)}{\sum_{j=1}^{c} \left(\frac{d_{ik}}{d_{jk}} \right)^2} + \alpha f_{ik} b_k \right] \qquad (5.5)$$

Moving on to the computations of the prototypes, the necessary condition for the minimum of Q taken with respect to the prototypes takes the form $\frac{\partial Q}{\partial v_{st}} = 0$, $s = 1, 2, \ldots, c; t = 1, 2, \ldots, n$. Calculating the respective partial derivatives, one arrives at

$$\frac{\partial Q}{\partial v_{st}} = \frac{\partial}{\partial v_{st}} \left[\sum_{i=1}^{c} \sum_{k=1}^{N} u_{ik}^2 \sum_{j=1}^{n} (x_{kj} - v_{ij})^2 \right.$$

$$\left. + \alpha \sum_{i=1}^{c} \sum_{k=1}^{N} (u_{ik} - f_{ik}b_k)^2 \sum_{j=1}^{n} (x_{kj} - v_{ij})^2 \right] \qquad (5.6)$$

$$= \frac{\partial}{\partial v_{st}} \left[\sum_{i=1}^{c} \sum_{k=1}^{N} [u_{ik}^2 + \alpha(u_{ik} - f_{ik}b_k)^2] \sum_{j=1}^{n} (x_{kj} - v_{ij})^2 \right]$$

Let us introduce the following shorthand notation:

$$\psi_{ik} = [u_{ik}^2 + \alpha(u_{ik} - f_{ik}b_k)^2] \qquad (5.7)$$

This leads to the optimality condition of the form

$$\frac{\partial Q}{\partial v_{st}} = 2 \sum_{k=1}^{N} \psi_{sk}(x_{kt} - v_{st}) = 0 \tag{5.8}$$

Finally, we derive

$$\boldsymbol{v}_s = \frac{\displaystyle\sum_{k=1}^{N} \psi_{sk}\boldsymbol{x}_k}{\displaystyle\sum_{k=1}^{N} \psi_{sk}} \tag{5.9}$$

5.4. EXPERIMENTAL EXAMPLES

For illustrative purposes, we consider a small, synthetic two-dimensional data set in the (x_1, x_2) plane, as shown in Figure 5.2. Partial supervision comes with the classification results of several patterns; their labels are shown in Figure 5.2 as well. These are patterns 5 and 7, with the membership grades [0.5 0.5] and pattern 10 with [0.0 1.0], respectively. These hints indicate two classes, so we set up two clusters; $c = 2$.

Clustering was completed for several increasing values of α, giving a detailed picture of the impact the classification hints have on the revealed structure of the patterns. This is shown in two different ways: by visualizing the partition matrices (Figure 5.3) and by locating the prototypes (Figure 5.4). In both cases, we note that by changing α, the discovered structure tend to conform to the classification

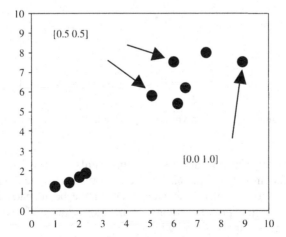

Figure 5.2. A two-dimensional synthetic data set; visualized are the classification hints — labeled patterns that are used to guide the clustering process.

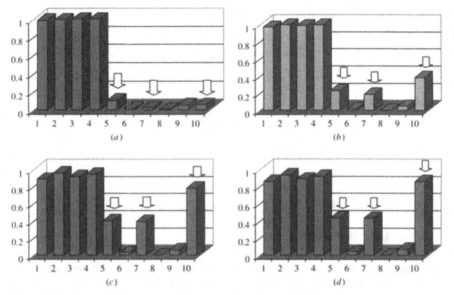

Figure 5.3. Membership grades of patterns for selected values of α: (a) $\alpha = 0.0$ (no supervision); (b) $\alpha = 0.5$; (c) $\alpha = 3.0$; (d) $\alpha = 5.0$. The arrows point at labeled patterns.

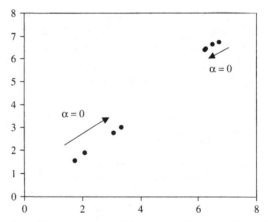

Figure 5.4. Migration of the prototypes of the clusters illustrated for selected values of α.

constraints. For reference, we have shown the results for $\alpha = 0$ when no supervision effect is taken into consideration. The prototypes (Figure 5.4) migrate from the original position when no labeled patterns have been made available. The movement of the first prototype tends to be more "plastic," while that of the second prototype is far more limited.

The alternative version of the objective function (5.3) produced very similar results; obviously, there could be some calibration of the weight factor, as its contribution to the membership functions being formed there is different from the one occurring in the previous objective function.

5.5. CLUSTER-BASED TRACKING PROBLEM

The effect of partial supervision can emerge in different ways. The one we discussed in the previous sections concerns a small subset of labeled patterns whose usage impacts the discovery of the overall structure in the data. While this is the most intuitively straightforward scenario, there are several alternatives reflecting the problem at hand. The structure-tracking problem is one of them. Figure 5.5 shows a series of snapshots of data taken in a sequence of discrete time moments. At the beginning of the tracking, a cloud of patterns is located at the origin and then starts moving along some trajectory. The shape of the migrating cloud is also affected. At some observation points the cloud becomes quite compact; at others, it expands significantly by changing from a circular to a more elongated and rotated shape. Our intent is to discover the structure of this data set. As there is a steady influx of data, clustering should encounter the dynamics of the data and react accordingly. The conceptual model we can envision is based upon the moving clusters that wander in the data space, tracking the data and capturing the main trends observed in the patterns.

What happens if the moving cloud of patterns (data) is affected by noise? To avoid the impact of noise, the consecutive clusters should "talk" to their predecessors. The intuitive requirement we adopt here is to avoid sudden jumps between two successive clusters. Smooth trajectories formed by the centers of the clusters could be taken as feasible development guidance in this clustering. The cluster-based tracking problem is a generalization of the standard tracking encountered in control, signal filtering, and the like. The partial supervision occurs through the interaction at the level of the prototypes of the clusters.

Before formulating the detailed optimization problem that casts this data analysis problem in a context of fuzzy clustering, it is helpful to set up some notation.

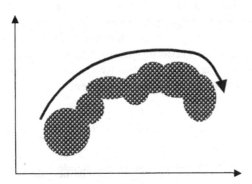

Figure 5.5. Dynamic accumulation of data: a concept of moving clusters that follow the data and communicate among themselves.

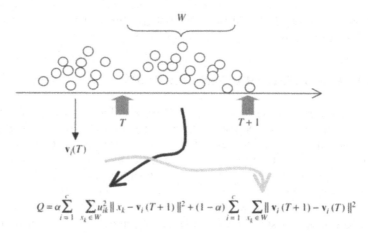

Figure 5.6. A schematic visualization of accumulation of data and their clustering.

This is of particular interest, as we are concerned with a global population-driven analysis. The data stream occurs on a continuous basis. We monitor the data and explore the structure at discrete time moments T, $2T$, $3T$, and so on. Within these intervals (windows of clustering) of length T, we see some accumulation of data (Figure 5.6).

In essence, we act on new data (those patterns accumulated during the window of clustering) while taking into consideration the structure we have developed so far during the previous window of clustering). This prudent method of using current data and relying on the structure (clusters) developed for the previous data has several benefits. First, the data collected during the current clustering window may be heavily biased by structural noise (so that a structure discovery based only on this data set could be questionable). Therefore, it is helpful to rely on the structure we designed for the current data and eventually use the previously discovered regularities (clusters) to help build those using the current data set. The simple mapping of data from two successive clustering windows is not an option because of the possible nonstationarity of data (and ensuing clusters). We can achieve a sound balance between eventual structural impurities in data structure (structural noise) and the nonstationary character of data by setting up a trade-off in the clustering problem treated as an optimization task.

As a result of our explanation, an objective function outlined below becomes a convex combination of the following components:

$$Q = \alpha \sum_{i=1}^{c} \sum_{x_k \in W} u_{ik}^2 \|\mathbf{x}_k - \mathbf{v}_i(T+1)\|^2 + (1-\alpha) \sum_{i=1}^{c} \sum_{x_k \in W} \|\mathbf{v}_i(T+1) - \mathbf{v}_i(T)\|^2$$

(5.10)

The first component determines the structure in the new data set and is a standard FCM objective function. The second component maintains the position of the prototypes. Here we rely on the "history" and try to keep the prototypes of the new data set close to the already existing prototypes constructed during the previous clustering window of the algorithm. The crucial parameter of the objective function is denoted by $\alpha \in [0,1]$. In boundary cases we end with complete adherence to the current data or a complete follow-up of the existing structure. The value of $\alpha = 1$ gives rise to the standard FCM being applied to the current data set. At the opposite extreme, the value of $\alpha = 0$ makes the prototypes $\mathbf{v}_i(T+1)$ equal to those computed for the past data. The most interesting scenarios occur for the values of α located between these two extreme situations.

Before deriving detailed formulas for the partition matrix and prototypes, it is educational to relate the tracking fuzzy clustering to the standard tracking algorithms that are well known in signal processing. In addition, note that the introduced objective function (5.10) resembles the smoothing expression applied in signal analysis and signal recovery. Consider a nonstationary discrete time series $\{x_k\}$. The intent is to "discover" the original signal on the basis of some readings $\{z_k\}$ whose values consist of the original signal affected by some noise. For instance, a typical additive model assumes that $z_k = x_k + \mu_k$, with μ_k denoting a noise component (commonly treated as a random variable that is normally distributed, with a zero mean value and a nonzero standard deviation). The estimation formula producing the recovery of the original signal is a weighted sum of the form

$$\hat{x}_{k+1} = \alpha z_{k+1} + (1 - \alpha)\hat{x}_k$$

with α playing the role of a smoothing factor. Here $\hat{x}_k$ is an estimate of the true signal, x_k. We note the same behavior of this relationship as described in clusters: high values of α place more emphasis on the current readings, which are considered to play a dominant role. A decrease in the values of α implies a stronger filtering effect and less impact of the current reading. The choice of α depends on the dynamics (nonstationarity) of the signal and the intensity of noise.

Evidently, there is a profound difference between signal processing and structure tracking. Whereas signal processing is concerned with a single signal (and individual samples), structure tracking deals with populations of data, which elevates the problem to the next higher level of sophistication.

Going back to the optimization of the objective function (5.10), we note that its second term does not explicitly involve the membership grades. In essence, we arrive at the same expression for the partition matrix as in the standard FCM. The computations of the prototypes are completed by setting up the necessary minimum condition for Q with respect to $\mathbf{v}_s$:

$$\frac{\partial Q}{\partial v_{st}} = 0, \quad s = 1, 2, \ldots, c; \ t = 1, 2, \ldots, n \tag{5.11}$$

In the sequel, this leads to the expression

$$\frac{\partial Q}{\partial v_{st}} = 2\alpha \sum_{k \in W} u_{sk}^2 (x_{kt} - v_{st}) + 2(1 - \alpha)(d_{st} - v_{st}) = 0 \tag{5.12}$$

Next, we obtain

$$\frac{\partial Q}{\partial v_{st}} = 2\alpha \sum_{k \in W} u_{sk}^2 (x_{kt} - v_{st}) + 2(1 - \alpha)(d_{st} - v_{st}) = 0 \qquad (5.13)$$

Finally

$$v_{st} = \frac{\alpha \sum_{k \in W} u_{sk}^2 x_{kt} + (1 - \alpha)d_{st}}{\alpha \sum_{k \in W} u_{sk}^2 + (1 - \alpha)} \qquad (5.14)$$

5.6. CONCLUSIONS

Supervision hints are an important component augmenting standard techniques of unsupervised learning. We have shown two typical cases of partial supervision. One involves a small subset of labeled patterns. The other concerns nonstationary data with supervision carried out at the level of prototypes of consecutive clusters. The proposed modifications of the objective functions result in their augmentation by some additive component, with a trade-off achieved by choosing a suitable value of the weight factor.

REFERENCES

J. Abonyi, F. Szeifert, Supervised fuzzy clustering for the identification of fuzzy classifiers, *Pattern Recognition Letters*, 24, 14, 2003, 2195–2207.

A.M. Bensaid, L.O. Hall, J.C. Bezdek, L.P. Clarke, Partially supervised clustering for image segmentation, *Pattern Recognition*, 29, 5, 1996, 859–871.

R. Coppi, P. D'Urso, Three-way fuzzy clustering models for LR fuzzy time trajectories, *Computational Statistics and Data Analysis*, 43, 2, 2003, 149–177.

P.R. Kersten, Including auxiliary information in fuzzy clustering, *Proc. 1996 Biennial Conf. of the North American Fuzzy Information Processing Society, NAFIPS*, 19–22 June 1996, 221–224.

H. Liu, S.T. Huang, Evolutionary semi-supervised fuzzy clustering, *Pattern Recognition Letters*, 24, 16, 2003, 3105–3113.

W. Pedrycz, Algorithms of fuzzy clustering with partial supervision, *Pattern Recognition Letters*, 3, 1985, 13–20.

W. Pedrycz, J. Waletzky, Fuzzy clustering with partial supervision, *IEEE Trans. on Systems, Man, and Cybernetics*, 5, 1997a, 787–795.

W. Pedrycz, J. Waletzky, Neural network front-ends in unsupervised learning, *IEEE Trans. on Neural Networks*, 8, 1997b, 390–401.

H. Timm, F. Klawonn, R. Kruse, An extension of partially supervised fuzzy cluster analysis, *Proc. Annual Meeting of the North American Fuzzy Information Processing Society, NAFIPS 2002*, 27–29 June 2002, 63–68.

6 Principles of Knowledge-Based Guidance in Fuzzy Clustering

As we have repeatedly stated, fuzzy clustering is synonymous with the unsupervised mode of learning used to determine structure in multidimensional data. The algorithms act on data while being directed by some predefined objective function (criterion) for which they discover a structure (clusters) that yields a minimal value of this criterion. In this chapter, we discuss the process of exploiting and effectively incorporating auxiliary problem-dependent hints that are part of the domain knowledge associated with the pattern recognition problem at hand. As such hints are usually expressed by experts/data analysts at the level of clusters (information granules) rather than individual data (patterns), we refer to them as knowledge-based indicators and allude to a set of them as knowledge-based guidance available for fuzzy clustering. The proposed paradigm shift in which fuzzy clustering incorporates this type of knowledge-based supervision is discussed and contrasted with the "pure" (that is, data-driven) version of fuzzy clustering. We introduce and discuss several types of guidance mechanisms, such as partial supervision, proximity-based guidance, and uncertainty-driven knowledge hints.

6.1. INTRODUCTION

Fuzzy clustering evolves around an objective function regarded as a optimization criterion. When such an objective function has been accepted, this criterion guides the process of forming groups of patterns. In spite of the diversity of the objective functions (e.g., distance functions, detailed algorithmic developments), all of them can be positioned in a general framework that emphasizes the underlying principle of predominant reliance on *data*, as visualized in Figure 6.1 (Hoppner et al., 1999; Jain and Dubes, 1988).

Figure 6.1 shows how the processing of data is realized by the clustering algorithm and indicates the character of communication (interaction) with the user. This communication is primarily unidirectional; the results are communicated in the form of the resulting partition matrix (or, equivalently, in a set of prototypes or centroids; obviously, the language of prototypes or partition matrices is equivalent, meaning that given one construct, we can infer the other). Note that

Knowledge-Based Clustering, by Witold Pedrycz
ISBN 0-471-46966-1   Copyright © 2005 John Wiley & Sons, Inc.

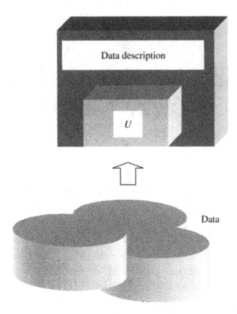

Figure 6.1. Fuzzy clustering as a data-driven search for data.

the resulting partition matrix is a direct consequence of the computing process carried out by the FCM algorithm.

The numeric data are the exclusive source of processing and guidance of the search process. In some cases, there is domain knowledge that could be incorporated to support clustering mechanisms. In a nutshell, we want to build a clustering environment based on numeric data and domain knowledge. This paradigm shift has to do with a way in which auxiliary knowledge can be injected into the clustering mechanisms. As visualized in Figure 6.2, some knowledge hints inserted by the user/analyst are accommodated at the level of results and start interacting with the FCM in an attempt to reconcile the data-driven optimization (the FCM itself) and the additional source of directing the mechanisms of clustering.

The notion of knowledge hints requires more attention. So far, we have not described them in detail (this will be done later on). We define knowledge hints as auxiliary information available at the time of data clustering and reflecting additional sources of problem domain knowledge. They may be very diversified. In general, they do not associate with all patterns but only with a small fraction of them. The hints may deal with a single pattern or pairs of patterns. The partition matrix is a reflection of information granules. Thus, any guidance is quantified and expressed in the language of fuzzy sets or fuzzy relations constructed at this level of generality.

This chapter is organized as follows. We start with several examples of knowledge-based guidance to illustrate the underlying concept and demonstrate their diversity. Then, we move on to the algorithmic issues, showing how these

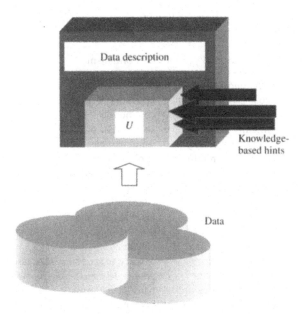

Figure 6.2. Knowledge-based paradigm shift in fuzzy clustering resulting in acceptance of various knowledge-based hints; note that the communication links point at the clustering results (e.g., partition matrix) rather than being linked with data.

hints impact the optimization algorithm and what optimization aspects requires special attention. The diversity of the hints reflects the variety of applications. To illustrate this, we discuss a specific application to the search for structure in Web pages.

6.2. EXAMPLES OF KNOWLEDGE-ORIENTED HINTS AND THEIR GENERAL TAXONOMY

Several formats of knowledge-based hints are provided by the user or data analyst. These can be problem-dependent. We elaborate on three of them that seem to be quite general and somewhat problem-independent. The first one is concerned with uncertainty of class membership. Using this format, we quantify our confidence or difficulty in the categorization (allocation) of a certain pattern. In the second category of guidance mechanisms, we encounter proximity-oriented, knowledge-based hints where we quantify knowledge about the proximity of pairs of patterns. The third is concerned with labeling of some patterns so that their class assignment is known and the hints are helpful in exploring the data and discovering their structure.

The taxonomy of knowledge hints is governed by several criteria concerning their generality, detailed knowledge, or required assumptions about the structure of the data. Table 6.1 elaborates on this in a more coherent manner.

TABLE 6.1. Knowledge-Based Hints and Their Characterization

Type of Knowledge Hint	Description	Formalism	Notes
Uncertainty	Reflects *uncertainty* about categorization of a pattern (e.g., the difficulty of assigning it to any category, borderline character of the pattern, assessing the pattern's class)	Entropy measure of fuzziness $H(.)$ commonly accepted as a suitable measure; applies to an individual pattern; no requirement concerning knowledge about the number of clusters	
Proximity	Reflects proximity between selected pairs of patterns and quantifies a subjective judgment about the closeness of some pairs of patterns	Proximity measure; applies to specified pairs of patterns; does not require any fixed number of clusters to be given in advance	Discussed in the context of Web mining; modification of the FCM to its P-FCM version (Pedrycz et al., 2004)
Labeling	Reflects the fact that some patterns are labeled (with classes assigned) and are part of the domain knowledge	Distance between provided membership grades and those contained in the partition matrix; requires the number of clusters to be specified in advance	Discussed in terms of clustering with partial supervision (Pedrycz and Waletzky, 1997)

While viewing these knowledge hints in a broader applied context, it is worthwhile to highlight the general rationale behind them.

Completeness of the Feature Space. The number of applications in which various knowledge hints become essential is directly implied by the effect of limited and incomplete feature spaces. A comprehensive feature space is an asset in any pattern recognition problem and implies potentially high recognition rates, particularly in classification tasks performed by or involving humans. Obviously, a number of essential features may not be available or cannot be easily quantified at the algorithmic end; however, these are the components that are implicitly used in human-based classification. The same argument holds in clustering: the available feature space cannot involve some of the critical features. In this case, any additional knowledge-based hints concerning the relationships between pairs of patterns or individual patterns start playing a pivotal role in enhancing the clustering activities. These hints compensate for the reduced character of the feature space. More formally, we envision the following model represented by Figure 6.3: the original feature space F ($\subset \mathbf{R}^n$) is available in its reduced version G ($\subset \mathbf{R}^m$), where usually $m \ll n$. The clustering realized in G is augmented by the logic predicates ϕ, ξ, and so on, whose active role compensates for the realization of classification activities in G. Ideally, one can anticipate achieving the close resemblance of results of clustering in F and knowledge-based clustering in G; refer again to Figure 6.3. We believe that the structures revealed in both cases are close enough, $S \approx T$.

Logic predicates (knowledge-based hints) can arise in different formats. In particular, the referential nature of the predicates can be indicated by proximity-based information about pairs of patterns (e.g., quantifications such as that two patterns are *similar*, with some level of closeness, or are very *different*). While the effect of availability of the reduced feature space is quite common, its impact is clear in the realm of Web-based exploration, classification, and organization. The crux of Web information lies in its inherently heterogeneous character: its textual layer (and resulting dictionaries) is a small fraction of what is really becomes available in the form of video, links, audio, graphics, etc. The formation of the complete associated feature space cannot be realized. This means that the navigation preferences of the user or designer about the proximity of selected Web information (pages) is essential. The studies in this involving fuzzy clustering with proximity-based hints are reported in Pedrycz et al. 2004. These papers are concerned with the development of taxonomies of Web pages based on textual information and exploiting information about proximities between selected pairs of the pages. These taxonomies are an alternative to other approaches existing in the literature (Boley et al., 1999; Guillaume and Murtargh, 2000).

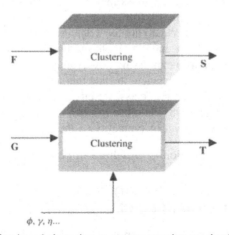

Figure 6.3. Knowledge-based clustering as a compensation mechanism for the use of the reduced feature space (G).

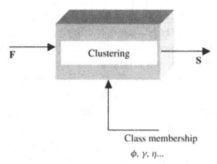

Figure 6.4. The functional aspects of knowledge-based guidance in clustering with implicit class information.

Knowledge-Based Guidance About Patterns. This category includes a variety of approaches where some information is available about class assignment of patterns. In general, the feature space becomes available in a complete form. Depending upon the form in which categories of patterns are incorporated in the knowledge-based guidelines, two essential directions are envisioned:

(a) *Explicit* information about class membership of selected patterns. For these patterns, we are given their membership grades (either by Boolean assignment or represented by some membership values). This assignment requires a fixed number of classes (categories). This situation is typical in the realm of fuzzy clustering under partial supervision (cf. Pedrycz and Waletzky, 1997a, b), where a relatively small subset of patterns has been fully labeled. This occurs, for example, in cases where labeling of all patterns is impractical but a subset of patterns can be handled quite effectively. For instance, we can label some OCR symbols, but classifying all of them is not feasible

(b) *Implicit* information about class membership. This type of information is less detailed than explicit information and focuses on the quantification of typicality of patterns. We do not have detailed information about allocation to classes; instead, we have a single numeric quantity (and thus implicit class characterization) expressing how *typical* or *relevant* a certain pattern is believed to be. For instance, to state that a pattern is a typical (representative) element of a certain class, we envision that its uncertainty measure (entropy) (Bezdek, 1981) is close to zero, $H \approx 0$. By stating that $H \approx 1$, we hint at the low level of typicality of the pattern.

The implicit class allocation is visualized in Figure 6.4.

6.3. THE OPTIMIZATION ENVIRONMENT OF KNOWLEDGE-ENHANCED CLUSTERING

Given the general observations in the previous section and the architectural investigations, we can now translate these into more operational details. This will lead us to the detailed algorithmic environment. First, we discuss a way in which ideas of uncertainty, proximity, and labeling can be incorporated into the general scheme of clustering. In what follows (as has already been envisioned in the communication scheme between data and knowledge-oriented processing), the knowledge hints are expressed at the level of the partition matrix (and specific membership grades). This becomes obvious in the following notation, with u_{ik} denoting the degree of membership of the kth pattern in the ith cluster.

Uncertainty. The typical model of uncertainty and its quantification comes in the form of an entropy function (Klir and Folger, 1988). Given a variable u assuming values in a unit interval, an entropy function $H(u)$ is defined as a continuous function from [0,1] to [0,1] such that (a) it is monotonically increasing in $[0, \frac{1}{2}]$ and (b) monotonically decreasing in $[\frac{1}{2}, 1]$ and satisfies the boundary conditions $H(0) = H(1) = 0$ $H(1/2) = 1$ (as intuitively expected, here the entropy function attains its maximum). Given a collection of membership grades $\mathbf{w} = [w_1, w_2, \ldots, w_c]^T$, the entropy easily generalizes to the form

$$H(w) = \frac{1}{c} \sum_{i=1}^{c} H(w_i) \tag{6.1}$$

with $H(w_i)$ being the entropy defined for the ith coordinate (variable). The form of the specific function coming from the class formulated above could vary. A typical example is a piecewise linear function or a quadratic function such as

$$H(u) = \begin{cases} 2u & \text{if } u \in [0, 1/2] \\ 2(1 - u) & \text{if } u \in [1/2, 1] \end{cases} \tag{6.2}$$

and

$$H(u) = 4u(1 - u) \tag{6.3}$$

where $u \in [0, 1]$. The plots of these realizations of the entropy function for two variables are shown in Figure 6.5.

Proximity Hints. Proximity is a fundamental concept to use when assessing the mutual dependency between membership occurring in two patterns. Consider two patterns with their corresponding columns in the partition matrix denoted by k and l, that is, $\mathbf{u}_k$ and $\mathbf{u}_l$, respectively. The proximity between them, Prox($\mathbf{u}_k, \mathbf{u}_l$), is defined in the form

$$\text{Prox}(\mathbf{u}_k, \mathbf{u}_l) = \sum_{i=1}^{c} \min(u_{ik}, u_{il}) \tag{6.4}$$

Note that the proximity function is symmetric and returns 1 for the same pattern ($k = l$); however, this relationship is not transitive. Given the properties of any

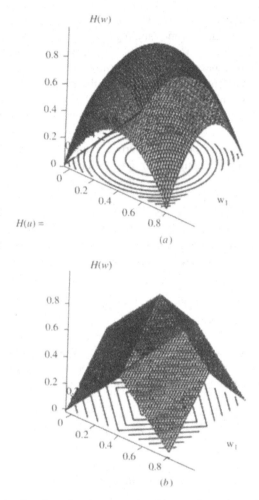

Figure 6.5. Entropy function for two membership grades: piecewise linear function (*a*) and quadratic function (*b*).

partition matrix, we immediately obtain

$$\text{Prox}(\mathbf{u}_k, \mathbf{u}_l) = \sum_{i=1}^{c} \min(u_{ik} u_{il}) = \text{Prox}(\mathbf{u}_l, \mathbf{u}_k)$$

$$\text{Prox}(\mathbf{u}_k, \mathbf{u}_k) = \sum_{i=1}^{c} \min(u_{ik} u_{ik}) = 1$$

(6.5)

Let us illustrate the concept of proximity for $c = 2$. In this case $u_{1k} = 1 - u_{2k}$, so we can confine ourselves to a single argument. The resulting plot (with the first coordinates of the patterns, u_{1k} and u_{1l}) is presented in Figure 6.6.

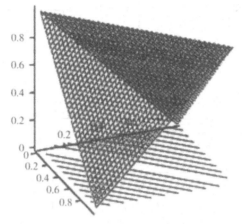

Figure 6.6. Proximity relationship as a function of entries of the partition matrix.

In addition to the proximity or uncertainty guidance expressed in terms of specific thresholds, we can envision their relaxed versions allowing for quantification in the form of containment predicates, say "*low* uncertainty level," "uncertainty not exceeding *about* 0.5," "*high* proximity," where the terms quantifying these constraints are regarded as fuzzy sets. Or, equivalently, we can relax the predicates "less than," and so on by allowing their truth values and consider these to be modeled by fuzzy sets. This makes these expressions more in keeping with the language of the user, and their use in forming the interfaces with the clustering environment contributes to the enhanced relevance and readability of the knowledge hints.

6.4. QUANTIFICATION OF KNOWLEDGE-BASED GUIDANCE HINTS AND THEIR OPTIMIZATION

As the hints are driven by preferences of the designer/data analyst, we can envision a situation in which the ensuing quantification can be more linguistic rather than purely numeric. This triggers our interest in linguistic-like quantification by admitting an interpretation of constraints in the language of fuzzy predicates. Thus the statement "x is less than λ," where both x and λ are positioned in the unit interval, can be regarded as a multivalued (fuzzy) predicate whose truth value is confined to the unit interval. Two pertinent logic models of predicates of inclusion and similarity are envisioned:

> The predicate "x is less than λ" (or, equivalently, "x does not exceed λ") returns a truth value $\tau(x$ less than $\lambda) \in [0,1]$ being a degree of inclusion of x in λ, $x \in \lambda$. In the language of fuzzy sets, this inclusion constraint is modeled by any implication that $x \rightarrow \lambda$, where this operator has three

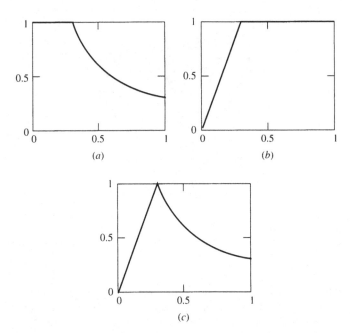

Figure 6.7. Example of the inclusion (a), dominance (b), and similarity (c) predicates (t-norm: product). The threshold level (λ) is equal to 0.3.

self-explanatory properties: (a) it returns 1 if $x \leq \lambda$; (b) it is a monotonically decreasing function of x, that is, for a fixed value of λ, the decrease in the value of x returns a higher value of the predicate. The logic-motivated model of this predicate is defined as $x \ \varphi \lambda l = x \rightarrow l = \sup(c \in [0, 1] | xtc \leq \lambda\}$, where t denotes a t-norm being used in fuzzy sets as a model of some *and* operator (*and* connective). Let us recall (see Chapter 2) that a t-norm is a two-argument function $t: [0, 1]^2 \rightarrow [0, 1]$ such that it is (a) symmetric, (b) commutative, (c) monotonically increasing, and (d) meets boundary conditions $0ta = 0$ and $1ta = a$, which are the ones satisfied for the *and* connectives in two-valued logic. The specific form of the t-norm depends upon the choice of the t-norm. For instance, for the t-norm treated as an algebraic product, we obtain $xt\lambda = x\lambda$. The plots of the predicates for some fixed values of λ are visualized in Figure 6.7. Interestingly, if x exceeds λ, the predicate reflects this effect of partial satisfaction of inclusion by returning lower truth values of the predicate.

With the inclusion predicate already defined, we can define a dominance predicate in which we determine the degree to which x dominates λ. In essence, the order of the variables is swapped and we have $\lambda \varphi x$.

The similarity predicate is built upon the already introduced inclusion predicate by combining these two, that is, $sim(x, a) = incl(x, a)t \ incl(a, x)$. The origin of this operation can be found in set theory; note that two sets A and

B are deemed equal (similar) if A is included in B *and* vice versa. With the inclusion operation already in place, we get $A \subset B$ and $B \subset A$.

6.5. ORGANIZATION OF THE INTERACTION PROCESS

As fuzzy clustering and incorporation of knowledge-based hints are the two streams of cooperating and intertwined activities, we can portray the following scheme of computing.

Fuzzy Clustering. We deal here with a general class of clustering techniques that return results of clustering arranged as a partition matrix U. Minimization of the underlying objective function induces the clustering to become a certain minimization problem, min $Q(U)$, where minimization is carried out for U as well as for the associated prototypes (centroids) of the clusters. The number of clusters (c) is specified in advance. The minimization of Q is a result of an iterative process: we cycle through computations of the partition matrix and the resulting prototypes.

Knowledge-Based Hints. Here we are concerned with optimization of the logic predicates leading to the maximization of the assumed performance index. Accommodation of the knowledge hints is achieved by maximization of the truth value of the corresponding predicates realized with respect to the membership grades (entries of the partition matrix), max $\mathcal{P}(U)$ where $\mathcal{P}$ stands for the general form of the predicate (inclusion or similarity) computed over the entire set of patterns and the associated maximum computed over the partition matrix.

Before proceeding with the detailed learning schemes, it is useful to build a general taxonomy of the logic predicates and the knowledge-based hints of entropy and proximity (see Table 6.2).

The condition that the entropy or proximity exceeds some threshold is given by repeating the first column of Table 6.2 with the reversed order of the arguments of the predicate so that we have

$$V = \sum_{k=1}^{N} \varphi[\lambda_k, H(\mathbf{u}_k),]b_k \tag{6.6}$$

and

$$V = \sum_{k_1=1}^{N} \sum_{\substack{k_2=1, \\ k_1 \geq k_{1_l}}}^{N} \varphi[\lambda_{k_1,k_2}, \text{Prox}(u_{k_1}, u_{k_2})]b_{k_1,k_2} \tag{6.7}$$

respectively.

TABLE 6.2. A General Taxonomy of Knowledge-Based Guidance and Types of Logic Predicates and Their Interpretation

Type of Hint and Logic Predicate	φ	ψ
Entropy (H)	$V = \sum\limits_{k=1}^{N} \varphi[H(\mathbf{u}_k), \lambda_k]b_k$	$V = \sum\limits_{k=1}^{N} \psi[H(\mathbf{u}_k), \mu_k]b_k$
	Entropy less than some specified threshold (λ_k)	Entropy similar to some specified value (μ_k)
Proximity (Prox)	$V = \sum\limits_{k_1=1}^{N} \sum\limits_{\substack{k_2=1. \\ k_1 \geq k_{1_l}}}^{N} \varphi[\text{Prox}(u_{k_1}, u_{k_2}), \lambda_{k_1,k_2}]b_{k_1,k_2}$	$V = \sum\limits_{k_1=1}^{N} \sum\limits_{\substack{k_2=1. \\ k_1 \geq k_{1_l}}}^{N} \varphi[\text{Prox}(u_{k_1}, u_{k_2}), \lambda_{k_1,k_2}]b_{k_1,k_2}$
	Proximity between pairs of patterns less than some predefined level given as $\lambda_{k_1 k_2}$	Proximity between pairs of patterns attains a predefined level given as $\lambda_{k_1 k_2}$

Considering the optimization (maximization) problem in the above form, the general gradient- based computing is governed by the expression

$$\mathbf{u}(\text{iter} + 1) = \mathbf{u}(\text{iter}) + \alpha \, \nabla_{\mathbf{u}} V \tag{6.8}$$

where $\nabla_{\mathbf{u}} V$ denotes the gradient computed with respect to the individual entries of the partition matrix. Note that we use the plus sign because of the maximization of the objective function.

The detailed formulas depend upon the specific details of the predicate and the form of the hint. We consider the entropy and assume that the uncertainty should not exceed 1. The starting point of our derivations is the maximized sum of the truth values of the predicates

$$V = \sum_{k=1}^{N} \varphi[H(\mathbf{u}_k), \lambda_k]b_k \tag{6.9}$$

where b_k is a binary variable (indicator) that assumes a value of 1 if the kth pattern is assessed with respect to the constraint under consideration. Otherwise, b_k is set to 0, meaning that there is no indication of this pattern. The calculations of the gradient of V taken with respect to u_{st} ($s = 1, 2, \ldots, c, t = 1, 2, \ldots, N$) lead to the expression

$$\frac{\partial V}{\partial u_{st}} = \frac{\partial}{\partial u_{st}} \sum_{k=1}^{N} \varphi[H(\mathbf{u}_k), \lambda_k]b_k = \sum_{k=1}^{N} \frac{\partial \varphi[H(\mathbf{u}_k), \lambda_k]}{\partial H(\mathbf{u}_k)} \frac{\partial H(\mathbf{u}_k)}{\partial u_{st}} b_k \tag{6.10}$$

If we assume that the implication is defined as shown in Figure 6.7 and that the entropy function is specialized in the form given by (6.3), then we obtain

$$\frac{\partial \varphi}{\partial H(\mathbf{u}_k)} = \begin{cases} 1 \text{ if } H(\mathbf{u}_k) \le \lambda \\ -\dfrac{\lambda}{H(\mathbf{u}_k)^2} \text{ otherwise} \end{cases} \qquad (6.11)$$

$$\frac{\partial H(\mathbf{u}_t)}{\partial u_{st}} = \frac{\partial}{\partial u_{st}} \left[\frac{1}{c} \sum_{i=1}^{c} 4u_{it}(1 - u_{it}) \right] = \frac{4}{c}(1 - 2u_{st}) \qquad (6.12)$$

As an illustrative example, we discuss the two-dimensional synthetic data set shown in Figure 6.8. There are three visible clusters. When running the standard FCM (with the fuzzification coefficient m equal to 2), the partition matrix (three fuzzy sets) reflects the structure of the data (Figure 6.8). With the optimized prototypes being equal to $\mathbf{v}_1 = [1.91\ 9.99]^T$, $\mathbf{v}_2 = [5.50\ 5.71]^T$, $\mathbf{v}_3 = [0.38\ 0.19]^T$, the computed membership functions are shown in Figure 6.9.

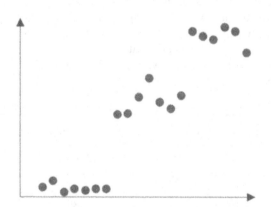

Figure 6.8. A two-dimensional synthetic data set.

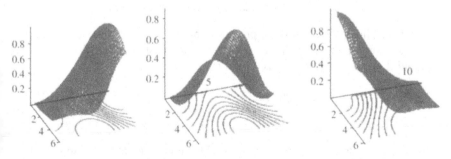

Figure 6.9. Plots of membership functions of the clusters.

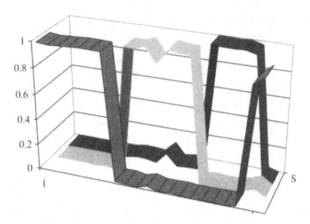

Figure 6.10. Membership functions of fuzzy sets.

Consider now six knowledge-based hints where the patterns have relatively high entropy values between 0.7 and 0.9 (Figure 6.10). Following the optimization of V (in which we use the product operator and the induced implication operator), we end up with the revised prototypes that are now equal to $[2.08\ 9.15]^T$, $[5.13\ 5.44]^T$, and $[0.40\ 0.50]^T$, with the ensuing partition matrix visualized in Figure 6.11. As a result of this optimization of the logic predicate, we end up with the adherence to the assumed values of λs. More specifically we get the following entries:

2162

Optimized Entropy	Knowledge-Based Constraint
0.90	0.89
0.70	0.76
0.80	0.81
0.90	0.89
0.70	0.77
0.80	0.82
0.80	0.83

In what follows, we consider proximity-based knowledge-based guidance, where, for some pairs of patterns, the proximity must not be below a certain threshold. Formally, the maximized performance index is

$$V = \sum_{i=1}^{N} \sum_{k=1}^{N} \text{Incl}(\lambda_{i,k}, \text{Prox}_{i,k}) b_{i,k} \qquad (6.13)$$

where $\lambda_{i,k}$ and $\text{Prox}_{i,k}$ denote the threshold values and the proximity values coming from the partition matrix. The Boolean indicator function $b_{i,k}$ identifies

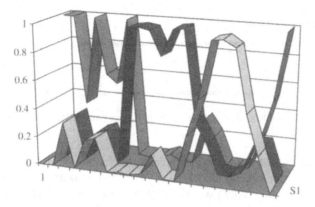

Figure 6.11. Membership functions resulting from knowledge-based clustering.

the pairs of patterns involved in knowledge-based guidance of the clustering process. The gradient of V taken with respect to u_{st} is equal to

$$\frac{\partial V}{\partial u_{st}} = \sum_{i=1}^{N} \sum_{k=1}^{N} \frac{\partial}{\partial \text{Prox}_{i,k}} \text{Incl}(\lambda_{i,k}, \text{Prox}_{i,k}) b_{i,k} \frac{\partial \text{Prox}_{i,k}}{\partial u_{st}} \qquad (6.14)$$

Let us define the inclusion in the form

$$\text{Incl}(a, b) = \begin{cases} 1 & \text{if } a \leq b \\ b/a & \text{otherwise} \end{cases} \qquad (6.15)$$

We obtain

$$\frac{\partial \text{Incl}(\lambda_{i,k}, \text{Prox}_{i,k})}{\partial \text{Prox}_{i,k}} = \begin{cases} 0 & \text{if } \lambda_{i,k} \leq \text{Prox}_{i,k} \\ \dfrac{1}{\lambda_{i,k}} & \text{otherwise} \end{cases} \qquad (6.16)$$

The second term can be calculated by plugging in the entries of the partition matrix

$$\frac{\partial \text{Prox}_{i,k}}{\partial u_{st}} = \frac{\partial}{\partial u_{st}} \sum_{j=1}^{c} \min(u_{ji}, u_{jk}) \qquad (6.17)$$

which leads to the following expression:

$$\frac{\partial \text{Prox}_{i,k}}{\partial u_{st}} = \begin{cases} 1 & \text{if } u_{si} \leq u_{sk} \text{ and } i = t \\ 1 & \text{if } u_{sk} \leq u_{si} \text{ and } k = t \\ 0 & \text{otherwise} \end{cases} \qquad (6.18)$$

Figure 6.12. General flow of cooperative computing in knowledge-based fuzzy clustering.

To focus our attention, we have concentrated on the FCM clustering scheme; however, any other iterative model based on the minimization of some objective function (performance index) would be of interest as well. The general organization of the computing flow of the clustering is displayed in Figure 6.12. The process is composed of the cycle consisting of a single iteration of the clustering computing followed by the sequence of iterations maximizing V.

Note also that as (6.9) does not return the membership grades whose sum is normalized to 1, we normalize these before returning the result to the iteration of the FCM computing. The other issue worth mentioning relates to the landscape of V treated as a function of the membership grades. The two-dimensional case (in which we encounter only two patterns, and therefore deal with $\mathbf{u}_1$ and $\mathbf{u}_2$) and two clusters (so that for u_{11} given, the second membership grade u_{12} is available as $1 - u_{11}$) are visualized in Figure 6.13 Here the entropy function is given by (6.2) and (6.3), the implication (inclusion) is expressed by (6.14), and $\lambda_1 = 0.4$, $\lambda_2 = 0.8$.

The maximized performance index V is multimodal, so potentially there could be several solutions. This nonuniqueness is a result of the nature of the fuzzy predicates expressing the knowledge hints. While one might argue that this lack of uniqueness could be detrimental to the optimization process, it is not that important. Note that as the clustering algorithm has already produced the partition matrix, its values form a starting point for the optimization of V (and it is very likely that we end up with the maximum that is the closest to the configuration of the membership grades returned by the fuzzy clustering).

Alluding to the general flow of computing visualized schematically in Figure 6.12, it is worth stressing that any overhead caused by the processing of knowledge-based hints that could become essential when dealing with large data sets is not critical. Even though the data sets could be large, the number of knowledge hints is quite limited. Consequently, the optimization phase involving these hints would not be very time-consuming.

6.6. PROXIMITY-BASED CLUSTERING (P-FCM)

In this section, we look more closely at proximity-based clustering (P-FCM) as an interesting and practically appealing clustering environment. Let us recall that knowledge-based hints are available as user-given experimental assessments

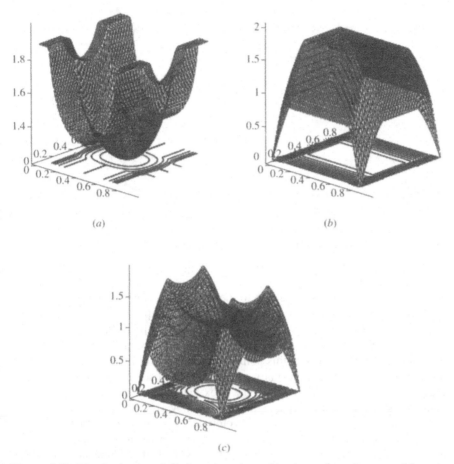

Figure 6.13. The landscape of V for selected combinations of predicates and knowledge-based guidance: (*a*) entropy and the inclusion predicate, (*b*) entropy and the dominance predicate, (*c*) entropy and the similarity predicate.

of proximity between selected pairs of patterns denoted here by $\text{Prox}^{\wedge}_{i,k}$. The generic objective function is the same as that encountered in the standard FCM. The gradient-based part of optimization minimizes the sum of squared differences (denoted by V) between the supplied proximity values and those computed on the basis of the already determined partition matrix. The overall minimization process cycles through the minimization of the objective function and the proximity performance index V.

Two illustrative experiments are helpful in presenting P-FCM.

Experiment 1. Here we are concerned with the small amount of two-dimensional synthetic data shown in Figure 6.14. Evidently, there is some structure with several visible but not necessarily clearly distinguishable clusters. The standard

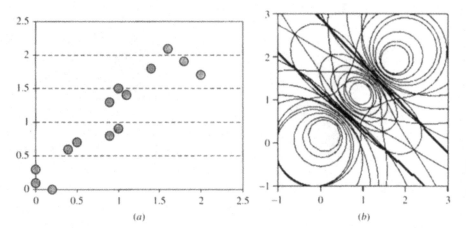

Figure 6.14. Two-dimensional data (a) and boundaries of the clusters resulting from the FCM algorithm (b).

FCM with $c = 3$ clusters gives rise to the cluster boundaries (Figure 6.14b) that distinguish between the produced groups. The prototypes of the clusters shown there are $\mathbf{v}_1 = [0.95\ 1.15]$, $\mathbf{v}_2 = [1.70\ 1.86]$, and $\mathbf{v}_3 = [0.15\ 0.25]$.

As we can observe, the proximity hints substantially affect the clusters and their boundaries. To illustrate this point more clearly, we added several proximity hints (constraints) to selected pairs of patterns. Two scenarios are presented in Figures 6.15 and 6.16. They show the proximity constraints and the resulting boundaries of the clusters. It is apparent that they are substantially affected by the proximities being a result of the prototypes being moved by them. Because of the low values of the proximities on the pairs of patterns, the region occupied by the second class has expanded compared to that of the first class (without any proximity constraints). The prototypes are equal to $\mathbf{v}_1 = [0.25\ 0.37]$, $\mathbf{v}_2 = [1.65\ 1.78]$, and $\mathbf{v}_3 = [1.06\ 1.36]$.

The situation in Figure 6.15 is completely remote from the previous cases. The proximity constraints (that are quite "decisive," assuming mostly binary values) have radically changed the landscape of the clustering. Notably, the proximity constraints throw some patterns with a visible neighborhood in the feature space into different groups (because of their low proximity values). The prototypes are now equal to $\mathbf{v}_1 = [0.67\ 0.77]$, $\mathbf{v}_2 = [0.58\ 0.79]$, and $\mathbf{v}_3 = [1.42\ 1.68]$. The resulting boundaries shown in Figure 6.16 are very distinct from those in Figures 6.14 and 6.15.

Experiment 2. Whereas in the previous case the proximity constraints were introduced freely, primarily to visualize how they impact the results of clustering by their interplay with the structure "discovered" within the original data, in this experiment the origin of the proximity values is very different. The general scheme we are using here is as follows: start with the original pattern positioned in an n'-dimensional space, pick up the subset (n) of the features (where $n < n'$),

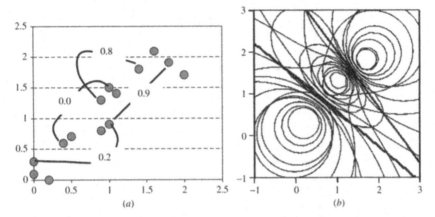

Figure 6.15. Proximity constraints (a) and resulting cluster boundaries (b).

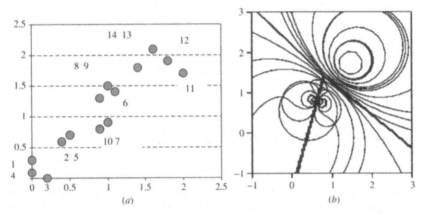

Figure 6.16. Proximity constraints—labeled patterns (a): (1 7 0.9), (2 6 0.0), (2 9 1.0), (7 12 0.9), (9 13 0.0), (8 14 0.0), (13 14 0.0), (1 2 0.0) and cluster boundaries (b).

and cluster them in this reduced feature space with some additional proximity constraints. These constraints are constructed in a systematic way: we cluster patterns in the n'-dimensional space and then build the proximities based on the resulting partition matrix. The overall setup of the experiment is portrayed in Figure 6.17.

The data set consists of three clusters of four-dimensional synthetic data with a Gaussian distribution:

$$\mathbf{m} = [0.0 \quad 0.0 \quad 3.0 \quad 0.0]^T \quad \Sigma = \begin{bmatrix} 1.0 & 0.0 & 0.0 & 0.0 \\ 0.0 & 1.0 & 0.0 & 0.0 \\ 0.0 & 0.0 & 1.0 & 0.0 \\ 0.0 & 0.0 & 0.0 & 0.5 \end{bmatrix}$$

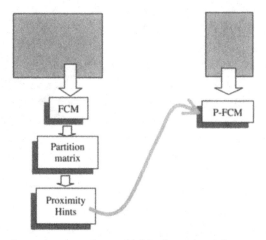

Figure 6.17. Experimental setup—from a highly dimensional feature space to its reduced version and proximity hints.

$$\mathbf{m} = [\,0.5 \quad 0.8 \quad 1.0 \quad 2.0\,]^T \quad \sum = \begin{bmatrix} 1.2 & 0.0 & 0.0 & 0.0 \\ 0.0 & 1.5 & 0.0 & 0.0 \\ 0.0 & 0.0 & 1.0 & 0.0 \\ 0.0 & 0.0 & 0.0 & 0.5 \end{bmatrix}$$

$$\mathbf{m} = [\,6.0 \quad 6.0 \quad 3.0 \quad 3.0\,]^T \quad \sum = \begin{bmatrix} 2.0 & 0.0 & 0.0 & 0.0 \\ 0.0 & 1.0 & 0.0 & 0.0 \\ 0.0 & 0.0 & 2.0 & 0.0 \\ 0.0 & 0.0 & 0.0 & 0.5 \end{bmatrix}$$

Each cluster consists of 100 patterns. The use of the FCM with $c = 4$ clusters leads to the prototypes $\mathbf{v}_1 = [0.04\ 0.13\ 3.00\ 0.33]$, $\mathbf{v}_2 = [5.90\ 6.01\ 3.34\ 3.08]$, and $\mathbf{v}_3 = [0.72\ 1.27\ 1.22\ 2.04]$, with the partition matrix illustrated in Figure 6.18.

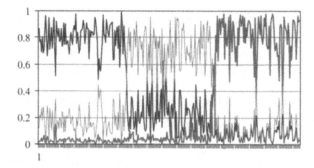

Figure 6.18. Membership functions produced by the FCM for the four-dimensional set of patterns (three clusters).

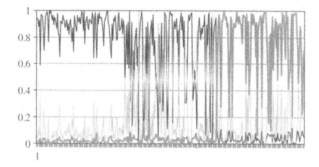

Figure 6.19. Membership grades of the patterns in three clusters (reduced two-dimensional data).

Now we take only the two first variables of the patterns. This leads to substantial overlap between the clusters (which becomes apparent by looking at the statistical parameters of the generated groups). This is also reflected in the more visible overlap between the clusters, as portrayed in Figure 6.19. The prototypes of the clusters are given as $v_1 = [0.16 - 0.03]$, $v_2 = [1.65\ 3.56]$, and $v_3 = [6.51\ 6.28]$.

Consider the P-FCM algorithm with randomly selected proximity hints; we consider 30 and 150 hints (that is, pairs of patterns for which we calculate the proximity level on the basis of the partition matrix obtained for the four-dimensional patterns). The resulting partition matrices (membership grades) are shown in Figure 6.20; the prototypes are:

No. of hints = 30 : $v_1 = [0.13 - 0.29]$, $v_2 = [0.76\ 2.07]$, $v_3 = [6.21\ 6.25]$

No of hints = 150 : $v_1 = [0.21 - 0.13]$, $v_2 = [0.96\ 2.25]$, $v_3 = [5.97\ 6.09]$

Another way to quantify the impact of the proximity constraints is to calculate the Hamming distance between the two partition matrices, namely, the one for the original four-dimensional patterns and the other corresponding to the reduced two-dimensional patterns without and with the proximity hints. The tendency is clear; if there are no hints, this distance is equal to 171.0 and is gradually reduced to 151. 6 and 146.4 for 30 and 150 hints, respectively. It becomes evident that the P-FCM can compensate for the unseen features by guiding the clustering process.

6.7. WEB EXPLORATION AND P-FCM

Considering the rapidly growing size of the Web, it is evident that any kind of manual classification and categorization of Web sources (sites, pages, etc.) is prohibitively time-consuming. Here is an important role for rapid, automatic, and accurate hypertext clustering algorithms. The main problem with the development of automated tools is related to finding, extracting, parsing, and filtering the user

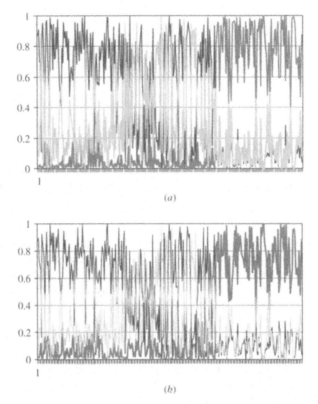

Figure 6.20. Membership grades for (*a*) 30 and (*b*) 150 proximity hints.

requirements from Web knowledge. A number of approaches help to automatically retrieve, categorize, and classify web documents. Clustering techniques
have been proposed by Hoppner et al. (1999) as generic information retrieval
tools. Two cluster-based approaches exploit graph partitions to induce clusters.
One is based on a hypergraph to define an association rule able to gather items
that appear together frequently in many transactions. The other method produces,
through recursive splitting, a binary tree of clusters in which the root is the starting document set, while each leaf node is a partition of the whole set. A recent
proposal of a fuzzy-based approach to Web mining concerns the use of medoids,
a form of relational fuzzy clustering. Some other clustering-oriented approaches
are more closely related to user interaction; the LOGSOM system was developed
in order to mine Web log data and provide a visual tool to guide the user during
the navigation, based on a SOM and organizing Web documents into a two-
dimensional map according to the user's navigation behaviors. Clustering can be
used to discover semantic relationship among specified concepts and organize
them into messages created during electronic meetings. There is a broad range
of existing approaches (see, e.g., Boley et al., 1999).

Owing to its concept, P-FCM plays an interesting role in Web mining. A few observations can place the discussion in a general setting. First, in assessing the proximity of two Web pages, we are faced with extremely heterogeneous information including text, images, video, and audio. Many factors play a significant role in our judgment of the proximity of Web pages, such as the layout of individual pages, form of the background, and intensity of links to other sites, as well as the origin in cyberspace that leads us to a specific page. Most importantly, many of these factors are difficult to quantify and translate into computationally meaningful features. The textual information is the most evident feature, and it is almost the only contributor to the feature space when determining structures in a collection of Web pages. This tendency is clear in the literature, and it is related to the extensive studies on information retrieval. Here we can envision a role for proximity hints, whose use can compensate for the consideration of a subset of the feature space. For instance, we can cluster Web pages in the subspace of textual information, and the proximity values provided by the user augment (and implicitly expand) this subspace by incorporating other features capturing the multimedia and layout portions of the description of the pages.

The schematic view of Web mining using P-FCM is presented in Figure 6.21; here we emphasize the origin of two sources of information: the data sets themselves and hints provided by the user/designer.

The usual approach to document categorization is based on analysis of its content, since information for categorizing a document is extracted from the document itself. Current (semi)-automatic attempts are too recent to be operative components of powerful Web search engines that prefer the work done by humans. ODP (Open Directory Project http://dmoz.org, informally known as *Dmoz*, e.g., Mozilla Directory) is the most widely distributed database of Web content classified by a volunteer force of more than 8,000 editors. ODP organization provides the means to manage the Web growth via editor/subeditor chains, whose integration covers all possible Web contents. This "collective"

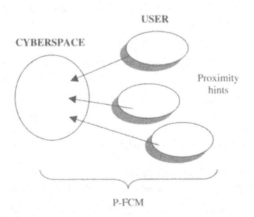

Figure 6.21. A schematic view of Web mining in the environment of P-FCM.

brain represents the human technology at the base of the most popular Web search engines and portals, including Netscape Search, AOL Search, Google, Lycos, HotBot, DirectHit, and hundreds of others.

For our testbed we considered three ODP categories:

1. Top:Computers:Software:Graphics:Image_Manipulation

   ```
   (www.dmoz.org/
       Computers/Software/Graphics/Image_Manipulation)
   ```

2. Top:Shopping:Gifts: Personalized:Photo_Transfers

   ```
   (www.dmoz.org/
       Shopping/Gifts/Personalized/Photo_Transfers)
   ```

3. Top:News:Media:Journalism:Photojounalism:Music_Photography

   ```
   (www.dmoz.org/News/Media/Journalism/Photojounalism/
       Music_Photography)
   ```

The open structure of ODP allowed us to analyze the structure of these three categories in order to acquire the necessary information to categorize the related Web pages. In fact, for each category, Dmoz provides a descriptive Web page, such as the one given in Figure 6.22, explaining the category in terms of related terms and keywords.

Table 6.3 presents the keywords used in the case study associated with each category.

The representation of each Web page is treated as a 14-dimensional vector, where each component is the frequency of occurrence (probability) of the term on the specific page. Although in our approach the features can be keywords, hyperlinks, images, animations, etc., in this experiment we consider just keyword-based features in order to comply with the Dmoz statistical classification (which is based on term frequency analysis).

TABLE 6.3. Selected Keywords Chosen as Features of the Data Set

Category	Keywords
Top:Shopping:Gifts: Personalized:Photo_Transfers	Transfer, gift, photo*, logo
Top:Computers:Software:Graphics:Image_Manipulation	Image, software, filter, digital, manipulat*
Top:News:Media:Journalism:Photojounalism: Music_Photography	Concert, music, journal, promot*, portrait

Note: Some keywords are stemmed to capture words with a common prefix denoted by an asterisk (for instance, *photo* or *manipulat*).

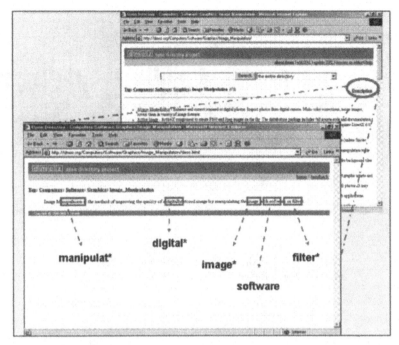

Figure 6.22. Selection of keywords from a descriptive Web page dealing with the category Top:Computers:Software:Graphics:Image_Manipulation.

Our test was performed on 20 Web pages per category, 60 pages in total. For the reference, we applied a standard FCM algorithm and partitioned the 60 pages into three clusters (see Figure 6.23). Figure 6.23 portrays the distribution of membership grades of the Web pages in each cluster: for the first 20 pages, the membership grades are highest in the 3 (white bars), while they become quite irrelevant in clusters 1 and 2 (denoted by black and gray bars, respectively). Analogous observations can be made for pages belonging to two other categories: pages 21 to 40 comprise cluster 1 (although some pages in this category assume higher membership values in cluster 2), while pages 41 to 60 form cluster 2. The prototypes of the clusters formed in this process of grouping are visualized in Figure 6.24.

Web navigation is in part a cognitive process that may be influenced by several factors. If the user wants to express proximity values between pairs of Web pages, these values can be established on the basis of personal judgment. In this situation, the P-FCM approach can be useful to capture the user's feedback and reflect its impact on reshaping the clusters.

In our test, the user identifies several pairs of Web pages and assigns to them some proximity values, as summarized in Table 6.4. For instance, the proximity

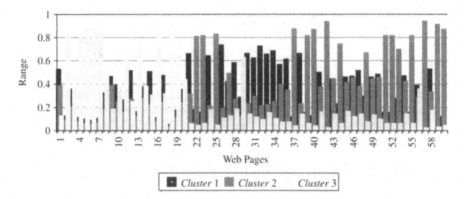

Figure 6.23. The results (membership grades) of the FCM clustering of Web pages.

TABLE 6.4. Proximity Values Between Selected Pairs of Web Pages Reflecting the User's Evaluation

Web Pages		Proximity
1	16	0.8
1	18	0.9
1	21	0.1
7	10	0.9
25	42	0.1
40	57	0
32	22	0.8
22	42	0.1
37	59	0.1
37	5	0.1
21	38	0.9
33	37	0.8

value for page 1 and 18 is equal to 0.9, and this value indicates that the pages are closely *analogous*. On the other hand, page 1 is very *different* from page 21, as reflected by the very low proximity value (e.g., 0.1) associated with this pair of pages (see Figure 6.25).

This user's feedback conveyed in terms of proximity values has an impact on the previous clustering results (refer to Figure 6.11). Figure 6.26 illustrates the results of P-FCM. It is evident that some pages have improve their membership in the right cluster. For instance, pages 1 and 21 are now in the right cluster, with higher values.

For comparative reasons, Figures 6.23 and 6.27 visualize the prototypes (highest membership values in corresponding clusters) produced by the FCM and P-FCM, respectively.

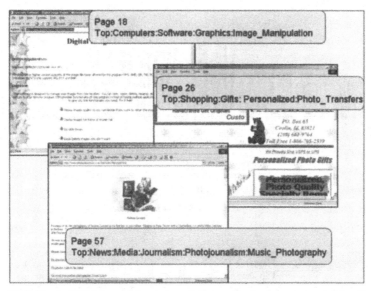

Figure 6.24. Typical Web pages (prototypes) for each cluster constructed by the FCM algorithm.

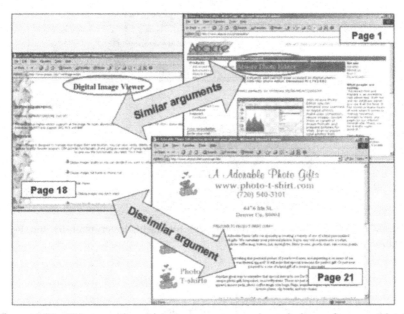

Figure 6.25. Web pages for which the user expressed proximity preferences (hints).

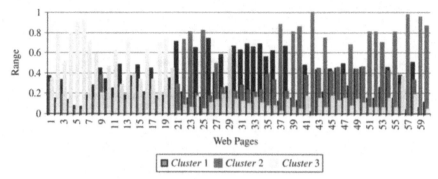

Figure 6.26. Results of P-FCM clustering.

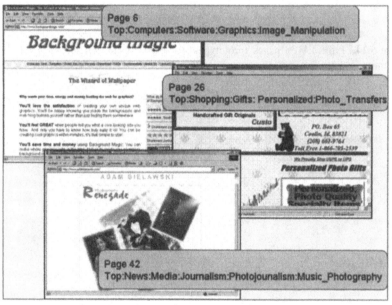

Figure 6.27. Typical Web pages (prototypes) for each cluster built by the P-FCM algorithm.

In many real-world cases, the user may disagree with a textual Web search classification. In our test, this situation may occur since the categorization has been based only on textual information; no other media (components of the hypertext) have been considered. Figure 6.28 shows this scenario; here the user defines the following proximity values between the pairs of the Web pages: (53 8 0.9) and (8 20 0.1).

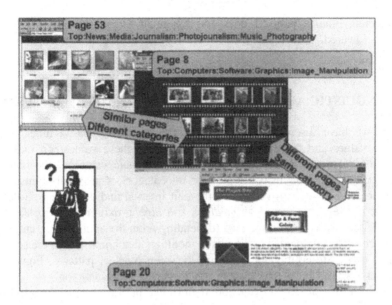

Figure 6.28. Subjective evaluation of the proximity between selected Web pages.

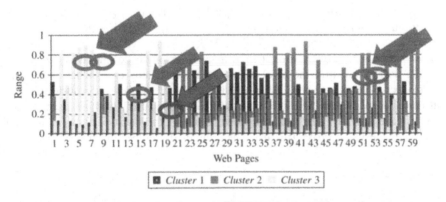

Figure 6.29. Results produced by P-FCM; note the shift in the values of the membership grades.

The user's intention is to assert that pages 53 and 8 are very similar, while page 20 is very different from page 8 (although both pages are in the same cluster); in fact, according to the user evaluation, pages 53 and 8 are similar because they show photo galleries, while page 20 is concerned with plugin tools.

After the use of P-FCM, see the results shown in Figure 6.29. Page 20 moves to cluster 1 (black bars), because it assumes the highest membership in that cluster (circled in the figure), while page 8 remains in cluster 3 (whitish bar), as expected. Finally, page 53 does not change its cluster, but its membership value is lowered. This effect follows the low proximity values introduced by the user.

6.8. LINGUISTIC AUGMENTATION OF KNOWLEDGE-BASED HINTS

So far, we have discussed a case where the knowledge-based hints assume numeric values and the designer/user must supply these values. To make this interaction more user-centric, we may admit linguistic values of the hints. To focus our discussion, let us concentrate on the category of proximity hints. These are modeled as fuzzy sets defined in the unit interval and describe well-defined semantic concepts such as, *low*, *medium*, and *high* proximity. Obviously, more linguistic terms could be introduced (depending upon the application), and those, in turn, could be augmented by linguistic modifiers that contribute to the semantic richness of the description of proximity between pairs of patterns. Figure 6.30 illustrates selected fuzzy sets used in this linguistic representation.

In light of this proposed extension, the original performance index introduced earlier, which takes the form of the sum $V = \sum_{k_{1_I}=1}^{N} \sum_{\substack{k_2=1, \\ k_1 \ge k_{1_I}}}^{N} \varphi[\lambda_{k_1,k_2}, \text{Prox}(u_{k_1}, u_{k_2})]$

b_{k_1,k_2} (where φ stands for the similarity predicate) must be revisited. Let us denote the linguistic terms (fuzzy sets) by $A_1, A_2, \ldots, A_r$, assuming that we are provided with an r level of quantification. The proximity value resulting from the fuzzy partition matrix U, $\text{Prox}^\wedge(.,.)$ is mapped through the linguistic term used by the user to quantify the level of proximity, so it tells us to what extent this preference is consistent with the proximity produced by the fuzzy clustering. For instance, suppose that the user has quantified the proximity as *high*, corresponding to fuzzy set A_i. The proximity computed on the basis of fuzzy clustering generates the value $\text{Prox}^\wedge(\mathbf{u}_k, \mathbf{u}_l)$. The computed membership degree $A_i(\text{Prox}^\wedge$

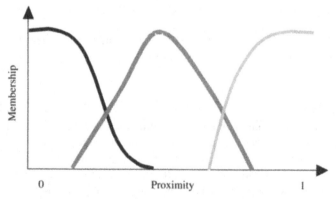

Figure 6.30. Examples of fuzzy sets used in the linguistic characterization of proximity.

$(\mathbf{u}_k, \mathbf{u}_l))$ reflects how much the results coincide. As we are interested in achieving the maximum level of agreement, the optimization focuses on maximum of the following sum:

$$V = \sum_{k=1}^{N} \sum_{\substack{l=1, \\ k \geq l_1}}^{N} A_i [\text{Prox}^{\wedge}(u_k, u_l)] b_{k,l} \qquad (6.19)$$

(As before, the Boolean indicator $b_{k,l}$ points at the pairs of patterns for which the proximity hints have been supplied.) As a standard, we use gradient-based optimization, that is, we update the values of the partition matrix following the formula $\mathbf{u}(\text{iter} + 1) = \mathbf{u}(\text{iter}) + \alpha \nabla_{\mathbf{u}} V$.

6.9. CONCLUSIONS

The knowledge-based guidance augmentation of unsupervised learning opens up some new and promising avenues of exploration of data structures. By building a unified optimization framework in which we seamlessly combine data and knowledge-based computing activities, we are able to address the fundamental matter of hybrid information processing. The interface between data and knowledge-based computing exploits models of logic optimization, where we develop a certain predicate and maximize its truth value by determining the underlying structure of the clusters (partition matrix). The fuzzy predicates are helpful in expressing linguistic relational constraints (such as *less than, approximately equal*, etc.) that are in line with the assessment of designers and data analysts. We have presented an array of possible scenarios arising in various applications ranging from Web exploration to uncertainty guidance in pattern classification to partial supervision involving labeling of selected patterns.

REFERENCES

J.C. Bezdek, *Pattern Recognition with Fuzzy Objective Function Algorithms*, Plenum Press, New York, 1981.

D. Boley et al., Partitioning-based clustering for Web document categorization, *Decision Support Systems*, 27, 1999, 329–341.

D. Guillaume, F. Murtargh, Clustering of XML documents, *Computer Physics Communications*, 127, 2000, 215–227.

F. Hoppner, F. Klawonn, R. Kruse, T. Runkler, *Fuzzy Cluster Analysis—Methods for Image Recognition*, John Wiley, New York, 1999.

A.K. Jain, R.C. Dubes, *Algorithm for Clustering Data*, Prentice Hall, Englewood Cliffs, NJ, 1988.

G.J. Klir, T.A. Folger, *Fuzzy Sets, Uncertainty, and Information*, Prentice Hall, Englewood Cliffs, NJ, 1988.

W. Pedrycz, V. Loia, S. Senatore, P-FCM: a proximity-based clustering, *Fuzzy Sets and Systems*, vol. 148 no.1, 2004, 21–41.

W. Pedrycz, J. Waletzky, Fuzzy clustering with partial supervision, *IEEE Trans. on Systems, Man, and Cybernetics*, 5, 1997a, 787–795.

W. Pedrycz, J. Waletzky, Fuzzy clustering in software reusability, *Software: Practice and Experience*, 27, 1997b, 245–270.

7 Collaborative Clustering

So far, knowledge-based clustering has operated on a single data set. In this chapter, we discuss a situation that arises when there are several data sets. A structure we are interested in revealing concerns all of them, but they need to be processed separately. This leads to the fundamental concept of collaboration: the clustering algorithms operate locally (namely, on individual data sets) but collaborate by exchanging information about their findings. We formulate two fundamental development scenarios referred to as vertical and horizontal collaboration. In the ensuing algorithms critical is a strength of collaboration and a way in which it is quantified within the network of clustering algorithms. The level of granularity at which this collaboration occurs reflects the requirement of data confidentiality and security to which one must adhere.

7.1. INTRODUCTION AND RATIONALE

Imagine a situation in which we have a collection of data sets existing at different organizations. These could be data describing customers of banking institutions, retail stores, and medical organizations. The data could include records of different individuals. They could also deal with the same individuals, but each data set may have different descriptors (features) reflecting the activities of the organization. The ultimate goal of each organization is to discover key relationships in its data set. These organizations also recognize that as there are other data sets, it would be advantageous to learn about the dependencies there occurring in order to reveal the overall picture of the global structure. We do not have direct access to other data, which prevents us from combining all data sets into a single database and carrying out clustering there. Access may be denied because of confidentiality requirements (e.g., medical records of patients cannot be shared and confidentiality of banking data has to be assured). There could also be some hesitation about the possibility of losing the identity of the data of the individual organization. We are more comfortable with revealing relationships in our own organization's data set. While appreciating the value of additional external sources of information, it is helpful to control how the findings there could

Knowledge-Based Clustering, by Witold Pedrycz
ISBN 0-471-46966-1   Copyright © 2005 John Wiley & Sons, Inc.

affect the results from the data within the company. In some cases, there could be technical issues; processing (clustering) of a single huge data set may not be feasible or sufficiently informative.

It is instructive to discuss two illustrative examples.

Security Issues and Discovery of Data Structures Across Different Data Sets. Consider the situation where information about the same group of clients is collected in different databases while an individual company (bank, store, etc.) builds its own database. Because of confidentiality and security requirements, the companies cannot share information directly. However, all of them are interested in deriving some associations that will help them learn about their customers (namely, their profiles and needs). As they are concerned with the same population of clients, we may anticipate that the basic structure of the population of such patterns, in spite of possible minor differences, should hold across all databases. The approach taken in this case would be to construct clusters specific to each database and exchange information at the level of the clusters, that is, information granules. In this way, security is not compromised and a sound mechanism of collaboration/interaction between the databases could be established.

Conceptual Split of Data as a Means of Structure Reconciliation. In clustering multivariable data, we can often split variables into distinct groups (e.g., in describing countries, we can distinguish between economic indicators, variables capturing cultural aspects, geography and climate). As the variables in each group are conceptually quite close (they deal with the same general concept), we may be interested in deriving the structure for each subset while maintaining links between the subsets in order to reconcile eventual structural differences and emphasize similarities. Here the data sets become disjoint, yet whatever we attempt to discover at the local level has to be consistent to some extent with the findings obtained at the level of other databases.

Collaboration in Processing Visual Data. Fuzzy clustering has been applied to a vast array of problems in image processing (e.g., image segmentation, edge detection). Given the multifaceted character of data, we can take into account various low- and high-end features of images (texture, color, gradient, etc.) and geometric features (location of objects or pixels). Naturally, we can envision clustering that focuses on these groups of features and operates on them while still completing active collaboration activities.

In this chapter, we refine the concept of collaboration at the level of information granules and distinguish between two fundamental models of collaboration, horizontal and vertical clustering (Pedrycz, 2002). We show that this approach stands in sharp contrast to the existing trends of clustering and data analysis (Bezdek, 1981; Duda et al., 2001; Hoppner et al., 1999). With the detailed notation in place, we design a suite of detailed clustering algorithms.

7.2. HORIZONTAL AND VERTICAL CLUSTERING

While collaboration can include a variety of detailed schemes, two of them are the most essential. We refer to them as horizontal and vertical modes of collaboration or simply horizontal and vertical clustering. More descriptively, given data sets $X[1], X[2], \ldots X[P]$ where P denotes their number and $X[ii]$ stands for the iith data set (we adhere to the practice of using square brackets to identify a certain data set), in *horizontal* clustering we have the same objects that are described in *different* feature spaces. In other words, these could be the same collection of patients whose records are developed within each medical institution. The schematic illustration of this mode of clustering portrayed in Figure 7.1 shows that every collaboration occurs at the structural level, viz., through the information granules (clusters) built over the data; the clusters are shown as an auxiliary interface layer surrounding the data. The net of directed links shows how the collaboration between different data sets takes place. The width of the links emphasizes the fact that the intensity of collaboration can differ, depending on the data set involved and the purpose of the collaboration (e.g., the willingness of an organization to accept findings from external sources).

Vertical clustering (Figure 7.2) is complementary to horizontal clustering. Here the data sets are described in the same feature space but deal with *different* patterns. In other words, we consider that $X[1], X[2], \ldots, X[P]$ are defined in the same feature space, while each of them consists of different patterns, $\dim(X[1]) = \dim(X[2]) = \ldots \dim(X[P])$, while $X[ii] \neq X[jj]$. We can show the data sets as being stacked on each other (hence the name of this clustering mode).

Collaboration involves mechanisms of interaction. While the algorithmic details are presented in the next section, it is instructive here to describe the nature of the possible collaboration.

> In horizontal clustering we deal with the same patterns and different feature spaces. The communication platform is based on through the partition

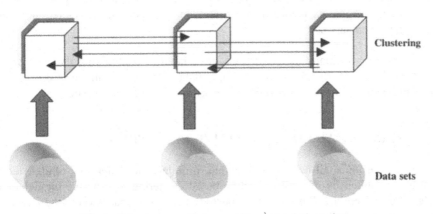

Figure 7.1. A general scheme of horizontal clustering.

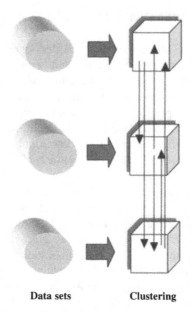

Data sets   **Clustering**

Figure 7.2. A general scheme of vertical clustering; note the stack of data sets communicating through a granular layer.

matrix. As we have the same objects, this type of collaboration makes sense. The confidentiality of data has not been breached: we do not operate on individual patterns but on the resulting information granules (fuzzy relations, that is, partition matrices). As this number is far lower than the number of data, the low granularity of these constructs moves us far from the original data.

In vertical clustering we are concerned with different patterns but the same feature space. Hence communication at the level of the prototypes (which are high-level representatives of the data) becomes feasible. Again, because of the aggregate nature of the prototypes, the confidentiality requirement has been satisfied.

There are also many hybrid models of collaboration involving data sets with possible links of vertical and horizontal collaboration. An example of collaborative clustering involving this hybrid scenario is presented in Figure 7.3.

7.3. HORIZONTAL COLLABORATIVE CLUSTERING

Here we introduce all necessary notation, formulate the underlying optimization problem implied by objective function-based clustering, and derive the detailed algorithm. There are P sets of data located in different spaces (viz., the patterns there are described by different features). As each subset deals with the same

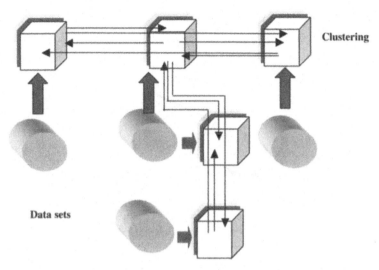

Clustering

Data sets

Figure 7.3. Example of hybrid collaboration involving horizontal and vertical clustering.

patterns (that is, each pattern is a concatenation of the corresponding subpatterns), the number of elements in each subset is the same and is equal to N. We are interested in partitioning the data into c fuzzy clusters. The clustering completed for each subset of data results in a partition matrix and a collection of prototypes. We use square brackets to point to the specific subset. That is, we use the notation $U[ii]$ and $\mathbf{v}[ii]$ to denote the partition matrix and the ith prototype produced by the clustering realized for the ii-set of data. Similarly, the dimensionality of the patterns in each data set could be different; to <u>show</u> this, we use a pertinent index, say $n[ii]$. The distance function between the ith prototype and the kth pattern in the same subset is denoted by $d_{ik}^{2}[ii]$, $i = 1, 2, \ldots, c, k = 1, 2, \ldots, N$.

The objective function guiding the formation of the clusters that is completed for each subset assumes a well-known form of FCM clustering:

$$\sum_{k=1}^{N}\sum_{i=1}^{c} u_{ik}^{2}[ii]d_{ik}^{2}[ii]$$

$ii = 1, 2, \ldots, P$. The collaboration between the subsets is established through a matrix of connections (or interaction coefficients or interactions; see Figure 7.4).

Each entry of the collaborative matrix describes the intensity of the interaction. In general, $\alpha[ii,kk]$ assumes nonnegative values. The higher the value of the interaction (collaboration) coefficient, the stronger the collaboration between the corresponding data sets. Note that the collaborative links along with the individual clustering algorithms form a directed graph where the nodes represent the clustering activities at the level of the individual data sets and the directed edges indicate the level of collaboration. Evidently $\alpha[ii,kk]$ need not be equal to $\alpha[kk,ii]$, as the impact from the kkth structure on the development of the iith

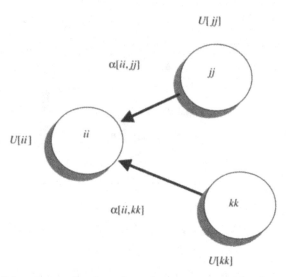

Figure 7.4. Collaboration in the clustering scheme represented by the matrix of collaboration levels between the subsets; the partition matrices generated for each data set are shown.

structure could be quite different from the one where the structure in the kkth data set is developed in collaboration with the iith structure.

To accommodate the collaboration mechanism in the optimization process, the objective function is expanded into the form

$$Q[ii] = \sum_{k=1}^{N}\sum_{i=1}^{c} u_{ik}^2[ii]d_{ik}^2[ii] + \sum_{\substack{jj=1 \\ jj \neq ii}}^{P} \alpha[ii, jj]\sum_{k=1}^{N}\sum_{i=1}^{c}\{u_{ik}[ii] - u_{ik}[jj]\}^2 d_{ik}^2[ii]$$

(7.1)

$ii = 1, 2, \ldots, P$. The second term in the above expression makes the clustering based on the iith subset "aware" of other partitions. It is obvious that if the structures in data sets are similar, then the differences between the partition matrices tend to be lower, and the resulting structures start becoming more similar.

As usual, we require the partition matrix to satisfy standard requirements of membership grades summing to 1 for each pattern and the membership grades contained in the unit interval. All in all, collaborative clustering converts into the following family of P optimization problems with the membership constraints

$$\min_{U \in U, \mathbf{v}_1, \mathbf{v}_2, \ldots, \mathbf{v}_c} Q$$

subject to

$$U \in \mathbf{U}$$

where $\mathbf{U}$ is a family of all fuzzy partition matrices. The minimization is carried out with respect to the fuzzy partition and the prototypes.

7.3.1. Optimization Details

The above optimization task splits into two problems, namely, determination of the partition matrix $U[ii]$ and the prototypes $\mathbf{v}_1[ii], \mathbf{v}_2[ii], \ldots, \mathbf{v}_c[ii]$. These problems are solved separately for each of the collaborating subsets of patterns.

To determine the partition matrix, we exploit a technique of Lagrange multipliers so that the constraint occurring in the problem becomes merged as a part of constraint-free optimization. This leads to the new objective function $V[ii]$:

$$
V[ii] = \sum_{i=1}^{c} u_{ik}^2[ii]d_{ik}^2[ii] + \sum_{\substack{jj=1 \\ jj \neq ii}}^{P} \alpha[ii, jj] \sum_{i=1}^{c} \{u_{ik}[ii] - u_{ik}[jj]\}^2 d_{ik}^2[ii]
$$

$$
- \lambda \left(\sum_{i=1}^{c} u_{ik}[ii] - 1 \right) \tag{7.2}
$$

for $k = 1, 2, \ldots, N$, where λ denotes a Lagrange multiplier. The necessary conditions leading to the local minimum of $V[ii]$ are

$$
\frac{\partial V[ii]}{\partial u_{st}[ii]} = 0 \qquad \frac{\partial V[ii]}{\partial \lambda} = 0 \tag{7.3}
$$

$s = 1, 2, \ldots, c$, $t = 1, 2, \ldots, N$. The derivative computed with respect to the partition matrix is

$$
\frac{\partial V}{\partial u_{st}} = 2u_{st}[ii]d_{st}^2[ii] + 2 \sum_{jj \neq ii} \alpha[ii, jj](u_{st}[ii] - u_{st}[jj])d_{st}^2[ii] - \lambda = 0 \tag{7.4}
$$

In other words,

$$
u_{st}[ii] = \frac{\lambda + 2d_{st}^2[ii] \sum_{jj \neq ii} \alpha[ii, jj]u_{st}[jj]}{2 \left(d_{st}^2[ii] + d_{st}^2[ii] \sum_{jj \neq ii} \alpha[ii, jj] \right)} \tag{7.5}
$$

Introduce the following shorthand notation:

$$
\varphi_{st}[ii] = \sum_{jj \neq ii} \alpha[ii, jj]u_{st}[jj]
$$

$$
\psi_{st}[ii] = \sum_{jj \neq ii} \alpha[ii, jj] \tag{7.6}
$$

In light of the constraint imposed on the membership values $\sum_{j=1}^{c} u_{jk}[ii] = 1$, the use of the above expression yields the result

$$\sum_{j=1}^{c} \frac{\lambda + 2d_{jk}^2[ii]\varphi_{jk}[ii]}{2d_{st}^2[ii](1 + \psi[ii])} = 1 \tag{7.7}$$

Next, the Lagrange multiplier computes in the form

$$\lambda = 2 \frac{1 - \dfrac{1}{1 + \psi[ii]} \sum_{j=1}^{c} \varphi_{jk}[ii]}{\sum_{j=1}^{c} \dfrac{1}{d_{jk}^2[ii]}} (1 + \psi[ii])$$

Plugging this multiplier into the formula for the partition matrix produces the final expression:

$$u_{st}[ii] = \frac{\varphi_{st}[ii]}{1 + \psi[ii]} + \frac{1 - \dfrac{1}{1 + \psi[ii]} \displaystyle\sum_{j=1}^{c} \varphi_{jt}[ii]}{\displaystyle\sum_{j=1}^{c} \dfrac{d_{st}^2[ii]}{d_{jt}^2[ii]}} \tag{7.8}$$

In the calculations of the prototypes, we confine ourselves to the Euclidean distance between the patterns and the prototypes. The necessary condition for the minimum of the objective function is of the form $\nabla_{v[ii]} Q[ii] = 0$. Let us rewrite $Q[ii]$ in an explicit manner to emphasize the character of the distance function:

$$Q[ii] = \sum_{k=1}^{N} \sum_{i=1}^{c} u_{ik}^2[ii] \sum_{j=1}^{n[ii]} (x_{kj} - v_{ij}[ii])^2 + \sum_{\substack{jj=1 \\ jj \neq ii}}^{P} \alpha[ii, jj]$$

$$\times \sum_{k=1}^{N} \sum_{i=1}^{c} \{u_{ik}[ii] - u_{ik}[jj]\}^2 \sum_{j=1}^{n[ii]} (x_{kj} - v_{ij}[ii])^2 \tag{7.9}$$

The patterns in this expression come from the iith data set. Computing the derivative of $Q[ii]$ with respect to v_{st} $[ii]$ $(s = 1, 2, \ldots, c, t = 1, 2, \ldots, N)$ and setting it to 0, we obtain

$$\frac{\partial Q[ii]}{\partial v_{st}[ii]} = -2 \sum_{k=1}^{N} u_{sk}^2[ii](x_{kt} - v_{st}[ii])$$

$$- 2 \sum_{k=1}^{N} \sum_{jj \neq ii}^{c} \alpha[ii, jj](u_{sk}[ii] - u_{jk}[jj])^2 (x_{kt} - v_{st}[ii]) = 0 \tag{7.10}$$

After some grouping of the terms, we derive the expression

$$
v_{st}[ii] \left\{ \sum_{k=1}^{N} u_{sk}^2[ii] + \sum_{k=1}^{N} \sum_{jj \neq ii}^{c} \alpha[ii, jj](u_{sk}[ii] - u_{sk}[jj])^2 \right\}
$$

$$
= \sum_{k=1}^{N} u_{sk}^2[ii]x_{kt} + \sum_{k=1}^{N} \sum_{jj \neq ii}^{c} \alpha[ii, jj](u_{sk}[ii] - u_{sk}[jj])^2 x_{kt} \qquad (7.11)
$$

We rewrite this in a compact format by introducing auxiliary notation:

$$
v_{st}[ii] = \frac{A_{st}[ii] + C_{st}[ii]}{B_s[ii] + D_s[ii]} \qquad (7.12)
$$

$$
s = 1, 2, \ldots, c, t = 1, 2, \ldots, n[ii], ii = 1, 2, \ldots P
$$

where

$$
A_{st}[ii] = \sum_{k=1}^{N} u_{sk}^2[ii]x_{kt} \qquad (7.13)
$$

$$
B_s[ii] = \sum_{k=1}^{N} u_{sk}^2[ii] \qquad (7.14)
$$

$$
C_{st}[ii] = \sum_{\substack{jj=1 \\ jj \neq ii}}^{P} \alpha[ii, jj] \sum_{k=1}^{N} (u_{sk}[ii] - u_{sk}[jj])^2 x_{kt} \qquad (7.15)
$$

$$
D_s[ii] = \sum_{\substack{jj=1 \\ jj \neq ii}}^{P} \alpha[ii, jj] \sum_{k=1}^{N} (u_{sk}[ii] - u_{sk}[jj])^2 \qquad (7.16)
$$

7.3.2. The Flow of Computing of Collaborative Clustering

The general clustering scheme consists of two phases:

Generation of clusters without collaboration. This phase involves the application of the FCM algorithm to each data set. Obviously, the number of clusters has to be the same for all data sets. During this phase we carry out an independent search for a structure in each subset of data.

Collaboration between the clusters. Here we start with the already computed partition matrices, set up the collaboration level (through the values of the interaction coefficients arranged in the collaboration matrix $\alpha[ii, jj]$), and proceed with a simultaneous optimization of the partition matrices.

Moving on to the formal algorithm, the computational details are arranged in the following fashion:

Given: subsets of patterns $X_1, X_2, \ldots, X_P$

Select: distance function, number of clusters (c), termination criterion, and collaboration matrix $\alpha[ii,jj]$.

Compute: initiate randomly all partition matrices $U[1], U[2], \ldots, U[P]$

Phase I
For each data

repeat

compute prototypes $\{v_i[ii]\}, i = 1, 2, \ldots, c$ and partition matrices $U[ii]$ for all subsets of patterns

until a termination criterion has been satisfied

Phase II

repeat

For the given matrix of collaborative links $\alpha[ii,jj]$, compute prototypes and partition matrices $U[ii]$ using (7.4) and (7.7)

until a termination criterion has been satisfied

The termination criterion relies on the changes to the partition matrices obtained in successive iterations of the clustering method; for instance, a Tchebyschev distance could serve as a sound measure of changes in the partition matrices. Subsequently, when this distance is lower than an assumed threshold value ($\varepsilon > 0$), the optimization is terminated.

The flow of optimization can be captured in terms of the values of the collaboration matrix. In the first phase, we consider that all values of $\alpha[ii,jj]$ are equal to 0 (implying no collaboration) and then, after some iterations, they assume nonzero values. In this sense, we can collapse the two phases into a single step (see Figure 7.5).

7.3.3. Quantification of the Collaborative Phenomenon of Clustering

The intensity of collaboration between the clusters, can be assessed at two levels: the level of data and the level of information granules (that is, fuzzy relations contained in the partition matrix). In this latter quantification, we use the results of clustering without any collaboration as a point of reference.

The *level of data* involves a comparison of the numeric representatives of the clustering, that is, the prototypes (centroids). The impact of the collaboration is then expressed in the changes of the prototypes occurring as a result of the collaboration.

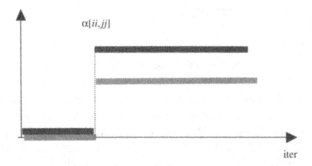

Figure 7.5. Values of the collaboration coefficients treated as functions of iterations of the algorithm (the noncollaborative phase comes with the 0 values of the $\alpha[ii,jj]$'s).

At the *level of information granules* (partitions and fuzzy sets), the effect of collaboration is expressed in two ways. The first one quantifies how much the collaboration impacts the clustering results. Using $U_{\text{ref}}[ii]$ to denote the partition matrix produced independently of other sources, we compute a distance $\|U[ii] - U_{\text{ref}}[ii]\|$ (the distance expressed in any way that is relevant when comparing two partition matrices) and treat it as a function of the intensity of collaboration:

$$\delta = \|U[ii] - U_{\text{ref}}[ii]\| \qquad (7.17)$$

The other measure expressing the effect of collaboration deals with the differences between the partition matrices. Intuitively, we can envision that once the level of collaboration increases, the structures within data sets start to exhibit smaller differences. The index (consistency measure)

$$\phi[ii, jj] = \|U[ii] - U[jj]\| \qquad (7.18)$$

indicates the structural differences between the partition matrices defined over two data sets (ii and jj, respectively). The level of collaboration can be adjusted, allowing for a certain maximal value of changes of the membership grades (entries of the partition matrix).

There is another option worth considering when selecting the collaboration level. Noting that the collaborative clustering is aimed at forming a consensus and that each external source of information should be used to refine the already developed structure within the given data set, we can consider entropy as a measure of enhancement of structure discovery. The process of setting up a level of collaboration is then performed in a stepwise manner: we start with a very low value of $\alpha[ii,jj]$ (which reflects a very loose form of interaction or collaboration) and increase it gradually by watching the values of the entropy. The increase is stopped once the entropy values start to increase, meaning that the collaboration tends to produce some mechanisms of competition rather than harmonious and supportive interaction.

7.4. EXPERIMENTAL STUDIES

To illustrate horizontal collaborative clustering, we carry out some experiments to get some insight into the performance of this method.

Experiment 1. We consider the data set distributed in two subspaces (see Figure 7.6). It is apparent that the structure of the data visible in these two subspaces is quite different. While we can eventually envision three clusters, these are positioned differently and their distinction is not clear. In space X_1, we note one cluster that is quite distinct, while the two others are far less clearly

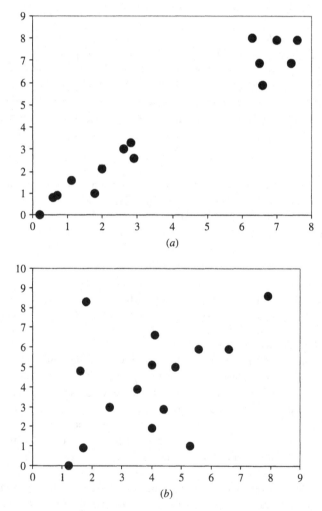

Figure 7.6. Synthetic data sets and their visualization in space X_1 (a) and X_2 (b).

delineated, forming almost a single elongated structure (which nevertheless could be split into spherical clusters). The data located in X_2 (Figure 7.6b) exhibit a far less visible structure.

We start with no collaboration by setting all the values of the collaboration matrix to 0; $\alpha[ii, jj] = 0$ for all values of ii and jj. The clustering is completed for $c = 3$ clusters. After clustering, the prototypes and the corresponding partition matrices are equal to

space X_1: $v_1 = [2.62 \ 2.79]^T$ $v_2 = [6.90 \ 7.28]^T$ $v_3 = [0.82 \ 0.85]^T$

$$
\begin{bmatrix}
0.002403 & 0.000220 & 0.997376 \\
0.073337 & 0.010147 & 0.916516 \\
0.006531 & 0.000642 & 0.992827 \\
0.139116 & 0.007933 & 0.852952 \\
0.217786 & 0.012458 & 0.769756 \\
0.767185 & 0.012808 & 0.220007 \\
0.993295 & 0.001160 & 0.005545 \\
0.963953 & 0.008303 & 0.027745 \\
0.981568 & 0.003061 & 0.015372 \\
0.009651 & 0.985867 & 0.004482 \\
0.016653 & 0.974473 & 0.008874 \\
0.021377 & 0.967951 & 0.010673 \\
0.008137 & 0.987674 & 0.004190 \\
0.009604 & 0.985596 & 0.004800 \\
0.065701 & 0.905631 & 0.028668
\end{bmatrix}
$$

space X_2: $v_1 = [3.81 \ 4.52]^T$ $v_2 = [2.09 \ 1.44]^T$ $v_3 = [6.29 \ 6.49]^T$

$$
\begin{bmatrix}
0.026211 & 0.965828 & 0.007961 \\
0.093903 & 0.871662 & 0.034434 \\
0.505755 & 0.388633 & 0.105613 \\
0.607660 & 0.080534 & 0.311806 \\
0.713268 & 0.055687 & 0.231046 \\
0.947006 & 0.019349 & 0.033645 \\
0.351095 & 0.601909 & 0.046996 \\
0.499791 & 0.229903 & 0.270306 \\
0.152238 & 0.063779 & 0.783983 \\
0.922959 & 0.052719 & 0.024322 \\
0.132055 & 0.024279 & 0.843666 \\
0.420602 & 0.363174 & 0.216224 \\
0.448355 & 0.450673 & 0.100972 \\
0.693879 & 0.190041 & 0.116080 \\
0.031356 & 0.008891 & 0.959753
\end{bmatrix}
$$

The plots of the membership functions formed on the basis of the prototypes are presented in Figure 7.7. This figure visualizes the class boundaries discriminates between the patterns belonging to different clusters.

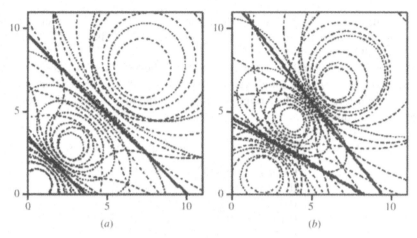

Figure 7.7. Contour plots of the clusters (dotted lines) and decision boundaries (solid lines) for space $\mathbf{X}_1$ (a) and $\mathbf{X}_2$ (b).

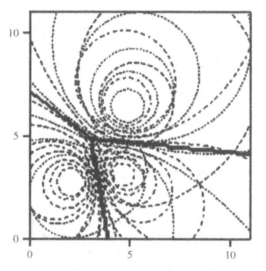

Figure 7.8. Contour plots of the clusters (dotted lines) and decision boundaries (solid lines) for space $\mathbf{X}_2$ with the prototypes given as $[4.84\ 6.25]^T$, $[4.55\ 3.20]^T$, and $[2.24\ 2.79]^T$.

We now introduce the collaboration involving the first data set, where we increase the values of $\alpha[2,1]$ (that entry of the collaboration matrix pointing at impact $\mathbf{X}_1$ starts exhibiting on $\mathbf{X}_2$). The results of clustering for $\alpha[2, 1] = 0.3$ are shown in Figure 7.8. The prototypes in $\mathbf{X}_2$ are quite different when compared with the previous clustering involving no collaboration.

The corresponding partition matrix produced in this space has the following entries:

$$U[2] = \begin{bmatrix} 0.052927 & 0.121152 & 0.825921 \\ 0.097542 & 0.167998 & 0.734460 \\ 0.118145 & 0.134497 & 0.747358 \\ 0.701552 & 0.065291 & 0.233157 \\ 0.544363 & 0.232398 & 0.223239 \\ 0.589833 & 0.261475 & 0.148691 \\ 0.239626 & 0.029546 & 0.730828 \\ 0.569272 & 0.185798 & 0.244930 \\ 0.678707 & 0.206891 & 0.114401 \\ 0.118356 & 0.633631 & 0.248013 \\ 0.664368 & 0.308076 & 0.027555 \\ 0.112732 & 0.736714 & 0.150554 \\ 0.060555 & 0.754855 & 0.184590 \\ 0.010806 & 0.976463 & 0.012730 \\ 0.522629 & 0.404842 & 0.072529 \end{bmatrix}$$

The measures of collaboration are plotted in Figure 7.9; obviously, the impact is visible on the entries of the partition matrix in X_2. As anticipated, the differences between the partition matrices start to be reduced over the increasing value of α. Interestingly the changes in the membership grades get lower when we increase the value of α. We note that there is some value of α above which any further changes in the values of δ or ϕ are quite limited. For instance, the value of α equal to 0.1 seems to be critical in this regard; higher values of α do not impact the values of δ, meaning that the structures in these two data sets are able to collaborate at the 0.1 level. Likewise, the differences between the two partition matrices are reduced visibly when we start applying nonzero values of α; however, stronger collaboration (with the values of α greater than 0.1) does not lead the differences between the partition matrices.

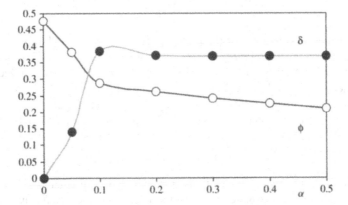

Figure 7.9. Indexes of structural differences treated as a function of collaboration (α).

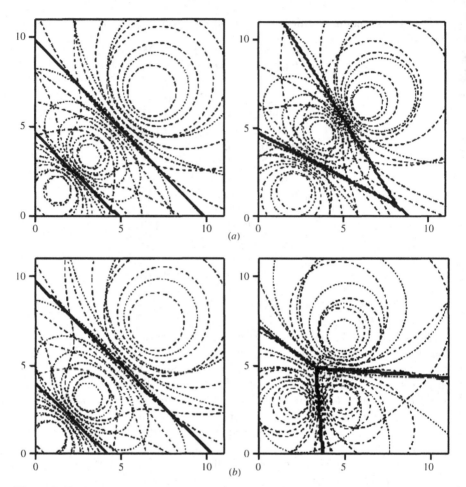

Figure 7.10. Contour plots of clusters and their classification boundaries for selected cases of collaboration: (a) $\alpha[1, 2] = 2.0; \alpha[2, 1] = 0.1$ and (b) $\alpha[1, 2] = \alpha[2, 1] = 0.8$.

By modifying the values of the collaboration matrix, we end up with different distributions of the prototypes. The plots presented in Figure 7.10 illustrate this effect for selected cases. First, $\alpha[1,2]$ set to 2.0 and $\alpha[2,1]$ set to 0.1 reflects a case where there is relatively high impact from the patterns in $\mathbf{X}_2$; thus, the search for structure in $\mathbf{X}_1$ is significantly affected by that impact (Figure 7.10a). The two values of $\alpha[1,2]$ and $\alpha[2,1]$, both set to 0.6, produce the prototypes visualized in Figure 7.10b. All these scenarios need to be contrasted with the prototypes constructed without any collaboration, as shown in Figure 7.8.

Experiment 2. In this experiment, we consider two synthetically generated data sets formed by the same patterns defined in two feature spaces $\mathbf{X}_1$ and $\mathbf{X}_2$. There are 2,000 patterns. In space $\mathbf{X}_1$ they form a clearly visible structure composed

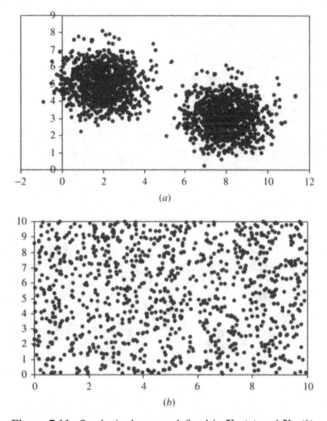

Figure 7.11. Synthetic data sets defined in $\mathbf{X}_1$ (a) and $\mathbf{X}_2$ (b).

of two well-delineated spherical clusters (Figure 7.11a). The patterns in $\mathbf{X}_1$ are governed by two Gaussian distributions $\mathbf{m}_1 = [2.0\ 5.0]^T$ and $\mathbf{m}_2 = [8.0\ 3.0]^T$, with the diagonal covariance matrices $\Sigma_1 = \Sigma_2 = \Sigma = \begin{bmatrix} 1.0 & 0.0 \\ 0.0 & 1.0 \end{bmatrix}$. The second pattern, defined in $\mathbf{X}_2$, is governed by a uniform distribution, so no structure becomes visible (Figure 7.11b).

The clustering is completed for $c = 2$. In the case of no collaboration, we obtain a structure where the prototypes are given in the form

$$\mathbf{X}_1 : \mathbf{v}_1 = [2.08\ 5.13]^T \quad \mathbf{v}_2 = [7.92\ 2.98]^T$$

$$\mathbf{X}_2 : \mathbf{v}_1 = [4.19\ 7.47]^T \quad \mathbf{v}_2 = [5.80\ 3.03]^T$$

and the partition matrix (one cluster shown) are visualized in Figure 7.12. It is apparent that the structure in $\mathbf{X}_1$ is fully revealed and quantified by the prototypes and the clear distinction between the groups. The membership grades are close to 1 in one group, and very few patterns in the second cluster exceed the membership grade of 0.6 of their allocation to the other group.

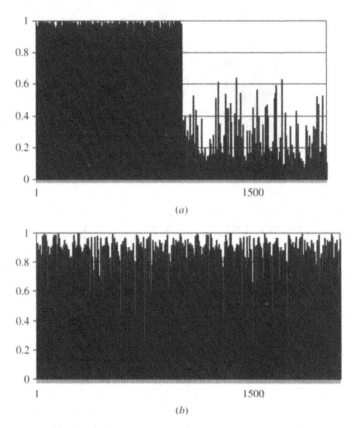

Figure 7.12. Results of clustering: partition matrices for patterns in $\mathbf{X}_1$ (a) and $\mathbf{X}_2$ (b).

Let us now consider a collaboration mechanism implemented via the link $\alpha[2,1]$ set to 1 (the other entries of the collaboration matrix were equal to 0). This affects the structure revealed in $\mathbf{X}_2$. The prototypes have changed their location and are equal to $\mathbf{v}_1 = [4.98\ 5.27]^T$ and $\mathbf{v}_2 = [5.00\ 5.31]^T$. More interestingly, the partition matrix shows a significant shift in comparison to the situation reported in Figure 7.13, where we see no structure and membership grades indicate clusters with a high level of overlap. We distinguish between two clusters. The first is formed by patterns with membership grades assuming values close to 0.8. The others started lowering their membership grades, thus discriminating between the clusters. In this way, the collaboration has helped build a structure by using some well-distinguished topology available in the other space.

To continue this experiment, we reverse the direction of the collaboration, allowing the patterns (and structure) in $\mathbf{X}_2$ to affect $\mathbf{X}_1$, so we set $\alpha[1,0]$ to 1 while keeping the remaining entries of the matrix equal to 0. As a result, the prototypes in $\mathbf{X}_1$ are shifted and now assume the location $\mathbf{v}_1 = [3.77\ 4.45]^T$, $\mathbf{v}_2 = [6.27\ 3.63]^T$. A more profound effect shows up when it comes to the membership

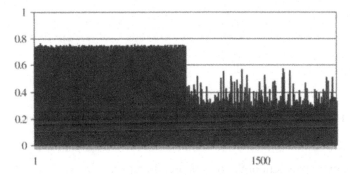

Figure 7.13. Partition matrix (one cluster shown) resulting from the collaboration defined by α[2, 1] = 1.0.

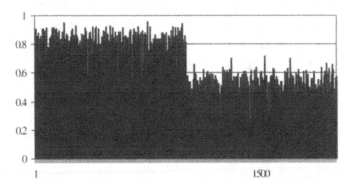

Figure 7.14. Partition matrix (one cluster shown) resulting from the collaboration quantified by α[1, 2] = 1.

grades. The patterns tend to be affected in a way so that they become less "declared" to belong to the specific cluster and start sharing their membership among the two of them. First, the patterns start losing their membership grades, originally equal to 1 (which is clear from Figure 7.14, especially the first patterns). Second, there is a group of patterns with membership degrees of about 0.5, in contrast to the clusters portrayed in Figure 7.13.

Experiment 3. Boston housing data are one of the data sets available in the Machine Learning Repository, ftp://ftp.ics.uci.edu/pub/machine-learning-databases/housing/. They consist of 506 patterns describing real estate in the Boston area. There are 14 features describing the patterns, including the crime rate, nitric acid concentration, and median value of the house. Several scenarios of collaborative clustering are envisioned (Figure 7.15).

X_1 is a data set including all features but the price of real estate. X_2 involves the patterns described by their price. We arbitrarily proceed with four clusters.

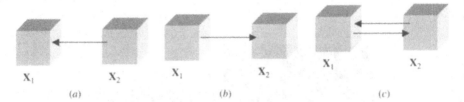

Figure 7.15. Several experimental setups of collaborative clustering.

Without any collaboration, the results are as follows:

$\mathbf{X}_1$:

$\mathbf{v}_1 = [2.21\ 3.59\ 12.37\ 0.08\ 0.57\ 6.14\ 79.42\ 3.11\ 7.00\ 385.16\ 18.68\ 361.36\ 13.92]$

$\mathbf{v}_2 = [1.02\ 6.23\ 8.26\ 0.05\ 0.51\ 6.31\ 61.78\ 4.12\ 5.32\ 313.75\ 18.32\ 380.00\ 10.35]$

$\mathbf{v}_3 = [10.04\ 0.55\ 17.97\ 0.03\ 0.67\ 6.03\ 91.17\ 2.09\ 21.98\ 639.04\ 19.96\ 322.80\ 18.68]$

$\mathbf{v}_4 = [0.58\ 39.36\ 5.16\ 0.04\ 0.45\ 6.62\ 38.61\ 6.08\ 4.85\ 313.26\ 17.07\ 383.91\ 7.39]$

$\mathbf{X}_2 : \mathbf{v}_1 = 31.45 \quad \mathbf{v}_2 = 21.46 \quad \mathbf{v}_3 = 47.65 \quad \mathbf{v}_4 = 12.86$

We consider collaboration where the search for the structure in $\mathbf{X}_1$ becomes affected by the topology discovered in $\mathbf{X}_2$ (that is, it is driven by the price of real estate). The collaboration link is set to $\alpha[1, 2] = 1.0$. The impact on $\mathbf{X}_1$ is profound: the prototypes there start shifting, yielding the following configuration:

$\mathbf{v}_1 = [0.60\ 30.24\ 5.36\ 0.07\ 0.47\ 6.75\ 48.16\ 5.10\ 5.15\ 303.99\ 17.05\ 385.30\ 7.13]$

$\mathbf{v}_2 = [1.29\ 8.78\ 10.36\ 0.06\ 0.53\ 6.11\ 63.00\ 4.18\ 7.15\ 366.82\ 18.64\ 379.20\ 11.57]$

$\mathbf{v}_3 = [1.73\ 17.86\ 8.78\ 0.18\ 0.53\ 7.20\ 67.58\ 3.53\ 7.48\ 347.13\ 16.75\ 377.51\ 6.74]$

$\mathbf{v}_4 = [10.38\ 0.59\ 17.35\ 0.03\ 0.67\ 5.92\ 92.09\ 2.22\ 18.46\ 587.79\ 19.79\ 294.12\ 20.41]$

One can easily identify patterns whose membership grades are shifted because of the collaboration; a snapshot of the results is shown in Figure 7.16, which demonstrates the differences in membership grades for selected clusters produced with and without collaboration. These changes to the membership grades of the patterns are candidates for a thorough analysis as potential outliers.

In the next experiment, we allow $\mathbf{X}_1$ to impact $\mathbf{X}_2$ by setting the collaborative link $\alpha[2,1]$ equal to 1. This allows us to affect the clusters built on the real estate price by the properties of the real estate. The prototypes in $\mathbf{X}_2$ are affected and are now $\mathbf{v}_1 = 20.61$, $\mathbf{v}_2 = 24.45$, $\mathbf{v}_3 = 14.61$, and $\mathbf{v}_4 = 30.93$ (Figure 7.17).

In the next scenario, we consider nonzero values of $\alpha[1,2]$ and $\alpha[2,1]$. As we envision different combinations of the collaborative links, it may be instructive to quantify the effect of collaboration on the consistency of the resulting information granules, that is, their partition matrices. A series of experiments was carried out for combinations of the collaborative links, with the results reported in Table 7.1.

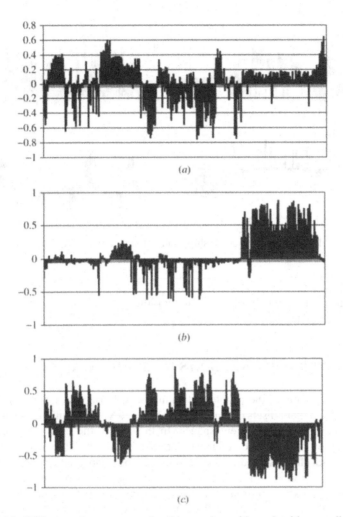

Figure 7.16. Difference between membership grades with and without collaboration: (*a*) first cluster, (*b*) third cluster, and (*c*) fourth cluster.

TABLE 7.1. Consistency Measure ϕ Computed for
Different Values of the Collaborative Links

$\alpha[1,2]$				
$\alpha[2,1]$	0.0	0.5	1.0	1.5
0.0	0.328	0.164	0.114	0.091
0.5	0.156	1.112	0.089	0.074
1.0	0.115	0.093	0.073	0.063
1.5	0.091	0.077	0.064	0.054

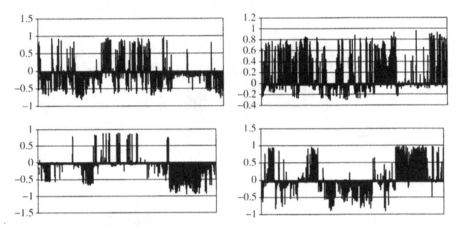

Figure 7.17. Differences in membership grades for patterns in $\mathbf{X}_2$ reported for successive clusters.

Note, that as the intensity of the collaboration increases, the revealed structures move close to each other.

7.5. FURTHER ENHANCEMENTS OF HORIZONTAL CLUSTERING

As the formulation of the problem demonstrates, horizontal clustering relies on the assumption that all sets of patterns $\mathbf{X}_1, \mathbf{X}_2, \ldots, \mathbf{X}_P$ deal with the same patterns (objects) described in different spaces. In this sense, the established collaboration hinges on the existence of these patterns so that the knowledge about their allocations to clusters in different spaces can be shared and used in a collective manner. There could be cases where some patterns may not be present in every data set. As a consequence, they do not contribute to the effect of collaboration. We have to exclude them from the optimization process. To do so, we introduce a Boolean matrix $B[ii, jj] = [b_k[ii, jj]]$, $k = 1, 2, \ldots, N$, with the two-valued valuation characterizing the allocation of the patterns (k) to two data sets.

The original objective function minimized at the level of the iith data set is refined to take into account the patterns not shared between the sets. We augment the second term of the function in the form

$$Q[ii] = \sum_{k=1}^{N} \sum_{i=1}^{c} u_{ik}^2[ii]d_{ik}^2[ii] + \sum_{\substack{jj=1 \\ jj \neq ii}}^{P} \alpha[ii, jj]$$

$$\times \sum_{k=1}^{N} \sum_{i=1}^{c} \{u_{ik}[ii] - u_{ik}[jj]\}^2 b_k[ii, jj]d_{ik}^2[ii] \qquad (7.19)$$

so only the relevant patterns are involved in the optimization. We can eventually relax the binary requirements by allowing partial involvement in the collaboration; in this case, we assume the values of $b_k[ii, jj]$ to be in the unit interval.

7.6. THE ALGORITHM OF VERTICAL CLUSTERING

The concept of vertical collaborative clustering comes into play when we are dealing with different data sets where all patterns are described in the same feature space. Bearing this in mind, we cannot establish communication at the level of partition matrices (as their dimensions would vary from set to set). The other alternative consists of prototypes of the data sets. They are defined in the same feature space, so in contrast to the previous model of collaboration, they could be a viable option. The proposed objective function governing a search for structure in the iith data set is

$$Q[ii] = \sum_{k=1}^{N[ii]} \sum_{i=1}^{c} u_{ik}^2[ii] d_{ik}^2[ii] + \sum_{\substack{jj=1 \\ jj \neq ii}}^{P} \beta[ii, jj] \sum_{i=1}^{c} \sum_{k=1}^{N[ii]} u_{ik}^2[ii] \|\mathbf{v}_i[ii] - \mathbf{v}_i[jj]\|^2$$

(7.20)

where $\beta[ii, jj]$ is a collaboration coefficient supporting an impact of the jjth data set and affecting the structure to be determined in the iith data set. The number of patterns in the iith data set is denoted by $N[ii]$. We use different letters to distinguish between horizontal and vertical collaboration. The interpretation of (7.20) is quite obvious: the first term is the objective function used to search for the structure of the iith data set, and the second term articulates the differences between the prototypes (weighted by the partition matrix of the iith data set) which have to be made smaller through the refinement of the partition matrix (or, effectively, the movements of the prototypes in the feature space).

The optimization of $Q[ii]$ involves the determination of the partition matrix $U[ii]$ and the prototypes $\mathbf{v}_i[ii]$. As before, we solve the problem for each data set separately and allow the results to interact, forming a collaboration between the sets. The minimization of the objective function with respect to the partition matrix requires the use of Lagrange multipliers because of the existence of the standard constraints imposed on the partition matrix. We form an augmented objective function V incorporating the Lagrange multiplier λ and deal with each individual pattern ($t = 1, 2, \ldots, N[ii]$):

$$V = \sum_{i=1}^{c} u_{it}^2[ii] d_{it}^2[ii] + \sum_{\substack{jj=1 \\ jj \neq ii}}^{P} \beta[ii, jj]$$

$$\times \sum_{i=1}^{c} u_{it}^2[ii] \|\mathbf{v}_i[ii] - \mathbf{v}_i[jj]\|^2 - \lambda \left(\sum_{i=1}^{c} u_{it} - 1 \right)$$

(7.21)

Taking the derivative of V with respect to $u_{st}[ii]$ and making it 0, we have

$$\frac{\partial V}{\partial u_{st}} = 2u_{st}[ii]d_{st}^2[ii] + 2\sum_{\substack{jj=1 \\ jj\neq ii}}^{P} \beta[ii,jj]u_{st}[ii]\|v_s[ii] - v_s[jj]\|^2 - \lambda = 0$$

(7.22)

For notational convenience, let us introduce the shorthand expression

$$D_{ii,jj} = \|v_s[ii] - v_s[jj]\|^2$$

(7.23)

From (7.22) we derive

$$u_{st}[ii] = \frac{\lambda}{2\left(d_{st}^2[ii] + \displaystyle\sum_{\substack{jj=1 \\ jj\neq ii}}^{P} \beta[ii,jj]D_{ii,jj,s}\right)}$$

(7.24)

Given the standard normalization condition $\displaystyle\sum_{j=1}^{c} u_{jt}[ii] = 1$, one has

$$\frac{\lambda}{2} = \frac{1}{\displaystyle\sum_{j=1}^{c} \frac{1}{d_{jt}^2[ii] + \displaystyle\sum_{\substack{jj=1 \\ jj\neq ii}}^{P} \beta[ii,jj]D_{ii,jj,j}}}$$

(7.25)

With the following abbreviated notation

$$\varphi[ii] = \sum_{\substack{jj\neq ii}}^{P} \beta[ii,jj]D_{ii,jj,j}$$

(7.26)

the partition matrix is

$$u_{st}[ii] = \frac{1}{\displaystyle\sum_{j=1}^{c} \frac{d_{st}^2[ii] + \varphi_s[ii]}{d_{jt}^2[ii] + \varphi_j[ii]}}$$

(7.27)

For the prototypes, we complete calculations of the gradient of Q with respect to the coordinates of the prototype $v[ii]$ and then solve the following equations:

$$\frac{\partial Q[ii]}{\partial v_{st}[ii]} = 0, s = 1, 2, \ldots, c; t = 1, 2, \ldots n$$

(7.28)

We obtain

$$
\frac{\partial Q[ii]}{\partial v_{st}[ii]} = 2\sum_{k=1}^{N} u_{sk}^2[ii](x_{kt} - v_{st}[ii]) + 2\sum_{\substack{jj \neq ii}}^{P} \beta[ii, jj]
$$

$$
\times \sum_{k=1}^{N} u_{sk}^2[ii](v_{st}[ii] - v_{st}[jj]) = 0 \qquad (7.29)
$$

Next,

$$
v_{st}[ii]\left(\sum_{\substack{jj \neq ii}}^{P} \beta[ii, jj]\sum_{k=1}^{N[ii]} u_{sk}^2[ii] - \sum_{k=1}^{N[ii]} u_{sk}^2[ii]\right)
$$

$$
= \sum_{\substack{jj \neq ii}}^{P} \beta[ii, jj]\sum_{k=1}^{N[ii]} u_{sk}^2[ii]v_{st}[jj] - \sum_{k=1}^{N[ii]} u_{sk}^2[ii]x_{kt} \qquad (7.30)
$$

Finally, we get

$$
v_{st}[ii] = \frac{\displaystyle\sum_{\substack{jj \neq ii}}^{P} \beta[ii, jj]\sum_{k=1}^{N[ii]} u_{sk}^2[ii]v_{st}[jj] - 2\sum_{k=1}^{N[ii]} u_{sk}^2[ii]x_{kt}}{\displaystyle\sum_{\substack{jj \neq ii}}^{P} \beta[ii, jj]\sum_{k=1}^{N[ii]} u_{sk}^2[ii] - \sum_{k=1}^{N[ii]} u_{sk}^2[ii]} \qquad (7.31)
$$

An interesting application of vertical clustering occurs when dealing with huge data sets. Instead of clustering them in a single pass, we split them into individual data sets, cluster each of them separately, and reconcile the results through the collaborative exchange of prototypes.

7.7. A GRID MODEL OF HORIZONTAL AND VERTICAL CLUSTERING

In addition to the horizontal and vertical clustering that constitute two generic modes of collaboration, we can envision a variety of intermediate situations where patterns from various sources give rise to common subsets of data as well as being positioned in the same feature space. In other words, one can invoke mechanisms of horizontal and vertical collaboration at the same time. This leads to the grid mode of clustering, with examples of collaboration shown in Figure 7.18.

The objective function formulated for the iith pattern as a subject of minimization is an aggregation (sum) of the components used in the previous modes of collaborative clustering. In general, we have

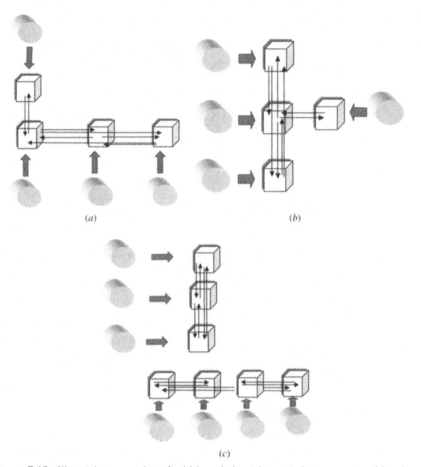

Figure 7.18. Illustrative examples of grid-based clustering: (*a*) data structure with a dominant component of horizontal clustering and some linkages of vertical clustering; (*b*) data structure with a dominant component of vertical clustering; (*c*) data structure with balanced mechanisms of collaborative activities where vertical and horizontal clustering become equally visible.

$$Q[ii] = \sum_{k=1}^{N}\sum_{i=1}^{c}u_{ik}^2[ii]d_{ik}^2[ii] + \underbrace{\sum_{\substack{jj=1\\jj\neq ii}}^{P}\alpha[ii,jj]\sum_{k=1}^{N}\sum_{i=1}^{c}\{u_{ik}[ii]-u_{ik}[jj]\}^2d_{ik}^2[ii]}_{D_1}$$

$$+ \underbrace{\sum_{\substack{jj=1\\jj\neq ii}}^{P}\beta[ii,jj]\sum_{i=1}^{c}\sum_{k=1}^{N[ii]}u_{ik}^2[ii]\|\mathbf{v}_i[ii]-\mathbf{v}_i[jj]\|^2}_{D_2} \qquad (7.32)$$

using the same notation used earlier. Note that the summation points at the corresponding data sets (operating in either mode of collaboration), that is, D_1 involves all data sets that operate in the horizontal mode of clustering, whereas D_2 concerns those using horizontal collaboration.

7.8. CONSENSUS CLUSTERING

In this model of clustering, which nicely complements the vertical mode of collaborative clustering, we consider data sets defined in the same feature space, $X[1], X[2], \ldots, X[P]$.

The clustering is carried out without any interaction between the data sets and the results produced there. In contrast to collaborative clustering, the grouping here can involve different numbers of clusters, say $c[1], c[2], \ldots, c[P]$. The results are provided in terms of prototypes and partition matrices. Because of the same feature space, the prototypes are of interest here. With P sets of data and the same number of clusters, the total number of prototypes is then equal to $\sum_{i=1}^{P} c[i]$. The prototypes are the aggregates summarizing the individual data sets. They, in turn, are treated as patterns to be clustered at the next higher level. This implies two-level clustering (Figure 7.19).

The fundamental question concerns the structure revealed at the higher end and what this implies with respect to the structures relevant to the data sets at the lower end. As the prototypes are clustered at the second level, we start building a *metastructure* that in the sequel gives rise to *metaclusters*. A convenient way to visualize and understand the structural dependencies (especially how the clusters

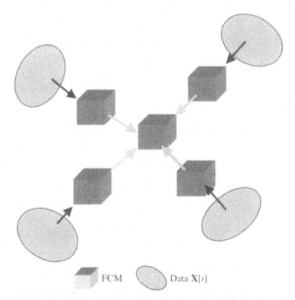

Figure 7.19. Two-level clustering—the development of structural consensus.

relate to each other) is to stress that the partition matrix built at the higher level of clustering is formed on the basis of the prototypes of the data sets, that is, $\mathbf{v}_1[1], \mathbf{v}_2[1], \ldots \mathbf{v}_{c[1]}[1], \ldots, \mathbf{v}_1[P], \mathbf{v}_2[P], \ldots \mathbf{v}_{c[P]}[P]$.

Recalling that each prototype at the lower level comes with a fuzzy relation formed there, each row of the above partition matrix is in essence built on top of the family of fuzzy relations, say $A_1[1], A_2[1], \ldots, A_{c[1]}[1], \ldots, A_1[P]$, $A_2[P], \ldots, A_{c[P]}[P]$. As an example, consider a partition matrix with the entries ($P = 4$, $c[1] = 2$, $c[2] = 2$, and $c = 3$ at the upper end).

$$U = \begin{bmatrix} 0.7 & 0.5 & 0.0 & 0.1 & 0.9 & 0.3 & 0.0 & 0.0 \\ 0.1 & 0.4 & 1.0 & 0.7 & 0.0 & 0.6 & 0.1 & 0.0 \\ 0.2 & 0.1 & 0.0 & 0.2 & 0.1 & 0.1 & 0.9 & 1.0 \end{bmatrix}$$

It produces three fuzzy sets of order -2 (Kandel, 1986; Pedrycz, 1995), as visualized in Figure 7.20.

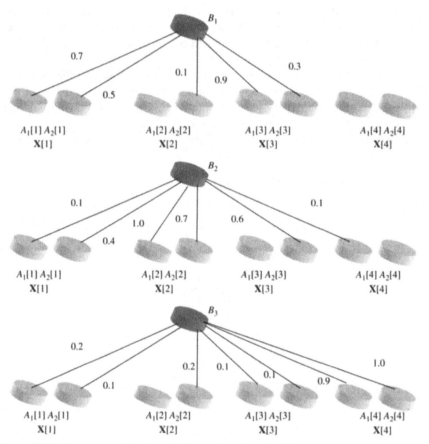

Figure 7.20. Fuzzy sets of order -2 formed on the "local" structure developed within the individual data sets (only nonzero membership grades are shown).

Interestingly, each fuzzy set B_1, B_2, and B_3 captures information about the complexity of the overall structure and how each of the metaclusters is formed as a mixture of fuzzy sets built in $\mathbf{X}[1], \mathbf{X}[2], \ldots, \mathbf{X}[P]$. We note that B_3 is homogeneous, being expressed (described) primarily over $A_3[3]$ and $A_4[3]$. The most heterogeneous one is the first cluster (B_1).

7.9. CONCLUSIONS

Collaborative clustering is useful to achieve interaction between different sources of information for the purpose of revealing underlying structures and regularities within data sets. It can be treated as a process of consensus building where we attempt to reveal a structure that is common across all sets of data. The introduced models of horizontal and vertical clustering achieve an active form of collaboration. We highlighted the aspect of data security and confidentiality: the level of granularity at which the communication takes place is a useful and practical way of retaining these features. The principle of collaborative clustering could be of interest in the design of intelligent agents operating within limited domains and benefiting from various forms of interaction and collaboration; needless to say, this must occur in an *active* mode, and that is what happens in collaborative clustering. Vertical, horizontal, and hybrid clustering are essential mechanisms of communication at some granular level. Various collaborative activities provide insight into the structure of data and help identify patterns that require more attention or could be labeled as outliers.

The models of collaborative clustering were designed under the assumption that each data set has the same number of clusters. While not being particularly limiting, in some cases this requirement could be relaxed, especially if we are interested in forming granular mappings where the assumption we have made so far does not seem to be fully justified. We link collaborative clustering with conditional clustering, which does not support active interaction (relying more on passive constraints) but allowing for various levels of granularity and different number of clusters in each data set, say $c[1], c[2], \ldots, c[P]$.

REFERENCES

J.C. Bezdek, *Pattern Recognition with Fuzzy Objective Function Algorithms*, Plenum Press, New York, 1981.

R.O. Duda, P.E. Hart, D.G. Stork, *Pattern Classification*, 2nd edition, John Wiley, New York, 2001.

F. Hoppner et al., *Fuzzy Cluster Analysis*, John Wiley, Chichester, England, 1999.

A. Kandel, *Fuzzy Mathematical Techniques with Applications*, Addison-Wesley, Reading, MA, 1986.

W. Pedrycz, *Fuzzy Sets Engineering*, CRC Press, Boca Raton, FL, 1995.

W. Pedrycz, Collaborative fuzzy clustering, *Pattern Recognition Letters*, 23, 14, 2002, 1675–1686.

8 Directional Clustering

Clustering is commonly regarded as a direction-free (or relational) activity aimed at the determination of structure in data (Duda et al., 2001). In this chapter, we develop methods that help endow clustering with some directional aspects of information granulation, recognizing that there may be some underlying mapping between information granules expressed in the corresponding spaces. We discuss various generalized objective functions aimed at striking a sound balance between the relational and directional optimization facets of fuzzy clustering.

8.1. INTRODUCTION

As we have emphasized, given multidimensional numeric experimental data, fuzzy clustering produces information granules (fuzzy sets or fuzzy relations). In the most generic version, we are concerned with a single multivariable data set for which a structure has to be revealed. Consider now a situation where we are given two separate data sets. Obviously, fuzzy clustering can discover the structure in each of these data sets treated separately or en bloc (assuming that they both have the same patterns with different features so that their feature vectors can be concatenated). In addition to the clustering itself, we are interested in forming a map between the information granules developed for these two data sets. Ideally, we would like to construct the granules in such a way that the mapping itself is optimized as well, that is, so that it transforms information granules defined in one space (domain) into granules in the other space (codomain) with no significant distortion (mapping error). Evidently, this problem statement goes far beyond the standard clustering optimization discussed so far. It exhibits some properties of collaborative clustering but contains new elements. Figure 8.1 highlights the crux of the problem: we cluster data set $X[1]$ (the notation will be clarified in Section 8.2) according to some performance criterion. The second data set $X[2]$ has to be clustered so that it reveals its granular structure, but here we also require these information granules to result from some logic mapping of information granules already constructed in $X[1]$. As the clustering in $X[2]$ is guided by two essential criteria, one of which pertains to the direction-sensitive

Knowledge-Based Clustering, by Witold Pedrycz
ISBN 0-471-46966-1   Copyright © 2005 John Wiley & Sons, Inc.

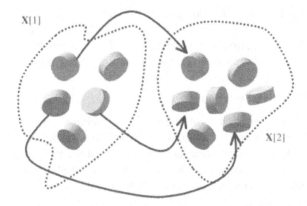

Figure 8.1. A general scheme of directional clustering: clusters in **X**[2] are directional constructs affected by the mapping between information granules in **X**[1] and **X**[2].

transformation from the information granules in **X**[1] to those in **X**[2], we refer to the ensuing optimization as directional clustering.

We start with the formulation of the objective function that helps capture the essence of the problem. Next, we describe its optimization and develop an overall algorithm, including a general flow of the corresponding optimization activities.

8.2. PROBLEM FORMULATION

Let us start by establishing a certain notation. Given are two data sets $X[1]$ and $X[2]$, where $X[1] = \{\mathbf{x}_1[1], \mathbf{x}_2[1], \ldots, \mathbf{x}_N[1]\}$ and $X[2] = \{\mathbf{x}_1[2], \mathbf{x}_2[2], \ldots, \mathbf{x}_N[2]\}$, where $\mathbf{x}_k[1] \in \mathbf{R}^{n_1}$ and $\mathbf{x}_k[2] \in \mathbf{R}^{n_2}$. Information granulation carried out in $X[1]$ involves $c[1]$ fuzzy clusters. For the second data set, the clustering gives rise to $c[2]$ clusters. The clusters constructed in $X[1]$ are denoted by $A_1, A_2, \ldots, A_{c[1]}$. For $X[2]$ the corresponding clusters are $B_1, B_2, \ldots, B_{c[2]}$. As is obvious, in the data as well as in granular constructs we use square indexes to distinguish between the data sets. We are interested in a mapping from $X[1]$ to $X[2]$ or, to be more precise, a correspondence between information granules defined in $X[1]$ and those developed in $X[2]$. Because of the level at which these mappings are formulated (they deal with information granules rather than numbers), we are interested in the logic–based form of transformations (aggregations). In other words, we express information granule B_i as a logic aggregation (ϕ) involving the information granules developed in $X[1]$:

$$B_i = \phi(A_1, A_2, \ldots, A_{c[1]}, \mathbf{w}_i)$$

$i = 1, 2, \ldots, c[2]$. More descriptively, B_i combines A_j's in a logic format to return the corresponding fuzzy relation (fuzzy set) defined in $X[2]$. The above expression comes with a weight vector (parameters) $\mathbf{w}_i$ that is used to calibrate the links between A_j and B_i. We discuss the details of the logic expression in the next section. At this point, it is worth mentioning that we are interested in the development of information granules $\{A_i\}$ and $\{B_j\}$ that satisfy the requirements of relational and directional character of constraints being imposed on data.

8.2.1. The Objective Function

The entire optimization starts with the objective function defined in such a way that it reflects the corresponding computing activities taking place in $X[1]$ and $X[2]$. They have been discussed separately because of the very different character of the required optimization. The activities in $X[1]$ are straightforward: we are concerned with the formation of information granules that capture the essence of the existing structure. Hence the minimized objective function assumes the well-known format

$$Q[1] = \sum_{i=1}^{c[1]} \sum_{k=1}^{N} u_{ik}^2[1] d_{ik}^2[1] \tag{8.1}$$

where

$$d_{ik}^2[1] = \|\mathbf{x}_k[1] - \mathbf{v}_i[1]\|^2$$

is a distance function between the pattern (that is, $\mathbf{x}_k[1]$) and the prototype (denoted here by $\mathbf{v}_i[1]$), with both of them being located in $X[1]$. At this point, it is instructive to emphasize that the partition matrix formed by this optimization yields a collection of fuzzy relations:

$$U[1] = [A_1 \; A_2 \; \ldots \; A_{c[1]}]^T = \begin{bmatrix} A_1 \\ A_2 \\ \ldots \\ A_{c[1]} \end{bmatrix}$$

Note that each A_i is defined in the finite space $X[1]$ (which implies that A_i has a discrete membership function).

In other words, the objective function in $X[1]$ can be regarded as a weighted sum of the distances between the prototypes and the data points, with the weights being the membership grades of the fuzzy relations:

$$Q[1] = \sum_{i=1}^{c[1]} A_i \bullet D_i \tag{8.2}$$

where

$$A_i^2 \bullet D_i = \sum_{k=1}^{N} u_{ik}^2 d_{ik}^2 \tag{8.3}$$

In more detail, we have

$$A_i = [u_{i1}^2 u_{i2}^2 \ldots u_{ic[1]}^2]$$

The minimization of (8.1) or, equivalently, (8.2) is standard as being completed with respect to the partition matrix and the prototypes.

In X[2], the optimization activities are more comprehensive, as we are faced with the two main objectives, that is, relational (finding a structure in X[2]) and directional (achieving granular mapping between information granules in X[1] and X[2]). Because there are two optimization components, the objective function has to reflect this situation. We take into consideration an additive form of $Q[2]$:

$$Q[2] = \sum_{i=1}^{c[2]} \sum_{k=1}^{N} u_{ik}^2[2] d_{ik}^2[2] + \beta \sum_{i=1}^{c[2]} \sum_{k=1}^{N} (u_{ik}[2] - \phi_i(U[1]))^2 d_{ik}^2[2] \qquad (8.4)$$

The first term is just a standard component encountered in the FCM that looks after the optimization of the structure in X[2]. The second term captures the differences between $U[2]$ and the mapping of the structure detected in X[1] (viz., the fuzzy partition $U[1]$) to X[2], that is, $\phi_i(U[1])$. In this sense, it characterizes the performance of the mapping between the information granules. The weight coefficient (β) is used to strike a sound balance between the relational and directional facets of the optimization.

The optimization of (8.1) is standard. The optimization of (8.4) requires detailed investigation. The minimization of the objective function $Q[2]$ is completed with respect to the partition matrix $U[2]$ (structure), the prototypes, and the parameters of the logic transformation (ϕ_i).

8.2.2. The Logic Transformation Between Information Granules

The granular mapping from X[1] to X[2] consists of a logic transformation between the information granules. It is worth stressing that there are many possible types of mappings. Our choice is implied by the transparency of the logic mapping that comes with the logic type of the spaces between which the mapping takes place. Two classes of mappings are discussed.

The first is *OR-based*. As the name indicates, we consider the information granule B_i to be an OR aggregation of the granules in the input space:

$$B_i = A_1 \ or \ A_2 \ or \dots \ or \ A_{c[1]} \qquad (8.5)$$

Not all fuzzy relations in the input space contribute to the formation of B_i, and not all of them have an equal impact on the membership of B_i. To gain this flexibility, we allow for a weight vector (connections) $\mathbf{w}_i$ whose role is to articulate (quantify) the contribution coming from A_j's. The following modification is made to (8.5):

$$B_i = (A_1 \ and \ w_{i1}) \ or \ (A_2 \ and \ w_{i2}) \ or \dots \ or \ (A_{c[1]} \ and \ w_{ic[1]}) \qquad (8.6)$$

where $\mathbf{w}_i = [w_{i1} w_{i2} \ldots w_{ic[1]}]$ are weights with values confined to the unit interval. The logic operations are realized by t- and s-norms, leading to the equivalent expression of (8.6) (s-t realization of the logic expression)

$$B_i = (A_1 \; t \; w_{i1}) \; s \; (A_2 \; t \; w_{i2}) \; s \ldots s (A_{c[1]} \; t \; w_{ic[1]})$$

and   (8.7)

$$B_i = \overset{c[1]}{\underset{j=1}{S}} (A_j t w_{ij})$$

Note that the above logic mapping concerns a single fuzzy relation in X[2]. In similar fashion, we can map the remaining information granules. After a careful examination of the mappings viewed together, we come up with the following concise notation. Arrange all weights in a matrix form

$$R = [r_{ij}] = \begin{bmatrix} w_{11} & \cdots & & w_{1c[2]} \\ \cdots & & & \\ & & w_{lj} & \\ w_{c[1]1} & \cdots & & w_{c[1]c[2]} \end{bmatrix}$$ (8.8)

Then the mapping from information granules in X[1] to information granules in X[2] is nothing but a fuzzy relational equation with a standard s-t composition (Di Nola et al., 1989) denoted here by a small dot

$$\mathbf{B} = \mathbf{A} \circ R$$ (8.9)

where $\mathbf{A} = [A_1 \; A_2 \; A_{c[1]}]$ and $\mathbf{B} = [B_1 \; B_2 \; B_{c[2]}]$.

The second type of mapping is *AND-based* aggregation of the information granules, meaning that we consider B_i to be a combination of A_j's aggregated AND-wise:

$$B_i = A_1 \; and \; A_2 \; and \; \ldots \; and \; A_{c[1]}$$ (8.10)

The straightforward generalization of this aggregation includes a weight vector and, subsequently, the combination of the t-s type

$$B_i = (A_1 \; or \; w_{i1}) \; and \; (A_2 \; or \; w_{i2}) \; and \ldots \; and \; (A_{c[1]} \; or \; w_{ic[1]})$$ (8.11)

$$B_i = (A_1 \; s \; w_{i1}) \; t \; (A_2 \; s \; w_{i2}) \; t \ldots t \; (A_{c[1]} \; s \; w_{ic[1]}) = \overset{c[1]}{\underset{j=1}{T}} (A_j s w_{ij})$$ (8.12)

These two aggregation mechanisms are dual in the sense of their logical functionality. Owing to the character of the AND and OR operations, their use depends on the number of information granules in the respective spaces. Intuitively, if $c[1]$ is greater than $c[2]$, we consider the OR type of aggregation (anticipating that the element in the output space is constructed as a union of several information granules in the input space; see Figure 8.2). Similarly, for $c[1]$ less than $c[2]$,

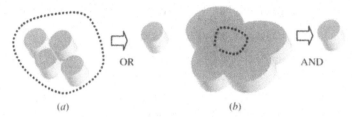

Figure 8.2. Logical character of mapping between information granules: (*a*) a generalization effect produced by the OR type of combination and (*b*) a specialization effect provided by the AND type of aggregation of information granules.

the AND type of aggregation is more appealing (as we project that B_i is made more specific in relation to the information granules existing in $X[1]$).

8.3. THE ALGORITHM

At this point we, are ready to proceed with the computational details for $Q[2]$ that lead to the complete algorithm. The objective function implies the following optimization task:

$$\min_{U[2],v_1[1],v_2[2],\ldots,v_{c[2]}]} Q[2]$$

subject to

$$U[2] \in \mathbf{U}$$

and

$$R \in \mathsf{R} \tag{8.13}$$

where the family of partition matrices $\mathbf{U}$ is defined in the usual manner (namely, we require that the elements in each column of $U[2]$ sum to 1 and that the sum of the elements in each row of R is nonzero and lower than N). R is an element of the family of the fuzzy relations R (viz., matrices with elements confined to the unit interval). The above optimization problem concerns a way of forming a structure in $X[2]$ with inclusion of the mapping properties. The clustering mechanisms in $X[1]$ follow the standard FCM and will not be discussed here.

The optimization of the partition matrix $U[2]$ in the objective function uses Lagrange multipliers (because of the constraint in the development of the partition matrix). For a given data point (k), we form an augmented objective function

$$V = \sum_{i=1}^{c[2]} \sum_{k=1}^{N} u_{ik}^2[2]d_{ik}^2[2] + \beta \sum_{i=1}^{c[2]} \sum_{k=1}^{N} (u_{ik}[2] - \phi_i(U[1]))^2 d_{ik}^2[2]$$

$$+ \lambda \left(\sum_{i=1}^{c[2]} u_{ik}[2] - 1 \right) \tag{8.14}$$

where λ is a Lagrange multiplier. Proceeding with the necessary conditions for the minimum of V

$$\frac{\partial V}{\partial u_{st}[2]} = 0 \quad \frac{\partial V}{\partial \lambda} = 0$$

we calculate

$$2u_{st}[2]d_{st}^2[2] + 2\beta(u_{st}[2] - y_{st})d_{st}^2[2] + \lambda = 0 \tag{8.15}$$

y_{st} stands for a logic-based mapping between the information granules:

$$y_{st} = \mathop{S}_{j=1}^{c[1]} (u_{st}[1]tr_{sj})$$

Computing $u_{st}[2]$ from (8.15), we obtain

$$u_{st}[2] = \frac{\beta y_{st} d_{st}^2[2] - \lambda}{2d_{st}^2[2](1 + \beta)}$$

As the membership grades sum to 1, this leads us to the expression

$$\sum_{j=1}^{c[2]} \frac{\beta y_{jt} d_{jt}^2[2] - \lambda}{2d_{jt}^2[2](1 + \beta)} = 1$$

and in the sequel

$$\frac{\lambda}{1 + \beta} = \frac{-1 + \dfrac{\beta}{1 + \beta} \displaystyle\sum_{j=1}^{c[2]} y_{jt}}{\displaystyle\sum_{j=1}^{c[2]} \frac{1}{d_{jt}^2[2]}}$$

We introduce the notation

$$\tilde{u}_{st}[2] = \frac{1}{\displaystyle\sum_{j=1}^{c[2]} \dfrac{d_{st}^2[2]}{d_{jt}^2[2]}}$$

Finally, we get

$$u_{st}[2] = \tilde{u}_{st}[2] + \frac{\beta}{1 + \beta}\left(y_{st} - \tilde{u}_{st}[2]\sum_{j=1}^{c[2]} y_{jt}\right) \tag{8.16}$$

$s = 1, 2, \ldots, c[2]$, $t = 1, 2, \ldots, N$.

The above formula has an interesting interpretation: if $\beta = 0$, then it reduces to the well-known formula for the partition matrix encountered in the FCM. When β increases, $u_{st}[2]$ is affected by the second term in (8.16).

The calculations of the prototypes do not come with any constraints, so we follow the necessary condition for the minimum of $Q[2]$, namely, $\dfrac{\partial Q[2]}{\partial v_{st}[2]} = 0$, $s = 1, 2, \ldots, c[2]$, $t = 1, 2, \ldots, n_2$.

In light of the weighted Euclidean distance governed by the expression

$$d_{ik}^2[2] = \sum_{j=1}^{n_2} \frac{(x_k[j] - v_{ij}[2])^2}{\sigma_j^2[2]} \tag{8.17}$$

(where $\sigma_j^2[2]$ denotes a variance of the jth variable), the above derivative is equal to

$$\frac{\partial Q[2]}{\partial v_{st}[2]} = 2\beta \sum_{k=1}^{N} u_{sk}^2[2] \frac{(x_k[t] - v_{st}[2])}{\sigma_t^2[2]} - 2\beta \sum_{k=1}^{N} \psi_{sk} \frac{(x_k[t] - v_{st}[2])}{\sigma_t^2[2]} \tag{8.18}$$

with the following notation:

$$\psi_{sk} = (u_{sk}[2] - y_{sk})^2$$

Bearing in mind the necessary condition for the minimum of $Q[2]$ with respect to the prototypes, they are equal to

$$v_{st}[2] = \frac{\displaystyle\sum_{k=1}^{N} x_k[t](u_{sk}^2 + \beta\psi_{sk})}{\displaystyle\sum_{k=1}^{N} (u_{sk}^2 + \beta\psi_{sk})} \tag{8.19}$$

Note that, when $\beta = 0$, we arrive at the standard expression for the prototypes that is identical to the one in the FCM method.

Finally, we optimize the fuzzy relation R describing the logic mapping between the spaces. In general, the solution is not expressed analytically and we have to use iterative optimization. The expression governing this optimization is

$$R(\text{iter} + 1) = R(\text{iter}) - \beta\nabla_{R(\text{iter})}Q \tag{8.20}$$

where the fuzzy relation is transformed on the basis of the gradient of the performance index Q. The learning rate shown above ($\beta > 0$) controls the rate of changes of the updates of the fuzzy relation. The gradient itself is computed for specific triangular norms. In what follows (and all experiments shown in Section 8.5 will exploit these assumptions), we consider two common models of the logic connectives, such as a product (t-norms) and a probabilistic sum (s-norm). On this basis, the gradient is

$$\frac{\partial y_{sk}}{\partial r_{st}} = (1 - A_{st})u_{tk}[1] \tag{8.21}$$

where A_{st} denotes an s-t composition that excludes the currently optimized element of the fuzzy relation

$$A_{st} = \mathop{S}_{\substack{j=1 \\ j \neq t}}^{c[1]} (u_{jk}[1]tr_{sj}) \tag{8.22}$$

8.4. THE DEVELOPMENT FRAMEWORK OF DIRECTIONAL CLUSTERING

The way in which the information granules are built stipulates a certain flow of optimization activities. These can be organized into two main phases (see Figure 8.3). The initial phase concentrates on the clustering completed independently for the two data sets X[1] and X[2]. The intent here is to establish some preliminary structure in the data so that we can have a reasonable starting point to proceed with the collaboration and further refine the initial relationships. During the second phase, the clustering processes start to collaborate through the mapping. At the same time, the fuzzy relation is subject to gradient-based optimization (as illustrated in Figure 8.3, this is an integral portion of the collaboration process and negotiation of the granular structures). Because of the direction of the mapping, the clustering in X[1] is not affected, while the relational and directional facets of the clusters emerge at the side of X[2].

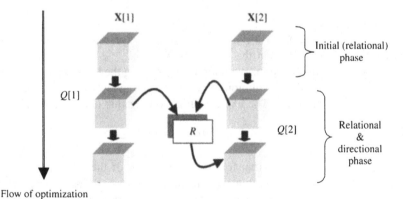

Figure 8.3. Relational and directional optimization of information granules—an overall development scheme.

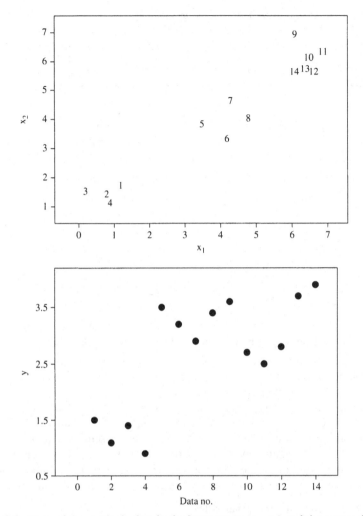

Figure 8.4. Plots of the synthetic data in the input space (x_1, x_2) and the output data (y).

8.5. NUMERICAL STUDIES

The proposed algorithmic framework is illustrated by numeric experiments. They include synthetic as well as real-world data available on the World Wide Web.

Synthetic Data. This experiment concerns three-dimensional data. The first data set includes input variables (x_1 and x_2), and the second involves one-dimensional data (y) (see Table 8.1 and Figure 8.4).

The same data points are shown in a three-dimensional space in Figure 8.5. This helps reveal the structure. The output variable has two clearly visible clusters. Moreover, the three clusters in the input space relate to the two clusters

TABLE 8.1. Synthetic Data Used in the Experiment

x_1	1.2	0.8	0.2	0.9	3.5	4.2	4.3	4.8	6.1	6.5	6.9	6.6	6.4	6.1
x_2	1.8	1.5	1.6	1.2	3.9	3.4	4.7	4.1	7.0	6.2	6.4	5.7	5.8	5.7
y	1.5	1.1	1.4	0.9	3.5	3.2	2.9	3.4	3.6	2.7	2.5	2.8	3.7	3.9

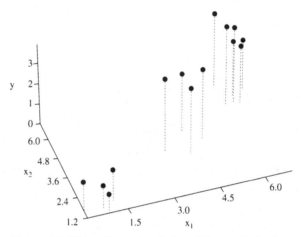

Figure 8.5. Three-dimensional plot of the synthetic data.

in the output space. In more detail, we note that the two clusters in the input space $X[1]$ map on a single cluster in $X[2]$.

Following these observations (which are easy to arrive at, as we are dealing with low-dimensional synthetic data), we set up $c[1] = 3$ and $c[2] = 2$. In the experiment, the learning rate of the gradient-based learning is equal to 0.1. This relatively low value helps avoid oscillations (which is more important than an eventual slowdown of the learning process itself). With no collaboration ($\beta = 0.0$), the obtained clusters are described in terms of the partition matrices:

Partition—space of input variables

$$
U[1] = \begin{bmatrix}
0.017434 & 0.005164 & 0.977402 \\
0.000068 & 0.000022 & 0.999910 \\
0.013630 & 0.004868 & 0.981502 \\
0.007460 & 0.002543 & 0.989997 \\
0.938615 & 0.030536 & 0.030849 \\
0.936182 & 0.032897 & 0.030921 \\
0.894294 & 0.082808 & 0.022897 \\
0.946818 & 0.039856 & 0.013326 \\
0.069128 & 0.914942 & 0.015930 \\
0.001119 & 0.998669 & 0.000212 \\
0.020588 & 0.974971 & 0.004441 \\
0.027533 & 0.967885 & 0.004582 \\
0.015457 & 0.982032 & 0.002511 \\
0.045319 & 0.948161 & 0.006520
\end{bmatrix}
$$

Partition—output variable

$$U[2] = \begin{bmatrix} 0.982731 & 0.017269 \\ 0.994259 & 0.005740 \\ 0.994833 & 0.005167 \\ 0.976867 & 0.023133 \\ 0.010273 & 0.989727 \\ 0.001394 & 0.998606 \\ 0.049299 & 0.950701 \\ 0.003565 & 0.996435 \\ 0.019316 & 0.980684 \\ 0.137239 & 0.862761 \\ 0.281143 & 0.718857 \\ 0.086491 & 0.913509 \\ 0.029929 & 0.970070 \\ 0.053702 & 0.946298 \end{bmatrix}$$

Subsequently, the prototypes are equal to

Input space: $\mathbf{v}_1[1] = [\,4.204584 \quad 4.015867\,]$

$\mathbf{v}_2[1] = [\,6.439176 \quad 6.117275\,]$

$\mathbf{v}_3[1] = [\,0.777533 \quad 1.524844\,]$

Output space: $v_1[2] = 1.265061 \quad v_2[2] = 3.272303$

The clusters emerging in both spaces are well delineated, with a very limited overlap. The clustering is carried out for several levels of collaboration (β).

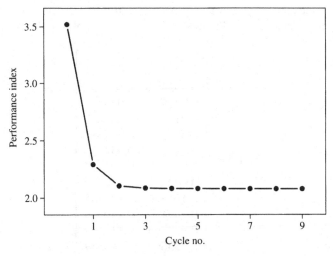

Figure 8.6. Performance index in successive development cycles ($\beta = 0.1$, $\alpha = 0.1$).

Following the general scheme (Section 8.4), we implement the collaboration after the initial clustering of the individual data. This requires five iterations. The performance index achieved throughout the clustering and learning of the relations is visualized in Figure 8.6 (note that the term "cycle" used there concerns the performance index recorded for a single clustering iteration and 20 learning epochs of the gradient-based learning). The optimization is efficient, as the values of the performance index are reduced from cycle to cycle.

Once the optimization has been completed, the fuzzy partition in the output space is

$$
U[2] = \begin{bmatrix}
0.980664 & 0.019336 \\
0.994456 & 0.005544 \\
0.993309 & 0.006691 \\
0.976525 & 0.023475 \\
0.010258 & 0.989742 \\
0.001312 & 0.998688 \\
0.045444 & 0.954556 \\
0.003853 & 0.996147 \\
0.023591 & 0.976409 \\
0.128656 & 0.871345 \\
0.259470 & 0.740530 \\
0.082898 & 0.917102 \\
0.033662 & 0.966339 \\
0.055437 & 0.944563
\end{bmatrix}
$$

We do not report the results of clustering in the input space, as in this model these fuzzy sets have not been affected. When comparing this partition matrix with the one obtained for the clustering without any collaboration, we conclude that there are no substantial differences. Obviously, the collaboration effect is quite limited, and this may be a reason for the evident coincidence in the information granules (conveyed in the respective partition matrices). The prototypes do not change when the collaboration effect comes into play at this level (namely, for $\beta = 0.1$).

$$
U[2] = \begin{bmatrix}
0.980664 & 0.019336 \\
0.994456 & 0.005544 \\
0.993309 & 0.006691 \\
0.976525 & 0.023475 \\
0.010258 & 0.989742 \\
0.001312 & 0.998688 \\
0.045444 & 0.954556 \\
0.003853 & 0.996147 \\
0.023591 & 0.976409 \\
0.128656 & 0.871345 \\
0.259470 & 0.740530 \\
0.082898 & 0.917102 \\
0.033662 & 0.966339 \\
0.055437 & 0.944563
\end{bmatrix}
$$

What becomes of interest is a fuzzy relation revealing the main relationships between the information granules (fuzzy sets) in the input and output spaces:

$$R = \begin{bmatrix} 0.000000 & 0.063547 & 0.738940 \\ 0.710587 & 0.889359 & 0.005974 \end{bmatrix}$$

There is a strong dependency (relationship) between the granules quantified by high membership grades. Denoting the fuzzy sets by A_1, A_2, and A_3 (input space) and B_1 and B_2 (output space), we translate the above fuzzy relation into two logic expressions:

$$B_1 = A_3 \ (0.73)$$

$$B_2 = A_1 \ (0.71) \ or \ A_2 \ (0.89)$$

(Note that we have included only the terms with high levels of association; the associations themselves are simply the corresponding entries of the fuzzy relation.)

Now we increase the value of β to 0.4. This change becomes reflected in the partition matrix, whose entries now start to diverge from the ones without any collaboration. Figure 8.7 illustrates these new membership grades of the partition matrices.

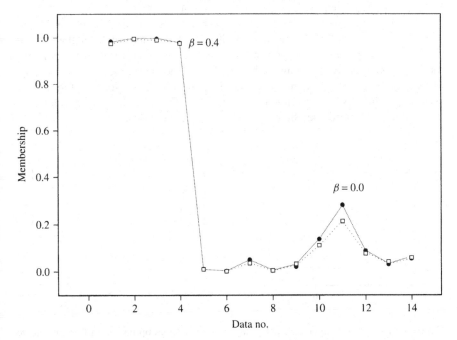

Figure 8.7. Changes in membership grades as a result of collaboration.

The differences between the prototypes are still negligible, as they are equal to $v_1[2] = 1.25503$ and $v_2[2] = 3.255817$, respectively. The fuzzy relation of the associations is now equal to

$$R = \begin{bmatrix} 0.000000 & 0.060559 & 0.947464 \\ 1.000000 & 0.908921 & 0.005793 \end{bmatrix}$$

Subsequently, the list of logical expressions is similar to the one obtained before:

$$B_1 = A_3 \ (0.95)$$

$$B_2 = A_1 \ (1.00) \ or \ A_2 \ (0.91)$$

However, now the strength of the associations between the information granules has been increased.

Higher values of β may lead to some instability, as the mechanisms used in the method tend to compete. This is visible in Figure 8.8; for higher β, the performance index tends to oscillate. These oscillations become more visible once the structures rely on each other more significantly (β increases). The lack of stability indicates that the structures no longer collaborate but tend to compete.

Auto-mpg dataset comes from the UCI repository of machine learning (http://www.ics.uci.edu/~mlearn/MLSummary.html) and concerns a collection of vehicles described in terms of their displacement, weight, country of origin, and so on. We consider all features but fuel efficiency (expressed in miles pcr gallon) as inputs. Fuel efficiency is treated as the output variable.

Clustering is carried out for different number of clusters in the input and output spaces. The level of collaboration (β) is maximized as much as the stability is retained. The results are summarized in the form of fuzzy relations (with the most essential links being highlighted) and the prototypes in the input and output spaces. Table 8.2 contains a sample of the findings. Notably, there are no significant changes to the information granules. The granules in the input space start to become more specific once their number increases.

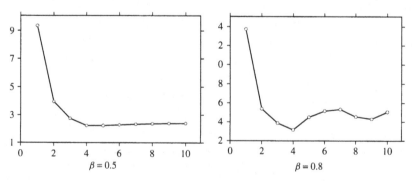

Figure 8.8. Objective function $Q[2]$ in successive cycles of optimization for two selected values of β.

TABLE 8.2. Results of Collaboration Between Clusters in the Input and Output Spaces for a Selected Number of the Clusters in the Input Space, $c[1] = 4$, 5, and 7 and $c[2] = 2$, and Maximal Values of the Collaboration Factors (β) and Obtained Prototypes of the Clusters

$\beta = 0.1$

$$R = \begin{bmatrix} 0.00 & \mathbf{1.00} & \mathbf{0.62} & 0.00 \\ \mathbf{0.85} & 0.00 & 0.00 & 0.36 \end{bmatrix}$$

Prototypes

INPUT SPACE

$v_1[1] = [\,4.08 \quad 103.92 \quad 76.34 \quad 2219.61 \quad 16.55 \quad 77.51 \quad 2.66\,]$
$v_2[1] = [\,7.94 \quad 347.37 \quad 160.89 \quad 4164.82 \quad 12.66 \quad 73.349 \quad 1.01\,]$
$v_3[1] = [\,5.88 \quad 218.63 \quad 99.789 \quad 3196.26 \quad 16.62 \quad 75.81 \quad 1.11\,]$
$v_4[1] = [\,4.23 \quad 127.19 \quad 83.87 \quad 2475.72 \quad 16.14 \quad 77.31 \quad 1.42\,]$

OUTPUT SPACE

$v_1[2] = 17.53 \quad v_2[2] = 31.19$

$\beta = 0.15$

$$R = \begin{bmatrix} \mathbf{1.00} & 0.00 & \mathbf{0.51} & 0.00 & 0.02 \\ \mathbf{0.63} & \mathbf{1.00} & 0.31 & \mathbf{0.52} & 0.35 \end{bmatrix}$$

Prototypes

INPUT SPACE

$v_1[1] = [\,4.08 \quad 102.48 \quad 73.39 \quad 2188.47 \quad 16.73 \quad 78.91 \quad 2.79\,]$
$v_2[1] = [\,7.96 \quad 350.07 \quad 162.00 \quad 4189.60 \quad 12.61 \quad 73.22 \quad 1.01\,]$
$v_3[1] = [\,4.27 \quad 137.07 \quad 85.29 \quad 2578.33 \quad 16.20 \quad 79.02 \quad 1.18\,]$
$v_4[1] = [\,6.068 \quad 230.51 \quad 102.26 \quad 3290.02 \quad 16.56 \quad 75.63 \quad 1.06\,]$
$v_5[1] = [\,4.16 \quad 113.299 \quad 84.54 \quad 2359.73 \quad 15.94 \quad 74.03 \quad 2.03\,]$

OUTPUT SPACE

$v_1[2] = 31.11 \quad v_2[2] = 17.78$

$\beta = 0.2$

$$R = \begin{bmatrix} 0.21 & 0.26 & 0.03 & 0.00 & 0.01 & \mathbf{0.75} & \mathbf{1.00} \\ 0.22 & \mathbf{1.00} & \mathbf{1.00} & \mathbf{0.62} & 0.18 & 0.13 & 0.32 \end{bmatrix}$$

Prototypes

INPUT SPACE

$v_1[1] = [\,4.09 \quad 108.24 \quad 83.89 \quad 2309.29 \quad 15.86 \quad 73.81 \quad 2.21\,]$
$v_2[1] = [\,7.97 \quad 362.46 \quad 168.89 \quad 4264.13 \quad 12.25 \quad 72.36 \quad 1.00\,]$
$v_3[1] = [\,7.79 \quad 309.14 \quad 138.76 \quad 3885.15 \quad 13.94 \quad 76.40 \quad 1.02\,]$
$v_4[1] = [\,5.95 \quad 226.26 \quad 99.46 \quad 3241.93 \quad 16.69 \quad 75.18 \quad 1.057\,]$
$v_5[1] = [\,4.33 \quad 130.68 \quad 84.46 \quad 2507.56 \quad 16.51 \quad 76.25 \quad 1.50\,]$
$v_6[1] = [\,4.217 \quad 135.32 \quad 84.60 \quad 2569.33 \quad 16.17 \quad 79.53 \quad 1.18\,]$
$v_7[1] = [\,4.04 \quad 99.76 \quad 71.44 \quad 2154.74 \quad 16.82 \quad 79.41 \quad 2.86\,]$

OUTPUT SPACE

$v_1[2] = 31.25 \quad v_2[2] = 17.65$

TABLE 8.3. Fuzzy Relations of Connections for $c[1] = c[2] = 2$, 3, and 6 with $\beta = 0.05$

$$R = \begin{bmatrix} 0.000000 & \mathbf{0.949054} \\ \mathbf{0.420400} & 0.000000 \end{bmatrix}$$

$$R = \begin{bmatrix} 0.110320 & 0.025040 & \mathbf{0.247469} \\ 0.000000 & \mathbf{0.877086} & 0.000000 \\ \mathbf{0.210435} & 0.000000 & 0.000000 \end{bmatrix}$$

$$R = \begin{bmatrix} \mathbf{0.090541} & 0.003166 & 0.000000 & 0.000788 & 0.000000 & 0.003596 \\ 0.004518 & 0.012588 & 0.076320 & 0.017965 & \mathbf{0.106748} & 0.006710 \\ 0.000000 & 0.051439 & \mathbf{0.156852} & 0.109466 & 0.000000 & 0.000000 \\ \mathbf{0.160533} & 0.000564 & 0.000000 & 0.000000 & 0.000000 & 0.037163 \\ 0.031722 & 0.004176 & 0.000000 & 0.003358 & \mathbf{0.117081} & 0.075881 \\ 0.000000 & \mathbf{0.706356} & 0.000000 & 0.000000 & 0.000000 & 0.000000 \end{bmatrix}$$

It is interesting to note that the collaboration can be made more vigorous without sacrificing stability when the number of clusters in the input space increases. This could have been expected, as the resulting information granules tend to be smaller (of higher granularity) and therefore could be moved around more freely, causing little distortion (and hence instability) during the collaboration process. The dependencies between the information granules, as expressed by the fuzzy relations, discriminate quite well between strong and weak links. In other words, the fuzzy relations start to contain values close to either 0 or 1. This shows that some information granules relate very strongly.

Now let us consider the same number of clusters in both spaces (Table 8.3). This arrangement helps us reveal how the granules relate in the two spaces. Because the number of fuzzy sets is the same, the logic formula may consist of one-to-one mapping, namely, mapping a single information granule in the input space to an information granule in the output space. Obviously, this happens at the level of information granules rather than numeric quantities. If we consider the entries of the fuzzy relations, the observation about this one-to-one mapping is completely legitimate. In each row of the fuzzy relation, there is only one dominant membership grade (indicated in boldface in Table 8.3). Notably, these are not necessarily high membership values. This is, however, justified, as the entries for the partition matrices start to lower once the number of clusters increases (recall that these membership grades must add up to 1).

The graph of links between the information granules for $c = 2$ and 3 is included in Figure 8.9 (we show only the most dominant connections).

8.6. CONCLUSIONS

In this chapter, we raised the issue of designing information granules (fuzzy sets or fuzzy relations) that take into consideration the structure in a data set, as well as addressing the mapping occurring at the level of these information granules.

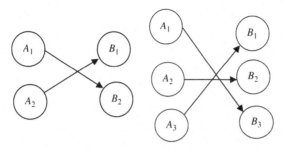

Figure 8.9. A graphical illustration of linkages between information granules in X[1] and X[2].

The novelty of this approach lies in this multifaceted aspect of information granulation. Fuzzy clustering itself is concerned with the structure of the data, not the nature of possible mappings. It is evident that fuzzy clustering, no matter which technique it involves, tackles the *relational* nature of data (so, when searching for a structure, no direction is taken into consideration). The augmented objective function includes an additional component to make the information granules consonant with the mapping requirements (which comes with a *directional* component). The additive form of the objective function with a modifiable component of directional activities makes it possible to model the level of influence attained at the level of the mapping formed by the granules in **X**[1]. In this way, it helps avoid potential competition in case of incompatible structures and the associated mapping.

Logic-based mapping (which relates to the use of fuzzy relational equations) is a consequence of the logic framework of information granules. One can, however, apply other types of mapping, including those implemented via neural networks. This generalizes the approach and promotes it as a general model of collaborative granular computing.

It is interesting to contrast directional clustering with collaborative clustering. While there is some similarity, primarily caused by the fact that a number of data sets are involved in the construction of information granules, the differences are quite significant. First, in collaborative clustering, we make an assumption about the equal number of clusters (so that collaboration can be invoked). Second, in collaborative clustering, we are considering the clustering algorithms operating at both data sets to send collaboration messages. In the directional approach, we emphasize the collaborative nature of the message generated within **X**[1] and transmitted to **X**[2] through logic granular mapping. The unidirectional nature of this communication is inherently associated with the mapping.

One can envision some extensions. While they do not affect the principles of directional clustering, they seem to be helpful in formulating the problem. The first extension deals with several data sets serving as inputs. The formalism is not modified, but the logic transformation can be established in the way it operates on input information granules formed for each dataset.

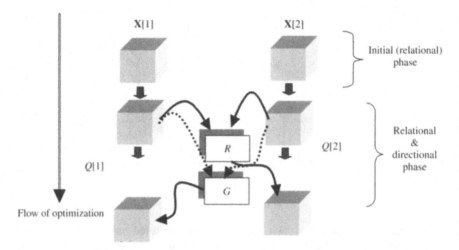

Figure 8.10. Bi-directional clustering: the concept and communication mechanism.

The second extension is concerned with the directional mapping achieved by two separate logic transformations. This evidently brings us closer to the model of collaborative clustering; the only exceptions are the different number of clusters in each data set and the message about the structure sent through the logic mapping. Through bidirectional clustering (this term is more reflective of the functionality of this algorithm), we construct a pair of mappings (R and G) whose domains and codomains are the reverse of each other. The computational concept is displayed in Figure 8.10. Both objective functions $Q[1]$ and $Q[2]$ involve the directional component in their formulation.

The directional model of information granulation is also in line with a broad range of techniques of fuzzy modeling; see Jang (1993), Delgado et al. (1998), Kandel (1986), Ma et al. (2000), Pedrycz and Vasilakos (1999), and Setnes (2000). The development of transparent models and building information granules consistent with the structure in the data and complying with the mapping requirements has been accomplished through the composite objective function of directional clustering.

REFERENCES

M. Delgado, A.F. Gomez-Skarmeta, F. Martin, A methodology to model fuzzy systems using fuzzy clustering in a rapid-prototyping approach, *Fuzzy Sets and Systems*, 97, 3, 1998, 287–302.

A. Di Nola, S. Sessa, W. Pedrycz, E. Sanchez, *Fuzzy Relational Equations and Their Applications in Knowledge Engineering*, Kluwer Academic Press, Dordrecht, 1989.

R.O. Duda, P.E. Hart, D.G. Stork, *Pattern Classification*, 2nd edition, John Wiley, New York, 2001.

J.-S.R. Jang, ANFIS: adaptive-network-based fuzzy inference system, *IEEE Trans. on Systems, Man, and Cybernetics*, 23, 3, 1993, 665–685.

A. Kandel, *Fuzzy Mathematical Techniques with Applications*, Addison-Wesley, Reading, MA, 1986.

M. Ma, Y.-Q. Zhang, G. Langholz, A. Kandel, On direct construction of fuzzy systems, *Fuzzy Sets and Systems*, 112, 2000, 165–171.

W. Pedrycz, A.V. Vasilakos, Linguistic models and linguistic modeling, *IEEE Trans. on Systems, Man, and Cybernetics*, 29, 6, 1999, 745–757.

M. Setnes, Supervised fuzzy clustering for rule extraction, *IEEE Trans. on Fuzzy Systems*, 8, 4, 2000, 416–424.

9 Fuzzy Relational Clustering

The patterns we have discussed so far are characterized by a vector of features and thus are considered to be points in a feature space. In practice, we can envision situations where, instead of individual patterns, specification and characterization are given for individual pairs of patterns. In this case, we say that such patterns are described in relational form, hence the name of the clustering algorithm.

9.1. INTRODUCTION AND PROBLEM STATEMENT

As their name indicates, relational data are concerned with structures that are formed on the basis of relationships (or relations) defined among original patterns. The typical predicate or relation we can define may deal with the concept of similarity or dissimilarity between pairs of patterns. The other predicate may be one of complementary character, such as a predicate expressing a degree of · difference or dissimilarity between pairs of patterns. Imagine that we have a set of N cities and compute the distances between pairs of them. Denote the result by d_{ij}. Now these distances are treated as new relational patterns. Within these patterns we characterize distance-oriented similarity clusters that could be discovered between the cities. This structure in the distance space is usually different from the one we could have expected to find in the original data (see Figure 9.1).

As another example, consider a typical biometric problem involving a collection of faces that need to be classified. It is natural to compare them in a pairwise manner. For instance, it makes sense to determine that face 1 is similar to face 4 to a certain degree. Next, we wish to cluster these proximity degrees. Clustering the original faces could have been more challenging for several reasons. First and foremost, constructing a meaningful feature space could be difficult. Second, even if this can be accomplished in a meaningful manner, the dimensionality of the resulting feature space may lead to a sparse distribution of patterns. The structure in such a data set may be quite unstable. A similar problem of sparse data arises when searching for structure in a collection of Web pages. A simple scenario in this search is to envision the use of textual information to form the feature space. In this sense, keywords could be plugged in as the entries of the feature vectors. Given the immense diversity of textual information, one can expect the

Knowledge-Based Clustering, by Witold Pedrycz
ISBN 0-471-46966-1   Copyright © 2005 John Wiley & Sons, Inc.

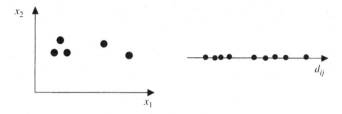

Figure 9.1. A collection of cities in two-dimensional space and the relational characteristics involving distances between them.

feature vectors to be very long. This high dimensionality could easily make the clustering of the data quite unreliable. The same patterns compared pairwise lead to the highly reduced feature space where the clustering takes place.

9.2. FCM FOR RELATIONAL DATA

The relational version of the FCM algorithm, that is, a clustering method applicable to relational data was introduced by Bezdek and Hathaway (1987), along with its extensions (Hathaway et al., 1996; Hathaway and Bezdek, 1994, 2002; Hathaway, et al., 1989; Runkler and Bezdek, 2003). The relational format of data assumes that there are degrees of dissimilarity between patterns rather than original patterns. Denote them by R_{ij}, where $i, j = 1, 2, \ldots, N$. They are conveniently organized in the matrix form $R = [R_{ij}]$. The following intuitively straightforward properties of R_{ij} are assumed to occur:

R_{ij} (nonnegativity)
$R_{ij} = R_{ji}$ (symmetry)
$R_{ii} = 0$ (dissimilarity of any pattern compared to itself is equal to zero)

Alternatively, one could define similarity as a complement of R_{ij} (assuming that these values are normalized), $D_{ij} = 1 - R_{ij}$. Clustering is then carried out on the basis of information about dissimilarities. Taking into account the above properties, we note that N patterns give rise to $N^2/2 - N = N(N-1)/2$ different pairs of dissimilarity values for the existing pairs of patterns. Those are now the entries (relational patterns) used to guide the relational clustering.

Like the standard FCM, its relational version is an iterative optimization scheme composed of the following phases. We first go through the process of computing the structure and then elaborate on several crucial components

Initialization. As in the standard FCM, we pick up the number of clusters (c), distance function $\| \ \|$, fuzzification factor (m), and stopping threshold ε. In addition, we set a spreading factor equal to 0, $\beta = 0$. Initiate a partition matrix $U = [u_{ik}], i = 1, 2, \ldots c, k = 1, 2, \ldots N$ to some random values (obviously, we assume that this initialization satisfies the set of requirements imposed on partition matrices).

Iterate. Update cluster centers $\mathbf{v}_i$, $i = 1, 2, \ldots, c$ in the following way:

$$\mathbf{v}_i = \begin{bmatrix} u_{i1}^m \\ u_{i2}^m \\ u_{iN}^m \end{bmatrix} \bigg/ \sum_{k=1}^{c} u_{ik}^m \tag{9.1}$$

Compute distances based on the given dissimilarity matrix R:

$$(d_{ik})^2 = (R^\beta \mathbf{v}_i)_k - \tfrac{1}{2}\mathbf{v}_i^T R^\beta \mathbf{v}_i \tag{9.2}$$

where

$$R_{ik}^\beta = \begin{cases} R_{ik} + \beta & \text{if } i \neq k \\ 0 & \text{if } i = k \end{cases} \tag{9.3}$$

and $(\ )_k$ denotes the kth entry of the given vector.

Check if the nonnegativity condition is satisfied for all entries of $(d_{ik})^2$, that is, do the following: If $(d_{ik})^2 < 0$ for any i and k, calculate

$$\Delta\beta = \max_{i,k}\left\{ -\frac{2(d_{ik})^2}{\|\mathbf{v}_i - \mathbf{e}_k\|^2} \right\} \tag{9.4}$$

and modify the values of $(d_{ik})^2$ as follows:

$$(d_{ik})^2 = (d_{ik})^2 + \frac{\Delta\beta}{2}\|\mathbf{v}_i - \mathbf{e}_k\|^2 \tag{9.5}$$

Subsequently, increase the value of the spread factor β by the increment defined above:

$$\beta = \beta + \Delta\beta \tag{9.6}$$

Above, $\mathbf{e}_k$ denotes the kth unit vector in $[0, 1]^N$.

Update partition matrix U in the usual way, that is, compute the membership grades:

$$u_{ik} = \frac{1}{\sum_{j=1}^{c}\left(\dfrac{d_{ik}^2}{d_{jk}^2}\right)^{\frac{1}{m-1}}} \tag{9.7}$$

until the given stopping criterion with the threshold value ε has been satisfied.

The above algorithm requires some explanation concerning the formation of the cluster centers. First and foremost, we learn that the prototypes are computed on the basis of the partition matrix U and do not directly involve the patterns.

This is easily explainable; we search the structure based on relational data (that is, dissimilarity or similarity values). The dimensionality of centers is equal to N, dim $(\mathbf{v}_i) = N$. As no assumption has been made with regard to the character of dissimilarities (with the exception of the properties listed above), there may be situations in which $(d_{ik})^2$ could assume negative values. In essence, these are not distanced values. The purpose of the spreading transformation guided by the values of β is to "spread" the nondiagonal values of R_{ij} so that the we end up with the nonnegative values of the distance. As discussed in Bezdek and Hathaway (1987), higher values of β yield a significant spread effect, which could affect the discovery of the structure. Being aware of that, we should avoid reaching high values of β and keep it at the lowest level that still leads to the nonnegative values. It has also been shown that if R satisfies the conditions listed above but is not Euclidean (meaning that its entries are not expressed as $R_{ij} = \|\mathbf{x}_i - \mathbf{x}_j\|^2$, with $\| \cdot \|$ being the Euclidean distance function), then there exists a positive value of β, say β_0, such that R_β is Euclidean for all $\beta \geq \beta_0$ and does not satisfy this property whenever $\beta < \beta_0$. This again suggests that the values of β should be kept as low as necessary.

While the determination of the prototypes (centers) in this clustering is quite different from that in data-driven FCM, the calculations of the partition matrix are done in the usual manner.

9.3. DECOMPOSITION OF FUZZY RELATIONAL PATTERNS

Relational data can be translated into the pattern-class membership dependency by using the idea of decomposition of fuzzy relations (Pedrycz, 1996). As before, we consider a collection of the similarity degrees r_{ij} collected into an N by N matrix (relation) $R = [r_{ij}], i, j = 1, 2, \ldots, N$, with N being the number of patterns (Zadeh, 1971). As usual, we assume reflexivity $(r_{ii} = 1)$ and symmetry $(r_{ij} = r_{ji})$. There are c categories (classes) of patterns. We are interested in designing a pattern-class relation $G = [g_{ij}], i = 1, 2, \ldots N, j = 1, 2, \ldots, c$ that captures the allocation of patterns to individual classes. The essence of relation decomposition (Di Nola et al., 1985) concerns the way in which information about class membership is inferred from R. We request that R is decomposed into G in the following manner:

$$R = G \circ G^T \tag{9.8}$$

where G^T denotes its transpose and "$\circ$" is a certain relation-relation composition operator. In general, we can treat it as an $s - t$ convolution, which translates the problem into the following format:

$$r_{ij} = \overset{c}{\underset{k=1}{S}} (g_{ik} t g_{jk}) \tag{9.9}$$

The solution to this problem (fuzzy relation G) is the one that decomposes R. Finding an exact solution seems optimistic and would not be manageable.

Treating this as an optimization task leading to the approximation of G is more manageable. The typical formulation along this line is the one using the sum of the squared performance index

$$Q = \|R - G \circ G^T\|^2 \tag{9.10}$$

that is,

$$Q = \sum_{i=1}^{N} \sum_{j=1}^{N} \left[r_{ij} - \underset{k=1}{\overset{c}{S}} (g_{ik} t g_{jk}) \right]^2 \tag{9.11}$$

Here the goal is to find G so that the values of Q become minimized, that is, $\mathrm{Min}_G Q$. Note that the values of G are confined to the unit interval, but this can be easily assured by clipping the values produced by applying any optimization algorithm. In what follows, we elaborate on two approaches. The first is the standard gradient-based technique that is guided by the values of the gradient of Q. The second approach maps the decomposition problem onto a certain fuzzy neural network and then translates the problem into the issue of supervised learning of this network.

9.3.1. Gradient-Based Solution to the Decomposition Problem

In the gradient-based approach, we compute the gradient of Q with respect to the fuzzy class assignment relation and update it following the scheme

$$G = G - \beta \nabla_G Q \tag{9.12}$$

with $\beta(>0)$ standing for the learning rate. In more detail, we get the updated expression

$$g_{ik} = g_{ik} - \beta \frac{\partial Q}{\partial g_{ik}} \tag{9.13}$$

The starting point of this iterative scheme is a fuzzy relation with random entries between 0 and 1.

Let us carry out detailed computations of the gradient

$$\frac{\partial Q}{\partial g_{lu}} = -2 \sum_{i=1}^{N} \sum_{j=1}^{N} \left[r_{ij} - \underset{k=1}{\overset{c}{S}} (g_{ik} t g_{jk}) \right] \frac{\partial}{\partial g_{lu}} \left(\underset{k=1}{\overset{c}{S}} (g_{ik} t g_{jk}) \right) \tag{9.14}$$

The inner derivative requires some attention, as there are four combinations of indexes that should be considered separately:

$$(\text{i}) \ i \neq l, j \neq l$$

$$(\text{ii}) \ i \neq l, j = l$$

$$(\text{iii}) \ i = l, j \neq l$$

$$(\text{iv}) \ i = l, j = l$$

Furthermore, as t- and s-norms are involved here, we have to specify their format. As an illustration, we consider the standard product (viewed here as a t-norm) and a probabilistic sum (recall that this computes as $a + b - ab$). For (i) the derivative is equal to 0, as there are no variables (entries of G) over which the calculations of the derivative take place. For (ii) we obtain

$$\frac{\partial}{\partial g_{lu}}\left(\underset{k=1}{\overset{c}{S}}(g_{ik}tg_{jk})\right) = \frac{\partial}{\partial g_{lu}}(A + g_{lu}g_{ju} - Ag_{lu}g_{ju}) \tag{9.15}$$

with the aggregate A defined in the form

$$A = \underset{\substack{k=1 \\ k \neq u}}{\overset{c}{S}}(g_{ik}tg_{jk}) \tag{9.16}$$

Then we get

$$\frac{\partial}{\partial g_{lu}}\left(\underset{k=1}{\overset{c}{S}}(g_{ik}tg_{jk})\right) = g_{ju}(1 - A) \tag{9.17}$$

Carrying out detailed computations for the two remaining cases, we finally arrive at the following derivative:

$$\frac{\partial}{\partial g_{lu}}\left(\underset{k=1}{\overset{c}{S}}(g_{ik}tg_{jk})\right) = \begin{cases} 0 & \text{if } i \neq l \text{ and } j \neq l \\ g_{ju}(1 - A) & \text{if } i = l \text{ and } j \neq l \\ g_{lu}(1 - B) & \text{if } i \neq l \text{ and } j = l \\ 2g_{lu}(1 - C) & \text{if } i = l \text{ and } j = l \end{cases} \tag{9.18}$$

The above expressions come with the auxiliary abbreviated notation

$$B = \underset{\substack{k=1 \\ k \neq u}}{\overset{c}{S}}(g_{ik}tg_{lk}) \qquad C = \underset{\substack{k=1 \\ k \neq u}}{\overset{c}{S}}(g_{lk}tg_{lk}) \tag{9.19}$$

Example 1. To illustrate the performance of this optimization approach, we take the two-dimensional data set shown in Figure 9.2.

Here we visualize three quite apparent clusters. The first one is formed by patterns 5-6-7 and 2-3-4. The other two patterns (8 and 1) can be treated as types of outliers; however, they exhibit a different character. Pattern 8 is located between the well-formed, condensed clusters, whereas pattern 1 is an outlier. To convert the two-dimensional patterns into their relational counterparts, we determine the level of similarity between the pairs of patterns using the following logic-driven format:

$$a \equiv b = 0.5[\min(a \to b, b \to a) + \min(\bar{a} \to \bar{b}, \bar{b} \to \bar{a})] \tag{9.20}$$

with a and b denoting the values of a certain feature of two patterns. The implication operator is taken as the one proposed by Lukasiewicz, namely,

$$a \to b = \min(1, 1 - a + b) \tag{9.21}$$

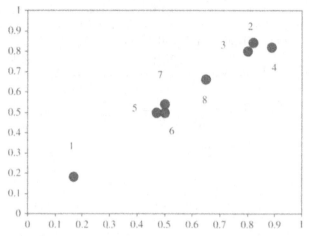

Figure 9.2. Two-dimensional synthetic patterns.

As the patterns are two-dimensional, we compute the similarity degree along each variable, average the results, and return the overall degree of similarity (match) between the patterns. We choose $c = 2$ classes and start the gradient-based optimization. The returned value of the performance index is equal to 0.3513. The fuzzy relation G reveals details about the structure of the classes:

$$G = \begin{bmatrix} 0.693 & 0.062 \\ 0.235 & \mathbf{0.973} \\ 0.295 & \mathbf{0.980} \\ 0.265 & \mathbf{0.953} \\ \mathbf{0.973} & 0.504 \\ \mathbf{0.967} & 0.542 \\ \mathbf{0.972} & 0.533 \\ 0.797 & 0.745 \end{bmatrix}$$

It is remarkable that the entries of G quantify well our intuitive observations. Patterns 5-6-7 (noted in boldface) are located in the same class, and all of them have high membership grades ($\sim$0.96). The same observation applies to patterns 2-3-4 (again, all the membership grades are over 0.95, marked in boldface). Pattern 8 shares its membership between two classes (again, as expected, as it serves as a bridging element between the two well-formed, highly condensed clusters). Finally, pattern 1 is classified as belonging to the second class but with a far lower membership degree. This could have been anticipated considering its limited proximity to the first class (cluster).

9.3.2. Neural Network Model of the Decomposition Problem

An alternative to the gradient-based learning discussed in the previous section is to map the problem onto a neural network and solve it via its training. As the

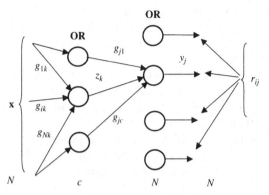

Figure 9.3. A topology of the fuzzy neural network formed with the use of OR neurons; connections between the entries of the unknown fuzzy relation G are indicated. Also shown are the dimensions of the successive layers of the network.

decomposition is inherently logical, the network has to comply with the character of the problem. We show that the decomposition can be translated into an input-output representation that is required to prepare a learning framework of the network and then develop its detailed topology.

We start with the structure of the logic network shown in Figure 9.3 and demonstrate that it delivers a network representation of the decomposition problem. Note the values of the connections of the network that form a distributed network-oriented model of the decomposition problem. The entries of the unknown relation G are treated as the connections of the network; the character of the problem dictates a structure with a single hidden layer.

Let us consider the following input $\mathbf{x}_l$ in the following binary format:

$$\mathbf{x}_l = [0\ 0\ 0\ \ldots\ 0\ 1\ 0 \ldots\ 0]$$

all of whose coordinates except the lth one are equal to 0. The lth coordinate is set to 1. We propagate this input through the network. The outputs of the OR neurons at the hidden layer are equal to

$$z_k = \overset{N}{\underset{i=1}{S}} (g_{ik}\, t x_i) = g_{lk} \tag{9.22}$$

$k = 1, 2, \ldots, c$. Then these intermediate signals generate the outputs computed as

$$y_j = \overset{c}{\underset{k=1}{S}} (g_{jk}\, t z_k) = \overset{c}{\underset{k=1}{S}} (g_{jk}\, t g_{lk}) \tag{9.23}$$

$j = 1, 2, \ldots, N$. Note that this output should be made equal to r_{jl} (refer to the formulation of the original decomposition problem). We can accomplish this by adjusting the connections of the network, that is, the components of G. as we have N outputs of the network, the input $\mathbf{x}_l$ produces the lth column of R. By varying index l from 1 to N, we keep concentrating on (learning) the successive columns of R. In essence, the entire learning set can be constructed by inputting Boolean vectors starting from [1 0 ... 0] and ending up at [0 ... 0 1]. Having formed the finite collection of input-output pairs of data, we are ready to carry out supervised learning.

While this learning is quite standard (referred to as error backpropagation), it is helpful to review the successive derivations that can help us follow the flow of computing. In what follows, we consider an on-line mode of learning in which the connections of the network are updated after the presentation of the individual input-output pair, which we denote as (**x, target**). For instance, when $\mathbf{x} = \mathbf{x}_l$, the output **target** occurs as the corresponding row of R. The performance index assumes a standard form of a sum of squared errors

$$Q = (\mathbf{target} - Net(\mathbf{x}, G))^T (\mathbf{target} - Net(\mathbf{x}, G))$$

that is,

$$Q = \sum_{j=1}^{N} (\text{target}_j - y_j)^2 \tag{9.24}$$

with Net denoting the vector of outputs of the fuzzy neural network while being activated by **x**. The derivations of the learning scheme invoke computations of the partial derivatives of Q with respect to the unknown connections (G):

$$\frac{\partial Q}{\partial g_{il}} = -2 \sum_{j=1}^{N} (\text{target}_j - y_j) \frac{\partial y_j}{\partial g_{il}} \tag{9.25}$$

$i = 1, 2, \ldots, n, l = 1, 2, \ldots, m$. The inner derivative consists of two components (note that the same connection occurs at two different locations of the network):

$$\frac{\partial y_j}{\partial g_{il}} = \left(\frac{\partial y_j}{\partial g_{il}} \right)_{\text{hidden-output}} + \left(\frac{\partial y_j}{\partial g_{il}} \right)_{\text{input-hidden}} \tag{9.26}$$

Further calculations are possible after the specification of the s- and t- norm. As before, with the product and probabilistic sum involved here, we derive

$$\left(\frac{\partial y_j}{\partial g_{il}} \right)_{\text{hidden-output}} = \begin{cases} 0 & \text{if } j \neq i \\ z_l(1 - A) & \text{if } j = i \end{cases} \tag{9.27}$$

and

$$A = \underset{u \neq l}{\overset{c}{S}} (g_{iu} t z_u) \tag{9.28}$$

For the connections occurring between the input and hidden layers we obtain

$$\left(\frac{\partial y_j}{\partial g_{il}} \right)_{\text{input-hidden}} = \sum_{u=1}^{c} \frac{\partial y_j}{\partial z_u} \frac{\partial z_u}{\partial g_{il}} \tag{9.29}$$

The corresponding derivatives are then computed in the form

$$\frac{\partial y_j}{\partial z_u} = g_{ju}(1 - B') \quad \frac{\partial z_u}{\partial g_{il}} = x_i(1 - C') \tag{9.30}$$

with the abbreviations describing the following compositions:

$$B' = \overset{c}{\underset{w \neq u}{S}} (g_{jw} t z_w) \quad C' = \overset{c}{\underset{w' \neq i}{S}} (g_{w'l} t x_l) \tag{9.31}$$

Example 2. Returning to this example, we decompose the fuzzy relation using the fuzzy neural network. Now the performance index attains the value of 0.3254 with the fuzzy relation G with the entries

$$G = \begin{bmatrix} 0.000 & 0.717 \\ \mathbf{0.979} & 0.334 \\ \mathbf{0.983} & 0.385 \\ \mathbf{0.971} & 0.348 \\ 0.417 & \mathbf{0.980} \\ 0.460 & \mathbf{0.979} \\ 0.450 & \mathbf{0.981} \\ 0.705 & 0.832 \end{bmatrix}$$

Example 3. This example comes from Tamura et al. (1971) and deals with the classification of portraits of members of three families ($c = 3$). The number of patterns is 16. The subjective similarities between the pairs of patterns collected in R are shown below (because of the symmetry, we show only the upper part of the relation).

The classification results are collected in the class membership matrix

$$G = \begin{bmatrix} 0.752 & 0.000 & 0.000 \\ 0.011 & 0.929 & 0.000 \\ 0.142 & 0.006 & 0.467 \\ 0.274 & 0.067 & 1.000 \\ 0.030 & 0.599 & 0.000 \\ 0.903 & 0.002 & 0.089 \\ 0.000 & 0.937 & 0.004 \\ 0.770 & 0.016 & 0.083 \\ 0.000 & 0.433 & 0.612 \\ 0.000 & 0.132 & 0.524 \\ 0.000 & 0.733 & 0.123 \\ 0.000 & 0.000 & 0.787 \\ 0.743 & 0.000 & 0.114 \\ 0.038 & 0.895 & 0.037 \\ 0.059 & 0.000 & 0.648 \\ 0.809 & 0.062 & 0.000 \end{bmatrix}$$

Its interpretation (by identifying the most dominant membership grades) returns the structure of three classes: class 1: {1, 6, 8, 13, 16}, class 2: {2, 5, 7, 11, 14}, class 3: {3, 4, 9, 10, 15, 17}.

Example 4. This experiment concerns the data set of Gowda and Diday, (1992), which deals with five-dimensional patterns concerning eight different types of oil. In, this experiment, the entries of the original similarity relation are normalized to 1. The result of the decomposition achieved with the use of the fuzzy neural network is presented in the following relation:

$$G = \begin{bmatrix} 0.609 & 0.007 \\ 0.767 & 0.237 \\ 0.810 & 0.432 \\ 0.822 & 0.361 \\ 0.768 & 0.368 \\ 0.763 & 0.478 \\ 0.069 & 0.905 \\ 0.114 & 0.921 \end{bmatrix}$$

We have highlighted in boldface the allocation of the patterns to the two classes; we conclude that class 1 involves all but the two last patterns. Class 2 is compact, with class membership grades over 0.90. Interestingly, the same data set, once clustered with the use of the relational form of the FCM (Hathaway and Bezdek, 1994), returns the membership grades in the first cluster equal to [0.704 0.818 0.935 0.924 0.816 0.834 0.036 0.028].

9.4. COMPARATIVE ANALYSIS

There are several distinctive features that distinguish direct and relational patterns. The patterns are different. In the first case, we use patterns directly. This implies that the clustering occurs in the feature space that comes with the patterns. In the second case, the pairs of patterns are formed and assessed with regard to their similarity, dissimilarity, or any other relative association, and the clustering takes place in this new relation-based induced feature space. The main advantage here starts to occur when the original (direct) patterns are defined in a highly dimensional space; in essence, we have far higher dimensionality of the feature space (on the order of hundreds or thousands) and a very limited number of patterns. This makes the set of patterns very sparse when they are located in the highly dimensional space. If the patterns themselves are difficult to describe, the feature space may not be able to fully capture the essence of the patterns and, subsequently, the structure of the data set.

Computationally, these two clustering methods are very different. Denoting by M and n the number of patterns and their dimensionality, respectively, we note that the relational clustering operates on $M^2 - M$ new patterns. In general, we assume that the relation capturing the pairs of patterns is not symmetrical, so we need information about the value of the relational constraint predicate defined

TABLE 9.1. Direct and Relational Clustering: A Comparative Overview

Direct Clustering	Relational Clustering
Number of patterns: M	Number of relational patterns: $M^2/2 - M > M$
Dimensionality of the feature space: n	Dimensionality of the feature space: r $r \ll n$
Design recommendation: ability to form a suitable feature space, $M \gg n$	Design recommendation: ability to form a relational space
Representation: prototypes and partition matrix	Representation: partition matrix

for the pairs (i, j) and (j, i). The value of the predicate on the same patterns (i, i) is not relevant, and such pairs do not enter into the clustering process. The dimensionality of the feature space of the relational patterns could be far lower than the original one. In particular, it could be equal to 1. Here let us assume that we have dimensionality equal to r. The main relationships are summarized in Table 9.1.

9.5. CONCLUSIONS

Clustering relational patterns is of significant interest when we are concerned with highly dimensional patterns in which no meaningful structure may be detected. We have discussed two main and conceptually diverse categories of algorithmic approaches. The first includes relational FCM and its generalizations (Bezdek). The second is based on decomposition of a relational matrix into pattern-class relations, and uses the calculus of fuzzy relational equations and ensuing gradient-based and fuzzy neural network learning schemes.

REFERENCES

J.C. Bezdek, R.J. Hathaway, Clustering with relational C-Means partitions from pairwise distance data, *Mathematical Modelling*, 9, 6, 1987, 435–439.

A. Di Nola, W. Pedrycz, S. Sessa, Decomposition problem of fuzzy relation, *Int. J. General Systems*, 10, 1985, 123–133.

K.C. Gowda, E. Diday, Symbolic clustering using a new similarity measure, *IEEE Trans. on Systems, Man, and Cybernetics*, 22, 1992, 368–378.

R.J. Hathaway, J.C. Bezdek, NERF C-Means: non-Euclidean relational fuzzy clustering, *Pattern Recognition*, 27, 3, 1994, 429–437.

R.J. Hathaway, J.C. Bezdek, J.W. Davenport, On relational data versions of the C-Means algorithm, *Pattern Recognition Letters*, 17, 1996, 607–612.

R.J. Hathaway, J.C. Bezdek, Clustering incomplete relational data using the non-Euclidean relational fuzzy C-Means algorithm, *Pattern Recognition Letters*, 23, 2002, 151–160.

R.J. Hathaway, J.W. Davenport, J.C. Bezdek, Relational duals of the C-Means clustering algorithms, *Pattern Recognition*, 22, 1989, 205–212.

W. Pedrycz, Classification of relational patterns as a decomposition problem, *Pattern Recognition Letters*, 17, 1996, 91–99.

T.A. Runkler, J.C. Bezdek, Web mining with relational clustering, *J. of Approximate Reasoning*, 32, 2003, 217–236.

S. Tamura, S. Higuchi, K. Tanaka, Pattern classification based on fuzzy relations, *IEEE Trans. on Systems, Man, and Cybernetics*, 1, 1971, 61–66.

L.A. Zadeh, Similarity relations and fuzzy orderings, *Information Sciences*, 3, 1971, 177–200.

10 Fuzzy Clustering of Heterogeneous Patterns

So far, we have discussed clustering of numeric patterns represented as points in R^n. It is of interest to discuss situations where the patterns are described as granular entities whose features take on nonnumeric values and therefore can be represented as, for example, intervals or fuzzy sets. In this chapter, our purpose is to revisit fuzzy clustering and extend the fundamental algorithm in such a way that it can deal with these augmented types of granular data. Because of the nature of the data, we refer to this process of structure determination as heterogeneous fuzzy clustering. We develop two fundamental modes of granular clustering by introducing parametric and nonparametric schemes of representation of granular data. Using on these models, we discuss the organization of the optimization process.

10.1. INTRODUCTION

So far, our primary emphasis in describing clustering has been to reveal structure in numeric data. The algorithmic diversity arose with regard to various ways of coping with the topology of numeric data. The issue of granular or heterogeneous data has not been a focal point of previous research and applications. Surprisingly, heterogeneous patterns are quite common. Four examples serve as an introduction to the subject.

(a) Imagine that we are given the results of polls where the respondents' replies are given in the form of information granules, say intervals (the individual indicates a certain range of possible outcomes, say [a, b], rather than specifying a single numeric value).

(b) A physical variable (e.g., temperature) can be measured in a precise way but can also be assessed subjectively and in this case represented by fuzzy sets (given their typical values along lower and upper bounds). In this way, we end up with a heterogeneous collection of data whose structure needs to be revealed.

(c) We may have incomplete data when some records come with missing values. As out data set is quite small and hence we are not allowed to remove incomplete patterns, the data set is carefully examined by an expert who provides estimates of the missing entries. These estimates are distinguished from other data in the sense that their granularity becomes lower. The expert may give assessments of the missing values that have to be treated as intervals or fuzzy sets. We end up with a heterogeneous data set where patterns are numeric as well as granular. The level of granularity of the data depends on various factors, especially the confidence of the expert in estimating the missing values. In some cases, the confidence level could be very high (hence the missing value is a narrow interval). In others, it could be very low and the results of data imputation may be speculative (and hence have broad intervals or highly spread fuzzy numbers). In an extreme case, the expert may hesitate to come up with any meaningful assessment and the result becomes totally *unknown* (modeled as a fuzzy set with the membership functions being identically equal to 1, that is, *unknown* $(x) = 1$ for all x in $\mathbf{X}$).

(d) In some cases, the experimental data are preprocessed, which gives rise to granular entities. These are more abstract constructs than the original physical variables. For instance, we may have histograms of pixels in image processing. To reveal a structure in the collection of images, we operate on histograms. In signal processing, we may encounter temporal or spatial aggregates of signals that are afterward a subject of clustering. For instance, temporal granulation of signals (Pedrycz et al., 2000) gives rise to fuzzy sets, and these, in turn, need to be clustered. When there are several sources of signals processes en bloc (a situation that is typical of sensor fusion or multi-lead biomedical signals), we aggregate these spatially distributed readings in the form of aggregate information granules that are later clustered.

Given such needs, in this chapter we discuss ways of augmenting the existing clustering environment so that granular data (patterns) can be accommodated. The two main approaches proposed here deal with so-called parametric and nonparametric representation of granular data. The representation mechanisms are important to the ensuing clustering mechanism invoked in this new representation space. We are concerned with understanding the main features of these two representation schemes and identifying their impact on the ensuing clustering process. We also elaborate on the format of clustering outcomes implied by the assumed representation model used for heterogeneous data

10.2. HETEROGENEOUS DATA

By heterogeneous data we mean experimental entities arising at different levels of granularity. Such patterns appear quite often. We estimate a value of the temperature and state that it is warm. Instead of making a precise measurement (which might not be of practical relevance), we express our perception and build its formal model in a form of a triangular fuzzy number (Figure 10.1a). We cannot give a numeric value of the time duration of a software project; we are convinced that it should take no more than 10 months, but it is very likely to take 6 to 8 months. Finally, no matter how many resources we allocate, its completion time will be no less than 4 months. Again, we develop a fuzzy number, now consisting of a trapezoidal fuzzy number (Figure 10.1b). The reading of the sensor has some variability, and in our experience this variability is 5%. In this sense, a sound model that could work in this scenario has an interval of a predefined width. These are examples of granular data. Another area with granular data is digital image processing. Here we have individual pixels, but in essence, all processing that occurs afterward attempts to compress the detailed data first and then work on these granular entities. In fact, each block (window) of pixels (typically 4 by 4 or 8 by 8) gives rise to the information granules of brightness levels, texture description, and so on.

Heterogeneity of data arises in two ways. First, data have different level of granularity. Our perception can vary when we switch between large, nonspecific information granules and numeric data (as we have discussed so far). They can occur as a unique mix of data in the problem under consideration. Second, the heterogeneous nature of data emerges because of the diversity in membership functions. There could be a immense mix of main categories of membership functions, including triangular, trapezoidal, Gaussian, parabolic, and other categories.

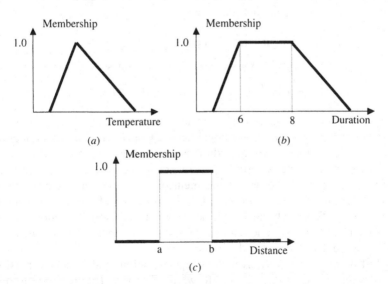

Figure 10.1. Examples of granular data: (a) a perception of warm temperature, (b) duration of a software project, (c) imprecise measurement of a sensor.

10.3. PARAMETRIC MODELS OF GRANULAR DATA

In this approach, information granules are represented in a parametric form, meaning that each of them is fully described in terms of a vector of parameters. The basic requirement is that all granular data come from the same family of membership functions so that we can create a homogeneous feature space (Hathaway et al., 1996). Consider trapezoidal fuzzy numbers. Each of them is uniquely described by a vector of four parameters of the corresponding membership function. In other words, if we confine ourselves to this family of fuzzy numbers (denote it by T), each element from T is given as a four-argument vector (a, m, p, b). Any fuzzy number in T comes with this representation and is uniquely described in this way. If we discuss Gaussian fuzzy numbers, each of them is fully described by a modal value and its spread (m, σ). We denote this family of fuzzy numbers by G. Each information granule can be expressed in the form of a two-argument vector. This representation fully justifies the name of the model, as each fuzzy number (or information granule) is fully captured by a vector of parameters. The information granules must come from the same family of membership functions (so that they can be uniquely described in this parametric manner). If some granules have different membership functions, we must choose one model (say, Gaussian fuzzy numbers) and approximate any other number outside this family by the two parameters of the Gaussian membership function. This mixed mode realization will come with some associated approximation error.

From the design standpoint, given a certain family of parametric representation, we should be confident that this form is rich enough to represent our granular data. For instance, trapezoidal fuzzy numbers seem quite convincing: by setting up the values of the four parameters, we can easily produce various subtypes of the fuzzy sets; for example:

$m = n$: triangular fuzzy number

$a = m, p = b$: interval (for different values of m and p)

$a = m = p = b$: single numeric entity (**R**)

Interestingly, a granular datum characterized by this type of membership function is equivalent to a four-dimensional vector. Considering information granules X, Y, and Z, all located in an n-dimensional feature space where along each dimension we encounter a certain level of granularity, the overall representation of X is a concatenation of the four-dimensional tuples contributing to its parametric description. We end up with $4 * n$ coordinates of the overall parametric representation. Obviously, there can be other forms of representation. The same trapezoidal membership function can be given in an incremental format of (da, m, db), where $da = m - a$ and $db = b - p$.

All of these forms are equivalent as representation models (and a transformation between them is pretty straightforward). The new feature space formed in this manner (either through the original cutoff points of membership functions or their incremental format) sets up an arena for fuzzy clustering. The multidimensional granular data are represented feature after feature, so that the final result becomes a vector of cutoff points of the membership functions defined for each dimension separately:

$$\mathbf{x} = [a_1 \ m_1 \ p_1 \ b_1 \mid a_2 \ m_2 \ p_2 \ b_2 \mid \ldots a_n \ m_n \ p_n \ b_n] \qquad (10.1)$$

Obviously, as a multidimensional construct, the data can be displayed as Cartesian products of fuzzy sets, that is, $X_1 \times X_2 \times \ldots \times X_n$ with the operation completed by the minimum operator (or any t-norm in general). A two-dimensional example involving triangular fuzzy numbers is shown in Figure 10.2. The minimum operation is noninteractive; the use of other t-norms introduces some interaction in the sense that the result is affected by the values of the two arguments involved in the calculations.

10.4. PARAMETRIC MODE OF HETEROGENEOUS FUZZY CLUSTERING

Once we have decided how information granules are to be represented (and this obviously depends upon the character of the information granules in the problem), this gives rise to a certain feature space. In comparison to the original space of dimensionality n, here we end up with $h * n$ variables, with h corresponding to the number of parameters required to represent fuzzy numbers. The minimization of the objective function in this new augmented feature space is standard. The results are in the generic format of a partition matrix (which now captures the relationships between information granules and the clusters). The prototypes are granular constructs (which is not surprising, considering that we started with information granules in the first place).

Example 1. The collection of two-dimensional granular data consists of a mixture of triangular fuzzy numbers (see Figure 10.3).

For clustering purposes they are represented as cutoff points, so we discuss their lower bounds, two boundary values representing the range of the modal values and the upper bound. This produces a six-dimensional feature space. The clustering is first completed for $c = 2$ clusters (Figure 10.4). The prototypes reflect the structure existing in the data; however, they try to compensate for the relatively small number of clusters allocated to this problem in comparison to the topology of the data set. What is more important is the shape of the prototypes along with their granularity. When we decided to move on to $c = 3$ clusters, this changes the form of the prototypes, which become far more compact and confined to a small portion of the data space. For $c = 2$ and the value of the fuzzification factor equal to 2 we get larger prototypes.

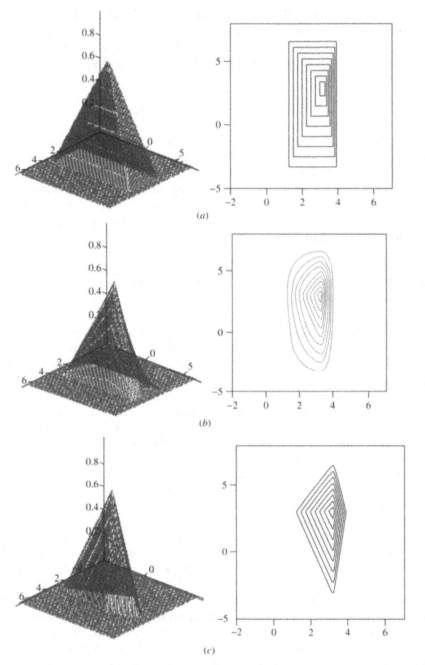

Figure 10.2. Examples of granular data treated as Cartesian products of fuzzy sets A_1 and A_2 defined in $\mathbf{X}_1$ and $\mathbf{X}_2$: $A_1 = (1.0, 1.5 \quad 3.2 \quad 4.0)$ and $A_2 = (-4.0 \quad -1.0 \quad 3.0 \quad 7.0)$: (a) minimum operator, (b) product, (c) Lukasiewicz *and* operator of the form max $(0, a + b - 1)$.

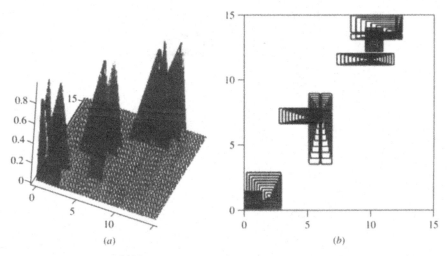

Figure 10.3. A collection of granular data represented as triangular fuzzy relations.

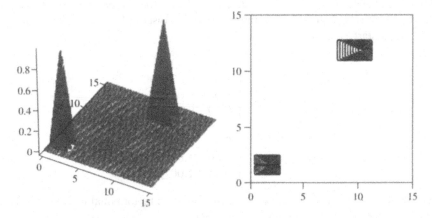

Figure 10.4. Prototypes of the two clusters (three-dimensional view and contour plot).

As expected, the partition matrix identifies several patterns with intermediate membership grades:

$$U = \begin{bmatrix} 0.978624 & 0.021376 \\ 0.989139 & 0.010861 \\ 0.997283 & 0.002717 \\ 0.616445 & 0.383555 \\ 0.483051 & 0.516949 \\ 0.012404 & 0.987596 \\ 0.021460 & 0.978540 \\ 0.005118 & 0.994882 \end{bmatrix}$$

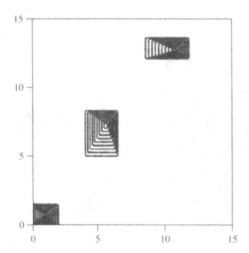

Figure 10.5. Granular prototypes formed in clustering completed for three clusters.

Now we change the fuzzification factor to 1.1. The prototypes start to "expand," while the entries of the partition matrix become closer to 0 and 1 and the "uncertain" data start vanishing:

$$
U = \begin{bmatrix}
1.0 & 0.00 \\
1.00 & 0.00 \\
1.00 & 0.00 \\
0.99 & 0.01 \\
0.99 & 0.01 \\
0.00 & 1.00 \\
0.00 & 1.00 \\
0.00 & 1.00
\end{bmatrix}
$$

For three clusters constructed with the fuzzification factor equal to 2, we produce the granular prototypes shown in Figure 10.5.

10.5. NONPARAMETRIC HETEROGENEOUS CLUSTERING

The second mode of representation of heterogeneous data is not concerned with any specific class of membership functions, meaning that we are not confined to a fixed format of the vector of parameters of these fuzzy sets, hence their nonparametric representation (Pedrycz et al., 1998).

10.5.1. A Frame of Reference

So far, we have discussed clustering of information granules that are represented in a properly chosen parameter space. This representation requires that the granules come from the same category of fuzzy sets (triangular, Gaussian, etc.) so that they can be uniquely described by means of the available parameters. The increased diversity in the information granules, say a mixture of triangular, exponential, and Gaussian fuzzy sets existing in the data, requires another type of treatment. One possible solution is to approximate membership functions so that they can be represented (approximated) in the format used to develop parametric heterogeneous clustering. If we are dealing with trapezoidal fuzzy numbers, a Gaussian membership function of a certain information granule with given values of the mode of the fuzzy set (u) and its spread (σ) has to be approximated by trapezoidal fuzzy sets. In other words, we seek values of the parameters (a, m, n, and b) such that the membership $G(x, u, \sigma)$ becomes approximated by $T(x; a, m, n, b)$ to the highest extent. An obvious optimization criterion would be

$$V(a, m, n, b) = \int_{-\infty}^{\infty} (T(x, a, m, n, b) - G(x, u, \sigma))^2 \, dx \qquad (10.2)$$

where the minimization is carried out over the parameters of the trapezoidal fuzzy number

$$\min_{a,m,n,b} V(a, m, n, b) \qquad (10.3)$$

Obviously, by approximating fuzzy numbers coming from one category by the elements of another category, we always end up with some approximation error. We must realize that the ensuing clustering has to deal with approximate information granules. The results of clustering, such as prototypes, are also directly impacted by such approximation error.

Another computationally appealing alternative would be to convert all fuzzy sets into the corresponding shadowed sets and perform clustering in this new augmented feature space. The attractiveness of this approach stems from the fact that the transformation from fuzzy sets into shadowed sets is quite apparent. In fact, for some classes of membership functions, the values of the thresholds have been derived in an analytic way and could be directly used. The representation scheme involving four parameters of each shadowed set is also fixed in advance so that it determines the dimensionality of this new feature space. Another attractive aspect of the clustering completed in this setting is that we can arrive at a transparent interpretation of the results of clustering, in particular its prototypes. Then the shadows of the prototypes become of particular interest, especially when it comes to their shadows and assessment of uncertainty regions.

The other alternative is to look for a so-called nonparametric representation of information granules that allows us to free the method from any assumption about the parametric format that is required to describe the granules. The central concept underlying the nonparametric model is a space of referential fuzzy granules. As the name implies, the proposed space in which clustering takes place consists of a collection of reference information granules—using which we characterize any information granule. The crux of such a frame of reference lies in its centrality in describing a certain variable or a collection of variables. In the first case, we are dealing with information granules such as intervals or fuzzy sets. In the second case, we are concerned with Cartesian products or fuzzy relations.

As an example, consider the speed of a vehicle. The frame of reference consists of a collection of generic information granules such as *low, medium, high,* and *very high.* The other frame of reference may involve terms like *low, safe,* and *dangerous* speed (obviously this one is conceptually more complex, as the granules of reference could be strongly affected by some other variable, such as driving conditions or weather). In either case, the frame of reference (or frame of cognition) serves as an important and practically sound setting within which we are interested in carrying out processing, moving on to future decision making, and so on. We can easily change the frame of reference, and this immediately modifies our view of the problem. The granularity of the frame of reference, which is strongly associated with the number of building blocks (conceptual constructs), defines the level of specificity of the problem. We may need a more detailed model, and this requires more information granules. A less detailed, more general look at the problem calls for a less specific frame of reference in which a few (say, five to seven) information granules are more than enough. The other option is to form a hierarchy of frames and start looking at data from different points of view. Given a frame of reference, any information granule can be expressed in the language of those granules. The form of the representation is not unique. In what follows, we discuss one alternative: a possibility-necessity transformation.

10.5.2. Representation of Granular Data Through the Possibility-Necessity Transformation

Any information granule can be represented through a frame of reference F consisting of r information granules. Denote them by $M_1, M_2, \ldots, M_r$. The specific character of F can vary from case to case: it can be formed by sets (intervals), fuzzy sets, rough sets, or shadowed sets. To focus our attention, we start with fuzzy sets and discuss the simplest scenario of a single variable ($\mathbf{X}$), that is, $M_i : \mathbf{X} \to [0, 1]$. Several fundamental and intuitively appealing requirements are formed concerning the elements of F (cf. Pedrycz, 1992, 1994; Pedrycz and Gacek, 2002; Pedrycz and Valente de Oliveira, 1995; Pedrycz and Vukovich, 2002):

Focus of attention (distinguishability criterion)—fuzzy sets of F should have a clear semantic meaning, implying that the corresponding reference fuzzy sets (linguistic terms) express a well-understood semantic range over the universe of discourse.

Justifiable number of referential fuzzy sets—the number of linguistic values (conceptual landmarks) should be compatible with the number of conceptual entities a human being can efficiently store and utilize during further processing. A rule of thumb in psychology suggests the number of such terms to be 7 ± 2. We may not insist on this requirements, but we should keep in mind that too many fuzzy sets defined in $\mathbf{X}$ may prevent their clearly defined semantic identity.

Coverage of the universe of discourse—the universe of discourse of the given variable should be "covered" by the supports of M_is. Formally we require that $\bigcup_{i=1}^{r} \text{supp}(M_i) = \mathbf{X}$, where supp(.) denotes the support of the corresponding fuzzy set. This requirement ensures that each element of $\mathbf{X}$ can be associated with at least one referential fuzzy set.

Unimodality and normality—each M_i must have at least one membership grade equal to 1 (these points can be treated as the prototypes of the fuzzy set). The unimodality reinforces the uniqueness of the prototype.

Now our objective is to represent any granular datum X in terms of elements of F. In a nutshell, we are concerned with the development of a matching scheme between X and M_i. The possibility and necessity measures come as an intuitive computing environment in this regard. As discussed earlier (see Chapter 2), the possibility measure expresses the degree of overlap between X and M_i. The necessity measure quantifies the extent to which X is included in M_i and as such is a measure of containment of X in the element of F.

When X and M_i are given in the Cartesian product of spaces (so that the information granules are effectively fuzzy relations), the calculations proceed in the same manner. Overall, the result of the possibility-necessity transformation becomes an element in the $2r$ unit hypercube, and the mapping itself can be described in the form

$$\mathbf{R}^n \rightarrow [0, 1]^{2r} \tag{10.4}$$

It is worth noting that from a computational perspective, this transformation is a nonlinear normalization of input information granules (and numeric data in particular). The nonlinear character of the reference fuzzy sets plays a pivotal role in the development of the nonlinear transformation; any changes made to the membership functions are reflected in the elements in the unit hypercube. The dimensionality of the data can be changed significantly, depending upon the number of referential fuzzy sets. As this number is quite limited (around seven) and as the original feature space can be highly dimensional (n being in the range of hundreds), this transformation contributes to the significant dimensionality reduction with the reduction rate of $2r/n$.

Several interesting observations can be made about the character of information granules, their diversity, and their relationship with the frame of reference. First, the possibility-necessity transformation is a general method of determining matching levels that does not impose any restrictions on the form of the membership function of the information granules to be clustered. Two "boundary" conditions are worth stressing. If $X = \{x\}$ is a numeric entity, then Poss $(\{x\}, M_i) = \text{Nec} (\{x\}, M_i) = M_i(x)$. If the specificity of X goes down, the possibility and necessity values start to reflect this (see Figure 10.6).

If we plot the results on the (Poss, 1-Nec) plane, we observe that with the lowered specificity (in limit X becomes *unknown*, that is, $X = \mathbf{X}$), we start moving to the right upper corner of the region. Here we get Poss $(X, M_i) = 1$, while the

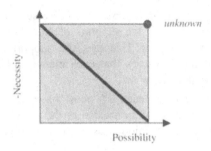

Figure 10.6. Results of possibility-necessity encoding vis-à-vis the granularity of X.

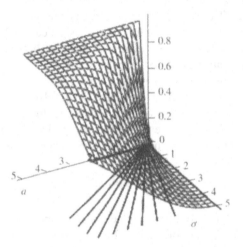

Figure 10.7. D as a function of a and σ.

necessity becomes equal to 0. The difference between the possibility and necessity values, $D = \text{Poss} - \text{Nec}$, is a useful indicator of the granularity of the data. When we have numeric data, $D = 0$. Less specific data produce higher values of D. As already indicated, for the *unknown* information granule, $D = 1$. Obviously, the level of granularity is relative to the granularity of the elements of F. An input X treated as an interval distributed around 0 and width of $2a$ matched with the referential fuzzy set M_i (with M_i being Gaussian and centered around 0, with the spread equal to σ) has a possibility value equal to 1. The necessity value is equal to $\exp(-a^2/\sigma^2)$ and depends on the values of a and σ. Hence the values of the difference are given by $D = 1 - \exp(-a^2/\sigma^2)$. The effect of the parameters capturing the granularity of X and M_i is illustrated in Figure 10.7.

At this point, we are ready to cluster data using their representation in the unit hypercube. The nonlinear normalization is a useful feature of this preprocessing. The FCM algorithm applied to the transformed patterns in the $[0,1]^{2r}$ space is quite typical. With the nonlinear transformation comes an interesting effect showing how the distance function in $[0,1]^{2r}$ is affected by the choice of

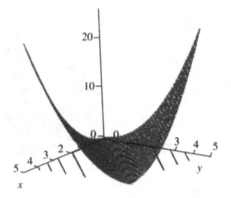

Figure 10.8. Distance function $d(x, y)$ treated as a function of its arguments.

membership function or changes in its parameters. Consider two data points x, y in **R**. The distance function $d(x, y) = (x - y)^2$ is smooth and exhibits the same shape across the entire space of its arguments (see Figure 10.8).

Now let us transform these two patterns through a collection of three reference fuzzy sets A_1, A_2, and A_3 that are described by Gaussian membership functions with modal values 1, 3, and 5 and the same spread σ. Compute the distance between x and y in the space of membership grades $D(x, y) = \sum_{i=1}^{3}(A_i(x) - A_i(y))^2$ (Figure 10.9). Obviously, if $x = y$, then $D(x, y)$ returns 0. In all remaining cases, the nonzero values of D are driven by the membership functions of the reference fuzzy sets. This effect is clearly visible when we start changing the value of σ. Low values give rise to a fairly rigged landscape with some regions that serve as regions (basins) of attraction. The increase in the spread values produces a smooth surface whose curvature is higher than the original quadratic function used in the original space of reals.

The interpretation of the results requires some attention, as these results are produced again in the unit hypercube and need to be brought back to the original space where the granular patterns are given. This transformation is referred to as granular dereferencing (an interesting analogy to the pointers and their usage in programming languages). To clarify the processing phases nonparametric heterogeneous clustering, refer to Figure 10.10. We have already discussed the representation (encoding) of granular data via the possibility-necessity transformation. The clustering in the unit hypercube is the second phase, and its calculations are standard.

The reference information granules can be constructed in three main ways:

(a) As a result of the of designer's preferences and domain knowledge. The reference fuzzy sets are the conceptual landmarks identified by the designer to establish a sound framework of data analysis. It is the designer who sets up the environment (specificity of granular information, number of information granules) within which the discovery of the data structure

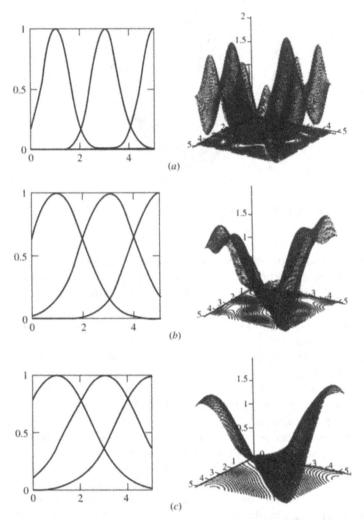

Figure 10.9. Plots of $D(x, y)$ for selected values of σ; membership functions of the reference fuzzy sets are also shown: $\sigma = 0.75$ (*a*), $\sigma = 1.5$ (*b*), $\sigma = 2.0$ (*c*).

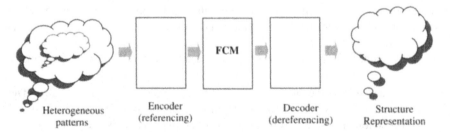

Figure 10.10. A flow of computing in the nonparametric fuzzy clustering of heterogeneous data.

takes place. As these results, especially prototypes of the clusters, are expressed in the language of the reference information granules, they must be of direct interest to the user so that their choice allows the user to become actively engaged in the search for the structure.

(b) By some other clustering process involving another previously available data set before the current data set was used. In this way, the previous data set builds an environment in which all further clustering techniques are "embedded."

(c) As a hybrid of (a) and (b). The initial membership values for some data are given by the designer, and the structure (partition matrix) is further refined by using patterns from the data set. We can envision a mode of clustering with partial supervision that contributes to this hybrid design mode.

10.5.3. Dereferencing

The results of fuzzy clustering in the unit hypercube need to be communicated back to the environment where the original data set was located. The clustering is completed for the logically transformed data (which contributed to its elevated level of abstraction), and now we must produce results that are "readable" in the original space. First, let us emphasize that there are no unique way of doing this. This is not surprising since we are concerned here with a higher level of conceptual abstraction. The prototypes are formed as a mixture of the reference information granules. If those were fuzzy sets, we can interpret the prototypes in $[0,1]^{2r}$ as an aggregate fuzzy set of order-2. More specifically, let us denote by $\mathbf{v}_i{}^{\sim}$ the ith prototype located in the unit hypercube. Its coordinates link to the possibility and necessity measures of the information granules $A_1, A_2, \ldots A_r$:

$$\mathbf{v}_i{}^{\sim} = [\text{Poss } (\mathbf{x}, A_1) \; \text{Poss}(\mathbf{x}, A_2) \ldots \text{Poss } (\mathbf{x}, A_r) \; \text{Nec}(\mathbf{x}, A_1) \; \text{Nec}(\mathbf{x}, A_2)$$
$$\ldots \text{Nec}(\mathbf{x}, A_r)] \quad (10.5)$$

Alluding to the graphic notation shown in Figure 10.11, we emphasize the fact that the prototype is formed on the basis of A_i's with the "activation" levels quantified by the corresponding possibility and necessity measures.

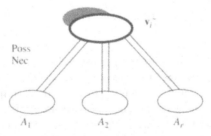

Figure 10.11. Prototype represented as a fuzzy set of order-2 with their membership grades expressed as possibility and necessity measures.

The representation shown in Figure 10.11 clearly emphasizes the abstract character of the prototype. If we want to translate it back to the original space, several interesting options are available. In view of the heterogeneous nature of data, the alternatives can be divided into three categories:

(a) Numeric representation of the prototype. This is the simplest version of the transformation where a single numeric realization of the prototypes is sought. Considering the modal values (or other representatives) of the information granules to be $z_1, z_2, \ldots, z_r$, the numeric description of the prototype is computed as a weighted sum of the form

$$\frac{\sum_{i=1}^{r} \text{Poss}(xA_i)z_i}{\sum_{i=1}^{r} \text{Poss}(xA_i)} \qquad (10.6)$$

(b) Interval-valued representation of the prototype. This is the second level of refinement of the prototype. As we encounter various classes of membership functions forming the frame of reference, a set-like form of the information is most preferred (as we cannot come up with a more specific form of the membership function). The lower and upper bounds of the interval are formed on the basis of the modes of the reference granules and the possibility and necessity measures.

(c) In this scenario, we construct an interval-valued fuzzy set of the prototype by solving an inverse problem of the necessity and possibility measures (which is regarded as a solution to some fuzzy relational equations).

The first set of alternatives, in which we generate a numeric prototype, poses an interesting question of reconstructability. Suppose that we get a numeric scalar datum x_0 (real number), transform it through the frame of reference, and derive possibility and necessity measures. Then we use these measures to complete dereferencing using the weighted sum of the form

$$\tilde{x} = \sum_{i=1}^{c} w_i z_i \qquad (10.7)$$

where z_i are the modal values of the corresponding reference fuzzy sets and w_i are the corresponding membership grades of these fuzzy sets. It is of interest to determine the conditions under which the result of dereferencing (denote it by $\tilde{x}_0$) is equal to x_0 and this result holds for any x_0 in $\mathbf{R}$. A surprising and interesting finding arises in the case of triangular membership functions with a one-half overlap between any two successive fuzzy sets (Pedrycz, 1994). In this case, x_0 is equal to x_0 for any value in $\mathbf{R}$. Because the possibility measures are equal to

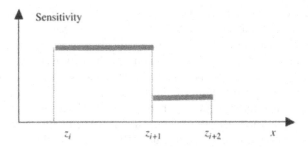

Figure 10.12. The sensitivity relationship for the reference fuzzy sets with triangular membership functions of one-half overlap. The stepwise character of the function reflects the distance between successive modal values of the fuzzy sets.

the necessity values, it is enough to keep only the possibility measures. Owing to this level of overlap (which is critical to the performance of the dereferencing), the sum of possibilities is equal to 1 and for each x_0 only two adjacent fuzzy sets contribute to the transformation formula. This yields $\tilde{x}_0 = w_i z_i + (1 - w_i)z_{i+1}$ The result reveals that the number of reference fuzzy sets does not matter, which may seem somewhat counterintuitive. A closer look at the formulation of the problem indicates that we made a strong assumption about the accuracy of the values of the possibility measure. Let us compute the sensitivity of x_0 with respect to w_i and take its absolute value. We have

$$\text{sensitivity} = \left| \frac{\partial \tilde{x}}{\partial w_i} \right| = |z_i - z_{i+1}| \tag{10.8}$$

which is simply the absolute difference between the modal values of the successive fuzzy sets; the more fuzzy sets, the smaller the distances and the lower the sensitivity. This dependency is a stepwise function, Figure 10.12 that assumes constant values (equal to the differences between z_{i+1} and z_i.

Thus, we can easily conclude that adding more reference fuzzy sets reduces sensitivity.

10.6. CONCLUSIONS

We have elaborated on the general clustering framework required to cope with heterogeneous data. We contrasted the two main representation approaches, discussing the advantages and shortcomings of parametric and nonparametric models of representation of information granules. We also emphasized the differences in the format of the clustering results. While the parametric approach is quite straightforward (in the sense that the result is in the same parametric format), nonparametric preprocessing results in several interesting opportunities that are not necessarily equivalent.

While we have concentrated on coping with information granules represented in fuzzy sets, a similar approach could be taken with respect to information granules represented in shadowed and rough sets. In general, any parametric approach would eventually increase the dimensionality of the new feature space, with the increase being related to the parametric representation scheme involved. The issue of coping with information granules represented in several fundamentally different conceptual settings has not been addressed and may require attention in the future exploration of this subject. This may eventually lead to interesting developments in frame-based architectures (Di Nola et al, 1994)

REFERENCES

A. Di Nola, W. Pedrycz, S. Sessa, Fuzzy information in knowledge representation and processing for frame-based structures, *IEEE Trans. on Systems, Man, and Cybernetics*, 6, 1994, 918–925.

R. Hathaway, J.C. Bezdek, W. Pedrycz, A parametric model for fusing heterogeneous fuzzy data, *IEEE Transactions on Fuzzy Systems*, 4, 1996, 270–281.

W. Pedrycz, Selected issues of frame of knowledge representation realized by means of linguistic labels, *Int. J. of Intelligent Systems*, 7, 1992, 155–169.

W. Pedrycz, Why triangular membership functions? *Fuzzy Sets and Systems*, 64, 1994, 21–30.

W. Pedrycz, J.C. Bezdek, R.J. Hathaway, G.W. Rogers, Two nonparametric models for fusing heterogeneous fuzzy data, *IEEE Trans. on Fuzzy Systems*, 6, 3, 1998, 411–425.

W. Pedrycz, A. Gacek, Temporal granulation and its application to signal analysis, *Information Sciences*, 143, 2002, 47–71.

W. Pedrycz, J. Valente de Oliveira, Optimization of fuzzy models, *IEEE Trans. on Systems, Man, and Cybernetics, Part B*, 26, 1995, 627–636.

W. Pedrycz, G. Vukovich, On elicitation of membership functions, *IEEE Trans. on Systems, Man, and Cybernetics, Part B*, 32, 2002, 761–767.

11 Hyperbox Models of Granular Data: The Tchebyschev FCM

In this chapter, we propose a model of granular data emerging through summarization and processing of numeric data. This model supports data analysis and contributes to further interpretation activities. The structure of data is revealed through the FCM equipped with the Tchebyschev (l_∞) metric. The chapter offers a novel contribution of gradient-based learning of the prototypes developed in the l_∞-based FCM. The l_∞ metric promotes the development of easily interpretable information granules, namely, hyperboxes. A detailed discussion of their geometry is provided. In particular, we discuss the deformation effect of the hyperbox shape of granules due to an interaction between the granules. We also show how the deformation effect can be quantified. Subsequently, we show how clustering gives rise to a two-level topology of information granules: the core part of the topology consists of hyperbox information granules. A residual structure is expressed through detailed, yet difficult-to-interpret, membership grades. Illustrative examples including synthetic data are studied.

11.1. INTRODUCTION

Clustering is widely recognized as one of the dominant techniques of data analysis. The variety of detailed algorithms and their underlying technologies (fuzzy sets, neural networks, heuristic approaches) is impressive. In spite of this diversity, the key objective remains the same: to understand the data. In this sense, clustering becomes an integral part of data mining (Cios et al., 1998; Maimon et al., 2001). Data mining is aimed at making the findings *transparent* to the end user. Transparency is accomplished through suitable knowledge representation mechanisms, namely, the ways in which generic data elements are formed, processed, and presented to the user. Information granularity is a basic concept that needs to be discussed in this context (cf. Bargiela, 2001; Zadeh, 1999).

The key idea we discuss here is that in any data set we can distinguish between a *core* part of the data structure, which is easily describable and interpretable in a straightforward manner, and a *residual* component, which has no clear pattern of regularity. The core part can be described in a compact manner through several

Knowledge-Based Clustering, by Witold Pedrycz
ISBN 0-471-46966-1   Copyright © 2005 John Wiley & Sons, Inc.

information granules, while the residual part exhibits no visible geometry and requires formal descriptors such as membership formulas. The approach proposed in this chapter focuses on the standard FCM method with a Tchebyschev distance that promotes geometry of the information granules (hyperboxes). Starting from the results of clustering, our objective is to develop information granules forming a core structure in the data set, characterize them, and discuss the interaction between the granules leading to their deformation.

The chapter consists of six sections. First, we formulate the problem and then move on to the modified clustering algorithm (Sections 11.2 and 11.3). Section 11.4 is concerned with the generation of granular prototypes. In Section 11.5 we analyze the geometry of information granules and quantify the deformation of regular hyperboxes. We then propose a general model of granular data description (Section 11.6) and present conclusions in Section 11.7. The concept and numerical studies are presented in parallel so that new ideas are made more tangible by illustrative material.

11.2. PROBLEM FORMULATION

Using the well-known objective function (Bezdek, 1981)

$$Q = \sum_{i=1}^{c} \sum_{k=1}^{N} u_{ik}^2 d_{ik} \qquad (11.1)$$

it is worth stressing that the choice of the distance function is critical to our primary objective of achieving transparency of the findings. We are interested in distances whose equidistant contours are "boxes" with the sides parallel to the coordinates. The Tchebyschev distance (l_∞ distance) satisfies this requirement (Bobrowski and Bezdek, 1991; Groegen and Jajuga, 2001; Jajuga, 1991; Pedrycz and Gomide, 1998). The boxes are easily decomposable, that is, the region within a given equidistant contour of the distance can be treated as a decomposable relation R in the feature space:

$$R = A \times B \qquad (11.2)$$

where A and B are sets (or, more generally, information granules) in the corresponding feature spaces. Note that the Euclidean distance does not lead to the decomposable relations in the above sense (as the equidistant regions in such constructs are spheres or ellipsoids). The decomposability property is illustrated in Figure 11.1.

The above clustering problem, known in the literature as an l_∞ FCM, was introduced and discussed by Bobrowski and Bezdek (1991) and Jajuga (1991) more than 14 years ago. Some recent generalizations can be found in Groegen and Jajuga (2001). The hyperbox aspects of clustering were discussed by Kersten (1999). This type of distance was introduced to handle data structures with

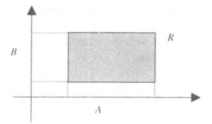

Figure 11.1. Decomposability property provided by the Tchebyschev distance; the region of equidistant points is represented as a Cartesian product of two sets in the corresponding feature space.

"sharp" boundaries (clearly, the Tchebyschev distance is more suitable in this regard than the Euclidean distance). The solution proposed in Bobrowski and Bezdek (1991) was obtained by applying a basis exchange algorithm.

In this chapter, as already noted, the reason for using the Tchebyschev distance is different. We want to describe the data structure and the related interpretability of the results of clustering so that the clusters can be viewed as basic models of associations existing in the data. Here we derive a gradient-based FCM technique enhanced with an additional convergence mechanism.

11.3. THE CLUSTERING ALGORITHM—DETAILED CONSIDERATIONS

The FCM optimization procedure is standard to a large extent and consists of two steps: determination of the partition matrix and calculation of the prototypes. The use of Lagrange multipliers converts the constrained problem into its constraint-free version. The original objective function (11.1) is transformed into the form

$$V = \sum_{i=1}^{c} \sum_{k=1}^{N} u_{ik}^2 d_{ik}^2 + \lambda \left(\sum_{i=1}^{c} u_{ik} - 1 \right) \tag{11.3}$$

with λ being a Lagrange multiplier. The problem is then solved with respect to each pattern separately, that is, we consider the following relationship for each data point ($t = 1, 2, \ldots, N$):

$$\frac{\partial V}{\partial u_{st}} = 0 \tag{11.4}$$

$s = 1, 2, \ldots, c$, $t = 1, 2, \ldots, N$. Straightforward calculations lead to the expression

$$u_{st} = \frac{1}{\sum_{j=1}^{c} \left(\dfrac{d_{st}}{d_{jt}} \right)^2} \tag{11.5}$$

The determination of the prototypes is more complicated, as the Tchebyschev distance does not lead to a closed expression (unlike the standard FCM with the Euclidean distance). Let us start with the situation in which the distance function is spelled out in an explicit manner:

$$Q = \sum_{i=1}^{c} \sum_{k=1}^{N} u_{ik}^2 \max_{j=1,2,\ldots,n} |x_{kj} - v_{ij}| \tag{11.6}$$

The minimization of Q with respect to the prototype (more specifically, its tth coordinate) follows a gradient-based scheme:

$$v_{st}(\text{iter} + 1) = v_{st}(\text{iter}) - \alpha \frac{\partial Q}{\partial v_{st}} \tag{11.7}$$

where α is an adjustment rate (learning rate) assuming positive values. This update expression is iterative; we start from some initial values of the prototypes and keep modifying them, following the gradient of the objective function. The detailed calculations of the gradient lead to the expression

$$\frac{\partial Q}{\partial v_{st}} = \sum_{k=1}^{N} u_{sk}^2 \frac{\partial}{\partial v_{st}} \{\max_{j=1,2,\ldots,n} |x_{kj} - v_{sj}|\} \tag{11.8}$$

Let us introduce the following shorthand notation:

$$A_{kst} = \max_{\substack{j=1,2,\ldots,n \\ j \neq t}} |x_{kj} - v_{sj}| \tag{11.9}$$

Evidently, A_{kst} does not depend on v_{st}. This allows us to concentrate on the term that affects the gradient. We rewrite the above expression for the gradient as follows:

$$\frac{\partial Q}{\partial v_{st}} = \sum_{k=1}^{N} u_{sk}^2 \frac{\partial}{\partial v_{st}} \{\max(A_{kst}, |x_{kt} - v_{st}|)\} \tag{11.10}$$

The derivative is nonzero if A_{kst} is less or equal to the second term in the expression

$$A_{kst} \leq |x_{kt} - v_{st}| \tag{11.11}$$

Next, if this condition holds, we infer that the derivative is equal to either 1 or -1, depending on the relationship between x_{kt} and v_{st}, that is, -1 if $x_{kt} > v_{st}$ and 1 otherwise. Putting these conditions together, we get

$$\frac{\partial Q}{\partial v_{st}} = \sum_{k=1}^{N} u_{sk}^2 \begin{cases} -1 \text{ if } A_{kst} \leq |x_{kt} - v_{st}| \text{ and } x_{kt} > v_{st} \\ +1 \text{ if } A_{kst} \leq |x_{kt} - v_{st}| \text{ and } x_{kt} \leq v_{st} \\ \quad\quad 0 \text{ otherwise} \end{cases} \tag{11.12}$$

The primary concern that arises about this learning scheme is not the piecewise character of the function (absolute value, a concern that could be easily raised from the formal standpoint) but the fact that the derivative zeroes for a significant number of situations. This may result in poor performance of the optimization method, so it could be easily trapped if the overall gradient becomes equal to 0. To enhance the method, we relax the binary character of the predicates (less or greater than) in (11.15). These predicates are Boolean (two-valued), as they return values equal to 0 or 1 (which translates into the expression "predicate is satisfied or it does not hold"). The modification consists of a degree of satisfaction of this predicate, meaning that we compute a multivalued predicate

$$\text{Degree}(a \text{ is included in } b) \tag{11.13}$$

that returns 1 if a is less than or equal to b. Lower values of the degree arise when this predicate is not fully satisfied. This form of augmentation of the basic concept was introduced in conjunction with studies of fuzzy neural networks and relational structures (fuzzy relational equations).

The degree of satisfaction of the inclusion relation is equal to

$$\text{Degree}(a \text{ is included in } b) = a \rightarrow b \tag{11.14}$$

where a and b are in the unit interval. The implication operation $\rightarrow$ is a residuation operation. Here we consider a certain implementation of such an operation where the implication is implied by the product t-norm:

$$a \rightarrow b = \left(\begin{array}{l} 1 \text{ if } a \le b \\ b/a \text{ otherwise} \end{array} \right. \tag{11.15}$$

Using this construct, we rewrite (11.12) as follows:

$$\frac{\partial Q}{\partial v_{st}} = \sum_{k=1}^{N} u_{sk}^2 \left\{ \begin{array}{ll} -(A_{kst} \rightarrow |x_{kt} - v_{st}|) & \text{if } x_{kt} > v_{st} \\ (A_{kst} \rightarrow |x_{kt} - v_{st}|) & \text{if } x_{kt} \le v_{st} \end{array} \right. \tag{11.16}$$

In the overall scheme, this expression will be used to update the prototypes of the clusters (11.7).

Summarizing, the clustering algorithm arises as a sequence of the following steps:

repeat

Compute partition matrix using (11.5).

Compute prototypes using the partition matrix obtained in the first phase. (Note that the partition matrix does not change at this stage, and all updates of the prototypes work with this matrix. This phase is more time-consuming than the FCM method with the Euclidean distance.)

until a termination criterion satisfied

Both the termination criterion and the initialization of the method are standard. The termination takes into account changes in the partition matrices at two successive iterations that should not exceed a certain threshold level. The initialization of the partition matrix is random.

As an illustrative example, we consider synthetic data involving four clusters (Figure 11.2). The two larger data groupings consist of 100 data points, and the two smaller ones have 20 and 10 data points, respectively.

Table 11.1 gives a representative set of clustering results for two to eight clusters. As expected, the two larger data groupings have a dominant influence on the outcome of the FCM algorithms. Both Euclidean and Tchebyschev distance–based FCM exhibit robust performance in that they find approximately the same clusters in their successive runs (within the limits of the optimization convergence criterion). While most of the identified prototypes fall within the large data groupings, the Tchebyschev distance–based FCM consistently manages to associate a prototype with one of the smaller data groupings (underlined in the table). This is clearly a very advantageous feature of our modified FCM algorithm and confirms our assertion that the objective of enhancing the interpretability of data through the identification of decomposable relations is enhanced with Tchebyschev distance–based FCM.

The above results are better understood if we examine the cluster membership function over the entire pattern space. The membership function for one of the two clusters, positioned in the vicinity of (0.2, 0.2) ($c = 2$), is visualized in Figure 11.3.

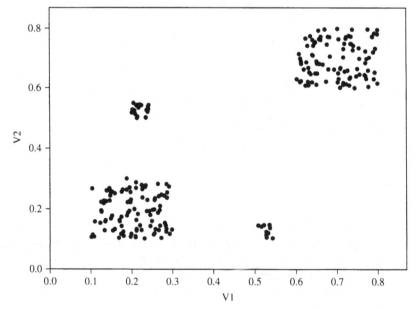

Figure 11.2. Two-dimensional synthetic data with four visible clusters of unequal size.

TABLE 11.1. Prototypes Identified by Two FCM Algorithms, with Euclidean and Tchebyschev Distance Measures, Respectively, for the Varying Number of Clusters (the Underlined Prototypes Correspond to the Smaller Data Groupings)

Number of Clusters	Prototypes for FCM with Euclidean Distance		Prototypes for FCM with Tchebyschev Distance	
2	0.6707	0.6706	0.2088	0.1998
	0.2240	0.2236	0.6924	0.6831
3	0.2700	0.3011	0.7000	0.6847
	0.6875	0.6841	<u>0.2440</u>	<u>0.4914</u>
	0.2302	0.2127	0.2124	0.1852
4	0.2255	0.2035	0.7261	0.7377
	0.2323	0.2479	<u>0.2278</u>	<u>0.5178</u>
	0.6872	0.6814	0.2092	0.1846
	0.6533	0.6588	0.6523	0.6498
5	0.2525	0.2784	0.2189	0.1451
	0.2282	0.2014	<u>0.2272</u>	<u>0.5188</u>
	0.6721	0.6757	0.1960	0.2258
	0.2343	0.2389	0.6568	0.6868
	0.6919	0.6841	0.7268	0.6593
6	0.2329	0.2562	0.7469	0.6650
	0.6809	0.6777	0.2151	0.1364
	0.6857	0.6830	<u>0.2278</u>	<u>0.5208</u>
	0.2272	0.2206	0.6570	0.6840
	0.2261	0.2008	0.2619	0.2648
	0.6447	0.6500	0.1945	0.2239
7	0.6646	0.6697	0.1967	0.2255
	0.7036	0.6619	0.2200	0.1450
	0.6993	0.7100	0.7278	0.6594
	0.2395	0.5019	<u>0.2277</u>	<u>0.5183</u>
	0.2382	0.1935	0.3976	0.4051
	0.2164	0.1955	0.6099	0.6117
	0.2271	0.2018	0.6588	0.6923
8	0.6962	0.6892	0.6607	0.7615
	<u>0.2398</u>	<u>0.5088</u>	0.2122	0.1327
	0.2360	0.1980	0.3209	0.3097
	0.2441	0.2203	0.6565	0.6830
	0.6962	0.6882	0.7267	0.6590
	0.6850	0.6756	0.6460	0.6492
	0.2385	0.1942	<u>0.2277</u>	<u>0.5191</u>
	0.2166	0.1965	0.2108	0.2249

It is clear that for higher values of the membership grades (e.g., 0.9), the shape of contours is rectangular. This changes for lower values of the membership grades, when we witness a gradual departure from this geometry of the clusters. This is an effect of the interaction between the clusters that manifests in a deformation of the original rectangles. The deformation depends on the

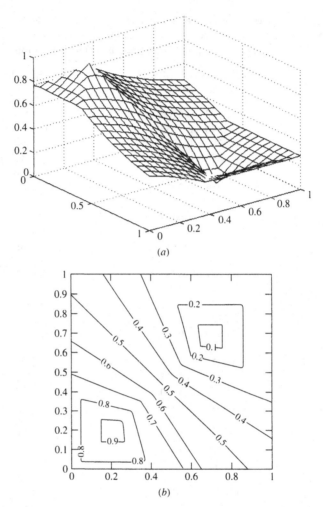

Figure 11.3. Visualization of the first cluster (membership function) centered around (0.2088 0.1998): (*a*) three-dimensional space and (*b*) contour plots.

distribution of the clusters, their number, and the threshold β selected. The lower the value of this threshold, the greater the departure from the rectangular shape. For higher values of β, such deformation is quite limited. This suggests that when high threshold level values are used, the rectangular (or hyperbox) form of the core part of the clusters is completely legitimate.

Let us contrast these results with the geometry of the clusters constructed using a Euclidean distance. Again, we consider two prototypes, as identified by the Euclidean distance–based FCM (see Figure 11.4). The results are significantly different: the clusters are close to the Gaussian-like form and do not closely approximate rectangular shapes.

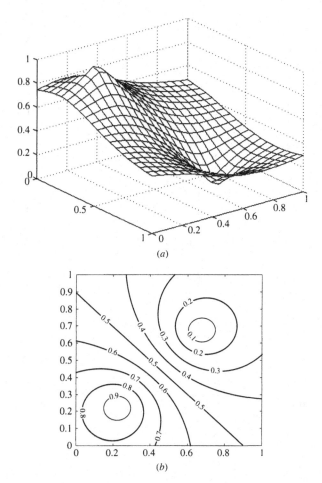

(a)

(b)

Figure 11.4. Visualization of the first cluster (membership function) centered around (0.2240 0.2236): (a) three-dimensional space and (b) contour plots. The Euclidean distance function was used in the clustering algorithm.

The above effect is even more pronounced when more clusters interact with each other. We consider eight prototypes identified by the two FCM algorithms (see Figure 11.5). In the case of the Tchebyschev FCM, it is clear that despite strong interactions between the clusters, the rectangular shape of the cluster membership function is preserved for a range of values of this function. These undistorted rectangles cover a good proportion of the original data, which is represented by the selected prototype. On the other hand, the Euclidean FCM results in contours of the membership function that are undistorted circles only in very close proximity to the prototype itself. Thus linking the original data with the prototype representing an association existing in the data is quite difficult for most of the data points.

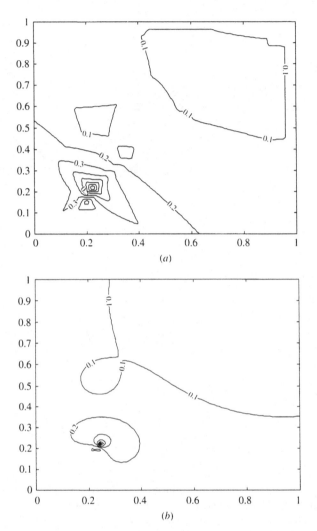

Figure 11.5. Contour plots for one of the eight clusters (membership function) centered around (0.2108 0.2248) for the Tchebyschev distance (*a*) and (0.2441 0.2203) for the Euclidean distance (*b*).

11.4. DEVELOPMENT OF GRANULAR PROTOTYPES

As we are dealing with the Tchebyschev metric, its underlying geometry promotes a rectangular shape of the information granules. Here we formalize how an information granule can be seen as a union of appropriately constructed hyperboxes. The essence of this construction is to move around the prototype by changing only a single feature. The moves are made separately toward higher and lower values (with the reference to the prototype) of the feature (see Figure 11.6).

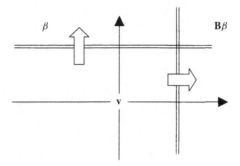

Figure 11.6. Generation of information granules (two-dimensional case); the region around the prototype is constructed by analyzing the membership grades of the clusters.

If we establish a certain threshold value (β), the resulting rectangle captures the corresponding part of the data set. The threshold values are in the unit interval. The development of the box uses the following method: move gradually from the prototype along one of the directions (as indicated in Figure 11.6) and compute the membership grade $u(\mathbf{x}, \mathbf{v})$ (see (11.8)) until it becomes equal to β. Note that as we move away from the prototype, this membership grade continuously decreases. Once we reach the threshold, the move stops and the corresponding $\mathbf{x}$ is treated as the face of the box (rectangle). The construction is carried out for all remaining directions. The final result is denoted by B_β.

Obviously, higher values of β confine the core part of the cluster and produce smaller information granules of the core of the data structure. The resulting hyperbox is tied to the threshold level through an evident relationship

$$B_\beta = \{\mathbf{x} \in [0, 1]^n | u(\mathbf{x}, \mathbf{v}) \geq \beta\} \qquad (11.17)$$

where B_β is a hyperbox with a threshold equal to β.

Second, for each β, the corresponding hyperbox is a relation in the feature space. When looking at these relations globally (considering varying values of β), we can represent them as a fuzzy relation, meaning that B_β is a fixed β-cut of it. We then have the following relationship:

$$B = \bigcup_\beta B_\beta \qquad (11.18)$$

where B is a fuzzy relation of the core of the data structure. To make the core meaningful, the threshold value should be high enough so that only the essential part of the data set becomes qualified as the core. There is another aspect of this design: deformation of the hyperboxes in the feature space caused by an interaction between the clusters. Again, with lower values of β, the interaction tends to be more critical, leading to more profound deformations of the hyperbox.

Following the discussion of the numeric example, a representative information granule (box), taken from the set of prototypes calculated for $c = 8$ clusters, is

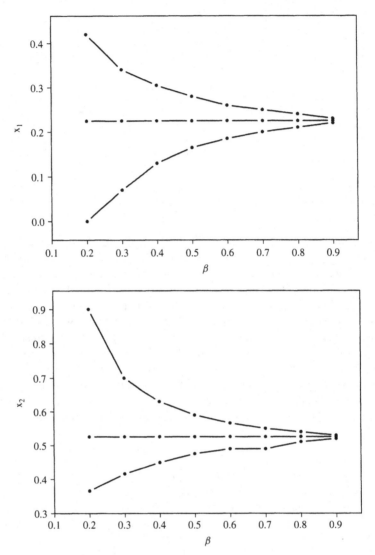

Figure 11.7. Coordinates of the information granules (x_1 and x_2) of a selected cluster (0.2277, 0.5191) for clustering with $c = 8$ clusters.

shown in Figure 11.7. Evidently, the box shrinks with increasing values of the threshold; at some point, the changes to the size of the granule become very small.

11.5. GEOMETRY OF INFORMATION GRANULES

As the contour plots of the clusters reveal (Figure 11.5), interaction between the clusters become responsible for the deformation of the hyperbox shape of the

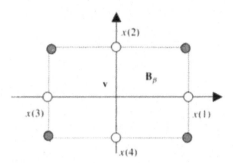

Figure 11.8. A concept of a deformation index—expressing departure from the hyperbox nature of cores implied by the Tchebyschev distance.

cores. This poses an interesting question concerning the size of the core structure of the data. The choice of the threshold level (β) needs to be controlled by an acceptable level of deformation of the equidistant lines of the Tchebyschev distance. We require quantification of this deformation effect. This can be done by finding the differences between the theoretical values of the membership (dictated by the Tchebyschev metric) and those resulting from the calculations of the membership grades based on the prototypes. The details follow the notation in Figure 11.8.

In the two-dimensional case, we identify four corner points of the box implied by the fixed threshold β. This means that all four of them belong to the information granule at the membership grade equal to β. Using (11.8) with $\mathbf{x} = \mathbf{x}(1), \mathbf{x}(2), \ldots, \mathbf{x}(4)$ and the prototype $\mathbf{v}$, the calculated membership grades could be different from this threshold, say, $u(1)$, $u(2)$, $u(3)$, $u(4)$. Let us consider the sum of differences

$$D = |\beta - u(1)| + |\beta - u(2)| + |\beta - u(3)| + |\beta - u(4)| \qquad (11.19)$$

Since $u = \beta$ was calculated only in "white" points (by virtue of construction (11.20)), ideally it should also be satisfied by "gray" points, which are the vertices spanned by the original vectors $\mathbf{x}(1)$, $\mathbf{x}(2)$, $\mathbf{x}(3)$, and $\mathbf{x}(4)$.

Therefore, (11.19) serves as a useful measure of deformation of the rectangular shape of the granules. The above construct easily expands to any dimension of the feature space; evidently, for n features, a search is completed for all corners of the hypercube, that is, 2^n.

Continuing the numeric example, we quantify the deformation of the boxes by means of (11.19). The approximation of the resulting dependency between the deformation measure D, viewed as a function of β, is done through a polynomial fit. A representative set of results is given in Figure 11.9. The assessment of information granules was carried out for every granule identified with two to eight clusters. It is evident that the deformation of the hyperboxes can be approximated by a low-order polynomial.

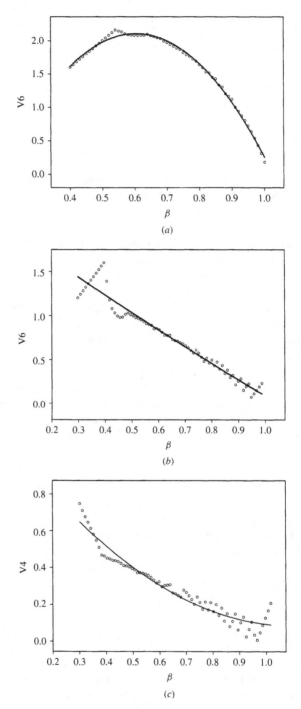

Figure 11.9. Quantification of the deformation effect D for (a) prototype $\mathbf{v}_6$ obtained with $c = 8$; (b) prototype $\mathbf{v}_6$ obtained with $c = 6$; and (c) prototype $\mathbf{v}_4$ obtained with $c = 4$.

11.6. GRANULAR DATA DESCRIPTION: A GENERAL MODEL

With the development of the granular prototypes guided by the clustering algorithm, we can concisely describe the data in the form

$$D = B_i \cup B_2 \cup \ldots \cup B_c \cup R \qquad (11.20)$$

where D is a data set under discussion, B_i are granular prototypes, and R is a residual structure of the data set. The brief characteristics of the this structure are summarized in Table 11.2.

Any element $\mathbf{x}$ in the data space of interest can be characterized as belonging to the core or being identified with the residual portion of the data. Membership in the core is binary—we identify the hyperbox to which this new pattern belongs. If $\mathbf{x}$ is identified to be a part of R, the degree of membership in the hyperbox is determined through the standard membership expression, in which we now consider the distance between a data point ($\mathbf{x}$) and a set (relation) $\mathbf{V}_i$:

$$u_i(\mathbf{x}) = \frac{1}{\displaystyle\sum_{j=1}^{c} \frac{d(\mathbf{x}, \mathbf{V}_i)}{d(\mathbf{x}, \mathbf{V}_j)}} \qquad (11.21)$$

11.7. CONCLUSIONS

In the description of data, we have developed two main components, namely, cores of the data that are well structured in the form of hyperboxes in the feature

TABLE 11.2. Main Structure Descriptors and their Features

Data Structure Descriptor	Description	Main Features
Hyperboxes	Capture the essence of the data. Serve as focal points of the structure. Can be easily interpreted as relations	Hyperboxes constructed on the basis of the Tchebyschev distance for a given threshold level. The choice of this level depends on an acceptable level of deformation of the hyperboxes allowed in the structure. The hyperboxes can be decomposed into intervals for individual features
Residual	Deals with the remaining data points and concentrates on patterns not clearly identified as focal points of the data structure	A lack of explicit description; more complex geometry of the regions. The degree of membership in the hyperbox is expressed through some formula

space and a far less regular structure that is described analytically through an expression for membership grades but has no clear geometric interpretation. The computing backbone of this approach is based on the well-known FCM technique using the Tchebyschev distance. We introduced a new way of optimizing the prototypes in this method that uses a gradient-based technique augmented by a logic-oriented mechanism of gradient determination. The geometry and the design of the hyperbox information granules were discussed, along with the important aspect of deformation of such granules. Quantification of this effect was also discussed.

The proposed approach to data analysis can be exploited in many different ways. A few options worth pursuing the following:

- Data mining. Considering the main pursuit of data mining articulated in the language of well-defined, semantically sound, and easily interpretable constructs, the information granules envisioned in this way are legitimate entities for data mining activities. They are easy to interpret, and thus cope with the underlying structure of data while leaving out the residual portion of data not exhibiting strong patterns of dependencies.

- In any modeling pursuit, the above data description helps concentrate on the design of local models assigned to the core parts. The residual part of data can be handled separately, anticipating that these data points may not lead to a model with a strongly manifested character.

- In classification problems, the core part of the data implies a collection of simple classifiers, while the residual part invokes more demanding and conceptually advanced classifiers such as neural networks.

REFERENCES

A. Bargiela, Interval and Ellipsoidal Uncertainty Models, in: W. Pedrycz (ed.), *Granular Computing, An Emerging Paradigm*, Springer-Verlag, Berlin, 2001.

J.C. Bezdek, *Pattern Recognition with Fuzzy Objective Function Algorithms*, Plenum Press, New York, 1981.

L. Bobrowski, J.C. Bezdek, C-Means clustering with the l_1 and l_∞ norms, *IEEE Trans. on Systems, Man, and Cybernetics*, 21, 1991, 545–554.

K. Cios, W. Pedrycz, R. Swiniarski, *Data Mining Techniques*, Kluwer Academic Publishers, Boston, 1998.

P.J.F. Groegen, K. Jajuga, Fuzzy clustering with squared Minkowski distances, *Fuzzy Sets and Systems*, 120, 2001, 227–237.

K. Jajuga, L_1-norm based fuzzy clustering, *Fuzzy Sets and Systems*, 39, 1991, 43–50.

P.R. Kersten, Fuzzy order statistics and their applications to fuzzy clustering, *IEEE Trans. on Fuzzy Systems*, 7, 6, 1999, 708–712.

O. Maimon, A. Kandel, M. Last, Information-theoretic fuzzy approach to data reliability and data mining, *Fuzzy Sets and Systems*, 117, 2001, 183–194.

W. Pedrycz, F. Gomide, *An Introduction to Fuzzy Sets*, MIT Press, Cambridge, MA, 1998.

L.A. Zadeh, From computing with numbers to computing with words—from manipulation of measurements to manipulation of perceptions, *IEEE Trans. on Circuits and Systems*, 45, 1999, 105–119.

12 Genetic Tolerance Fuzzy Neural Networks

In this chapter, we introduce a genetically optimized fuzzy neural network that reveals a hyperbox-based structure in numeric data. This type of network is developed around fuzzy tolerance neurons. Tolerance neurons produce a generalized version of intervals (sets) arising in the form of fuzzy intervals. The architecture of the network reflects a hierarchy of geometric concepts exploited in data analysis: fuzzy intervals combined *and*-wise give rise to fuzzy hyperboxes, and these, in turn, aggregated *or*-wise to generate a summary of the data as a collection of hyperboxes. We discuss an overall development process that consists of two main phases: (a) fuzzy clustering that initiates the search for the structure and (b) genetic optimization of the networks. We provide an in-depth view of the geometry of the individual hyperboxes as well as the overall topology of the network. Numerical experiments deal with two-dimensional synthetic data.

12.1. INTRODUCTION

The design of logic-driven and interpretable neural networks has been an ongoing challenge in the area of data analysis. The interpretability of such networks goes hand in hand with the geometry of the constructs identified and described in the highly dimensional space of data. The learning capabilities of the network have some parametric flexibility exhibited by the geometric constructs. Granular computing supports the first requirement (interpretability and transparency), with its broad agenda of building semantically sound information granules and forming environments for their processing (Gabrys and Bargiela, 2000; Ishibuchi et al., 1995; Pedrycz et al., 2000; Simpson, 1993; Sudkamp and Hammel, 1998). Neural networks help satisfy the second requirement (learning abilities). Evolutionary computing is an important optimization environment in which we can adapt the networks, aiming at global minimization of some performance measures. Together these three paradigms form a highly synergistic environment of computational intelligence (CI) (Pedrycz and Vasilakos, 1999).

The simplest interpretable information granules are those of intervals and hypercubes (cf. (Simpson, 1993)), with their long history of computing (e.g., interval mathematics) originating in interval analysis. The networks based on

Knowledge-Based Clustering, by Witold Pedrycz
ISBN 0-471-46966-1   Copyright © 2005 John Wiley & Sons, Inc.

these constructs were named Min-Max architectures by Simpson (1993) and enhanced by others (see Gabrys and Bargiela, 2000). We generalize this idea by introducing fuzzy hyperboxes developed on the basis of fuzzy tolerance neurons—basic logic processing units rooted in fuzzy logic and exhibiting parametric flexibility required for learning purposes. While the Min-Max networks dwell on hyperboxes, the approach taken here uses fuzzy hyperboxes, with their inherent flexibility in modeling smooth boundaries.

The material is arranged into five sections. We briefly revisit the logic operations of dominance, inclusion, and tolerance and discuss their geometry and parametric realization (Section 12.2). This leads us to the architectural considerations in Section 12.3, where we present an overall topology of the network and reveal the computing carried out at each of its layers. Section 12.4 elaborates on the issue of genetic optimization of the network. We show how genetic algorithms lead us to the minimization of some performance index (or maximization of the corresponding fitness function). Experimental examples are presented in Section 12.5 and conclusions in Section 12.6. In the experiments we use t- and s-norms represented as a product operation and a probabilistic sum.

12.2. OPERATIONS OF THRESHOLDING AND TOLERANCE: FUZZY LOGIC–BASED GENERALIZATIONS

The well-known Boolean operations (relational operators) "greater than" and "less than" are self-explanatory. The predicates $x > a$ or $x < b$ are binary in the sense that they become true or false, depending upon the relationship between the arguments in the expression. The numeric interval $[a, b]$ can be articulated in the language of these two predicates by building a compound statement (conjunction of the predicates)

$$x \text{ is in } (a, b) = (x > a) \,\&\, (x < b) \tag{12.1}$$

where *and* is a logic connective (&) encountered in two-valued logic. Proceeding with a multidimensional case, we build a hyperbox (or a box in a two-dimensional case) by defining the corresponding intervals for each variable treated separately and combining them *and*-wise.

The above Boolean predicates generalize to their continuous (fuzzy) logic counterparts. We will refer to them as inclusion and dominance operators, respectively. The operation of inclusion of x in a, denoted by $\text{incl}(x; a)$ returns a truth value (in [0,1]) expressing the extent to which x can be treated as being included in a. Likewise, an operation of dominance $\text{dom}(x; b)$ expresses the extent to which x dominates b (again, this level of dominance is quantified in the unit interval). Conceptually, these are intuitively appealing generalizations of the two-valued relational operators.

From now on, consider all variables to be confined to the unit interval. Furthermore, the crux of the ensuing definitions is in the logic operation of residuation (known as a φ-operator or a multivalued implication) used in fuzzy sets. For a continuous t-norm we define it in the form $a \rightarrow b = \sup\{c \in [0, 1] \mid atc \leq b\}$, where $a, b \in [0,1]$.

Definition 1. A degree of inclusion of x in a, $\mathrm{incl}(x; a)$ is expressed as

$$\mathrm{incl}(x; a) = a \rightarrow x \tag{12.2}$$

The operation of inclusion quantifies the extent to which x is included in a. Intuitively, it becomes obvious that if x is *less* than a, then this operation should return 1 (this condition is satisfied because of the nature of the multi-valued implication). The degree of inclusion decreases once x exceeds a. Again, this monotonicity property holds because of the implication used in the definition. Obviously, the way in which the inclusion operation quantifies the fact of inclusion depends upon the form of the t-norm. Every t-norm may affect the characteristics of the operation, yet it retains the required monotonicity property. Figure 12.1 illustrates inclusion for selected implications induced by two t-norms, namely, a product operation and the Lukasiewicz &-connective, that is, $xty = xy$, $xty = \max(0, x + y - 1)$:

$$\mathrm{incl}(x; a) = \begin{cases} 1 \text{ if } x \leq a \\ \dfrac{a}{x} \text{ if } x > a \end{cases} \tag{12.3}$$

$$\mathrm{incl}(x; a) = \begin{cases} 1 \text{ if } x \leq a \\ 1 - x + a \text{ if } x > a \end{cases} \tag{12.4}$$

In both cases, the functions monotonically decrease; for the Lukasiewicz implication, this decrease is linear.

Definition 2. A dominance of x over b, denoted by $\mathrm{dom}(x; b)$, is expressed as

$$\mathrm{dom}(x; b) = x \rightarrow b \tag{12.5}$$

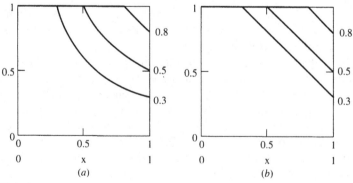

Figure 12.1. Inclusion operation treated as a function of x for selected values of a: (a) implication induced by the product operation; (b) implication induced by the Lukasiewicz t-norm.

Note that the position of the arguments in the above definition is interchanged in comparison with the previous operation. The dominance returns 1 if x dominates b, that is, x exceeds the value of b. Otherwise, the operation returns values lower than 1. The monotonicity property holds: the lower the value of x in comparison with b, the lower the value of the dominance.

The plots of the dominance for the two implications are presented in Figure 12.2. Note that the essence of the operation is retained, while the changes can be noted for the decreasing edge of the graph, whose form is affected by the t-norm used to quantify the dominance effect. In these two cases, we note a linear change in the dominance level.

The two definitions generalize the Boolean predicates to a continuous case by showing how the concept can be enriched by admitting the notion of partial satisfaction of the given concept.

Now we introduce the notion of a tolerance operation. It generalizes the idea of an interval used in set theory and interval analysis.

Definition 3. A tolerance operation, $\text{tol}(x; a, b)$ expresses the degree to which x is contained in an interval $[a, b]$

$$\text{tol}(x; a, b) = \text{incl}(x; a)\ t\ \text{dom}(x; b)$$

where t is a certain t-norm (in fact, the choice of the t-norm in the above expression does not really matter; we may take a t-norm that is computationally the simplest, say, a minimum operation or product). Plots of two examples of the tolerance operation are displayed in Figure 12.3.

The tolerance operation has an interesting interpretation. As indicated earlier, it generalizes a tolerance interval by giving rise to a fuzzy interval. If $a = b$, then

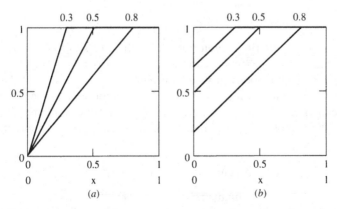

Figure 12.2. Dominance operation treated as a function of x for selected values of a: (a) implication induced by the product operation; (b) implication induced by the Lukasiewicz t-norm.

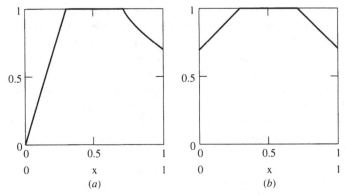

Figure 12.3. Examples of tolerance for two t-norms and several values of a: (a) product t-norm; (b) Lukasiewicz t-norm.

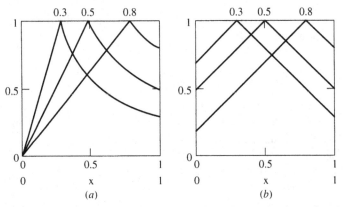

Figure 12.4. Examples of similarity for two t-norms and several values of a: (a) product t-norm; (b) Lukasiewicz t-norm.

we end up with a similarity operator; if $x = a$, then the similarity is equal to 1. It gracefully reduces to 0 when we start moving away from this point (Figure 12.4).

From the operational point of view, it is advisable to control the "edges" of the fuzzy interval and change their shape (i.e., sharpen or smooth them), depending upon the experimental data. This edge control is achieved by augmenting the previous definitions by a linguistic modifier. The modified inclusion and dominance operators are now

$$\text{incl}(x; a, p) = \text{incl}(x; a)^p \tag{12.6}$$

$$\text{dom}(x; b, r) = \text{dom}(x; b)^r \tag{12.7}$$

where p and r assume nonnegative values. This operation relates to what is known in fuzzy sets as a linguistic modifier (hedge). Let us recall that a fuzzy

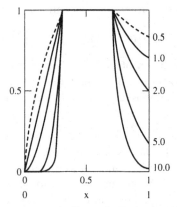

Figure 12.5. Tolerance operation and effect of dilution and concentration on selected values of p.

set (linguistic term) A can be transformed by linguistic modifiers that affect the semantics of the linguistic term

$$very\ A(x) = A(x)^2$$

$$more\ or\ less\ A(x) = A(x)^{0.5}$$

The values of $p(r)$ less than 1 dilute the original linguistic term, while $p(r) > 1$ gives rise to the concentration effect. The same effect is present here. Note that the edges of the operation are profoundly affected.

In a similar manner, we extend the definition of dominance and finally come up with a tolerance definition of the form

$$tol(x; a, b, p, r) = incl(x; a, p)\ t\ dom(x; b, r) \tag{12.8}$$

Refer to Figure 12.5. The above definition can be rewritten in a compact manner by arranging all parameters in a single vector $\mathbf{w} = [a\,b\,p\,r]^T$, that is, tol $(x;\mathbf{w})$. Those are the parameters that provide the tolerance operation with a significant level of parametric flexibility. In this sense we, can regard its implementation as a *tolerance* neuron—a processing element with well-defined logic properties and connections (parameters) that can be learned. In the next section we use this neuron in the design of the neural network and discuss its properties.

12.3. TOPOLOGY OF THE LOGIC NETWORK

The tolerance neuron presented in Section 12.2 generalizes the concept of an interval leading to a fuzzy interval. By combining several of them fuzzy intervals we can construct fuzzy hyperboxes—a generic and easily interpretable geometric entity located in a multidimensional data space. More specifically, a fuzzy hyperbox in an n-dimensional space $[0,1]^n$ is constructed by taking a Cartesian product of n fuzzy intervals formed in consecutive dimensions of the hypercube:

$$\text{tol}(x_1, \mathbf{w}_1) \times \text{tol}(x_2, \mathbf{w}_2) \times \cdots \times \text{tol}(x_n, \mathbf{w}_n) \qquad (12.9)$$

Moving on to the realization of the hyperbox H, the outputs of the tolerance neurons are combined *and*-wise using a certain t-norm. In other words, we form another processing layer composed of an *and* operation (AND neuron):

$$z = \mathsf{H}(\mathbf{x}; \mathbf{W}) = \text{tol}(x_1, \mathbf{w}_1)t\ \text{tol}(x_2, \mathbf{w}_2)t \ldots t\ \text{tol}(x_n, \mathbf{w}_n) \qquad (12.10)$$

The output of this AND neuron represents a degree of membership in a hyperbox. As the parameters of the tolerance neuron (denoted here collectively by $\mathbf{W}$) can be learned, the hyperbox can change its position in the hypercube and modify its size. The edges of the hyperbox can be easily adjusted as well.

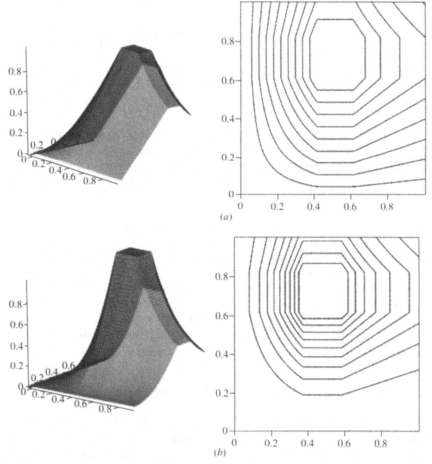

Figure 12.6. Characteristics of the AND neuron-hyperbox for several values of p and r: $p = r = 1$ (*a*); $p = r = 2.0$ (*b*); $p = r = 10$ (*c*); $p = r = 0.5$ (*d*).

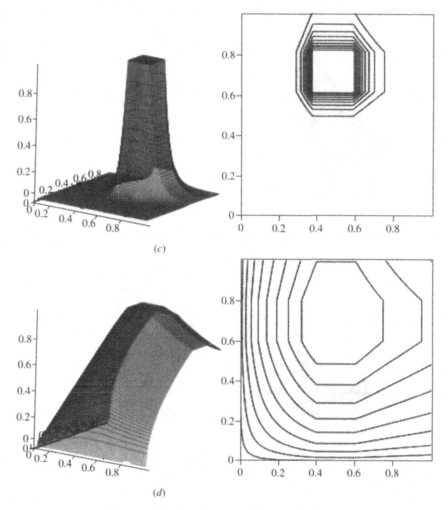

Figure 12.6. (*continued*)

The characteristics of the AND neuron-hyperbox $H(\mathbf{x}; \mathbf{W})$ in the case where $n = 2$ are shown in Figure 12.6. Note that with the increase of powers to around 5, the fuzzy boxes become practically (Boolean) boxes as the edges become very steep. Also, when the tolerance regions shrink to single points, the landscape becomes quite rugged (tolerance neurons convert into the matching neurons), with a number of focal points at which full matching takes place (Figure 12.7).

The geometry of patterns (data) can be captured by a collection of hyperboxes that again comes with a very clear interpretation: the patterns belonging to a given class are "covered" by a family (union) of c fuzzy hyperboxes:

$$H(\mathbf{x}; \mathbf{W}_1) \; or \; H(\mathbf{x}; \mathbf{W}_2) \; or \ldots or \; H(\mathbf{x}; \mathbf{W}_c) \qquad (12.11)$$

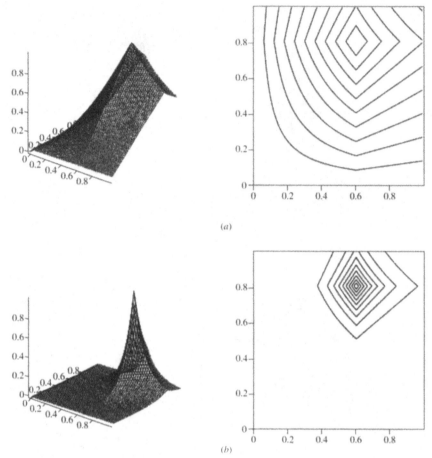

Figure 12.7. Characteristics of the matching neurons for $p = r = 1.0$ (a) and $p = r = 5$ (b).

In light of this, we require another OR neuron (realized by a certain s-norm) that aggregates *or*-wise the outputs of the hyperboxes (AND neurons):

$$y = \text{OR } (H(\mathbf{x}; \mathbf{W}_1), \ H(\mathbf{x}; \mathbf{W}_2), \dots, H(\mathbf{x}; \mathbf{W}_c))$$
$$= H(\mathbf{x}; \mathbf{W}_1) \ s \ H(\mathbf{x}; \mathbf{W}_2) \ s \dots s \ H(\mathbf{x}; \mathbf{W}_c) \tag{12.12}$$

where s is an s-norm (for instance, max or probabilistic sum). The output (y) describes a degree of class membership (as we are concerned about the binary 0–1 class assignment, these will be the values close to 1 or 0).

Summarizing, the overall architecture of the logic neural network is shown in Figure 12.8. Note that the first layer is constructed by means of the tolerance neurons, whose outputs are then aggregated *and*-wise. Then all hyperboxes are summarized *or*-wise by the OR neuron located in the output layer.

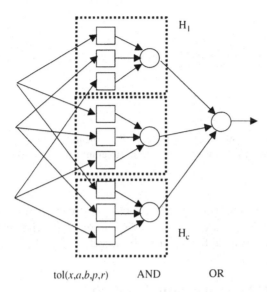

tol(x,a,b,p,r)   AND   OR

Figure 12.8. An overall topology of the neural network with tolerance neurons.

12.4. GENETIC OPTIMIZATION

The network is heterogeneous, consisting of neurons belonging to three different categories (tol, AND, and OR). For problems of high dimensionality that require a significant number of hyperboxes, the number of parameters also becomes quite high. This suggests the use of genetic computing as a suitable optimization vehicle. Genetic optimization has already been found useful in the development of rule-based systems (Ishibuchi et al., 1995). Because of the regular structure of the network, a standard genetic algorithm (GA) is worth considering. The parameters of the network are mapped onto a chromosome by grouping the parameters describing each hyperbox (Figure 12.9). Altogether, for n inputs and c hyperboxes, we end up with $4*n*c$ entries of the string (each tolerance neuron is fully characterized by four parameters). The standard floating point coding (instead of the binary one) is a preferred option, taking into account that the ranges of the connections are fixed: the bounds of the fuzzy intervals ($a, b, c \ldots$) are located in the unit interval, while the modifiers ($p, r, \ldots$) assume values from 0 to around 10–15 (it is clear that higher values of the modifiers make no visible difference to the characteristics of the neuron).

The fitness function that is an essential component of any environment of genetic computing reflects the performance of the network. In what follows, we consider the following form of the fitness function to be maximized:

$$\text{fit} = 1 - \frac{1}{N} \sum_{k=1}^{N} (\text{target}_k - NN(\mathbf{x}_k))^2 \qquad (12.13)$$

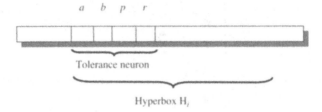

Figure 12.9. Chromosome of the tolerance fuzzy neural network; a detailed portion accommodating parameters of a fuzzy interval (tolerance neuron) and the ensuing single hyperbox are shown.

The learning is achieved in a supervised mode: we are provided with a training set $\{(\mathbf{x}_k, \text{target})\}$ $k = 1, 2, \ldots N$. Owing to the nature of the classification problem, the target values are made binary, namely, $\text{target}_k = 0$ or 1.

12.5. ILLUSTRATIVE NUMERIC STUDIES

In this section, we present four numeric examples that reveal the most essential properties of the networks and elaborate on the efficacies of the resulting geometric structure. The examples are two-dimensional, as they can be easily visualized (the topology and the ensuing optimization support multidimensional cases). In all experiments we use the following parameters of genetic optimization:

> population size: 200
> number of generations: 200
> crossover rate: 0.6
> mutation rate: 0.2
> selection strategy: elitist (the best individual is carried over from the previous
>   generation)

In the first three examples we construct a single hyperbox and analyze its geometrical aspects—in particular, the edges of this construct vis-à-vis a distribution of patterns occurring in the learning set.

Example 1. The data set comprising 600 patterns belonging to two classes (0–1) is shown in Figure 12.10.

The genetic optimization progressed as shown in Figure 12.11 and achieved a final value of the fitness function equal to 0.916156.

The values of the parameters of the tolerance neuron are collected in a single weight vector (cf. Section 12.3):

$$\mathbf{W}_1 = [0.300211 \quad 0.694968 \quad 7.182305 \quad 7.514652 \quad 0.559465 \quad 0.447005$$
$$9.805124 \quad 8.653694]^T$$

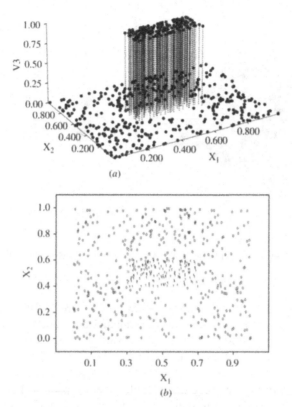

Figure 12.10. Synthetic two-class data set: (*a*) three-dimensional plot and (*b*) two-dimensional visualization.

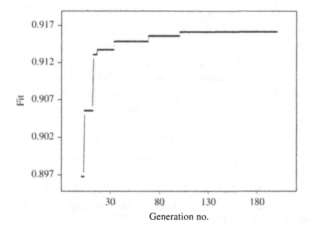

Figure 12.11. Fitness function (the best individual) in successive generations of GA.

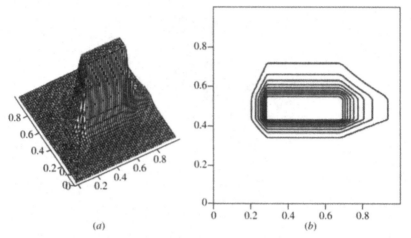

(a)   (b)

Figure 12.12. Genetically optimized hyperbox: (a) three-dimensional and (b) two-dimensional visualization.

They translate into a fuzzy box, as shown in Figure 12.12. It is worthwhile observing that high values of the power (around 7 and higher) lead to quite steep edges of the box; this effect is especially visible for the second variable, where the values of membership move down to zero very quickly.

Example 2. Here the data set exhibits a clear belt of patterns belonging to class '1' that goes across the entire range of the second feature (Figure 12.13).

The value of the fitness function after 200 generations is equal to 0.992209, with the parameters of the hyperbox equal to $\mathbf{W}_1 = [0.767083 \quad 0.505051 \quad 9.924768 \quad 9.238625 \quad 0.010102 \quad 0.819453 \quad 8.146714 \quad 4.389990]^T$. Again the result is very appealing, showing how the box was able to capture the geometry of the data (Figure 12.14).

Example 3. The structure of the two-class data is shown in Figure 12.15. We note a single cluster of patterns belonging to class 1 that are located in the middle of the unit square, while four very condensed clusters of data are coming from class 2 (0 patterns). In fact, these clusters were generated from a truncated Gaussian distribution with $\sigma = 0.05$.

The resulting fuzzy box (Figure 12.16), is no surprise (as it tends to spread across the unit square, with the exception of the corners occupied by the patterns belonging to class 0).

Now we modify the data (see Figure 12.17), by adding some clusters to the existing patterns (the new patterns being added belong to class 0) and analyze what happens to the optimized box.

The size of the box is reduced substantially by being "repelled" by the patterns in class 0. It now occupies a region where there are only patterns belonging to class 1 or no patterns at all. In this case, the box overflows freely into an empty

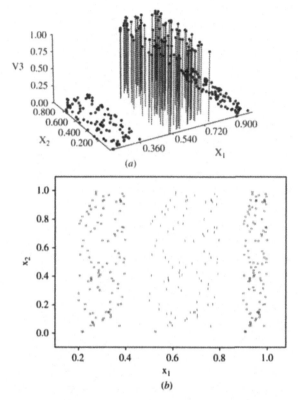

Figure 12.13. Synthetic two-class data: (*a*) three-dimensional and (*b*) two-dimensional visualization.

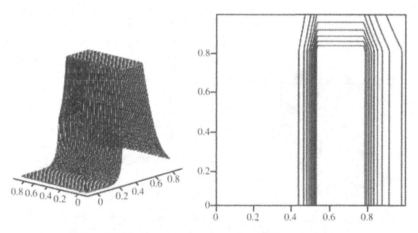

Figure 12.14. Fuzzy box as a result of GA optimization.

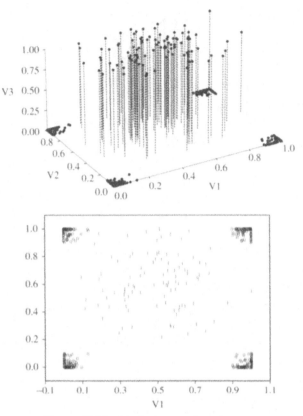

Figure 12.15. Synthetic two-class data set.

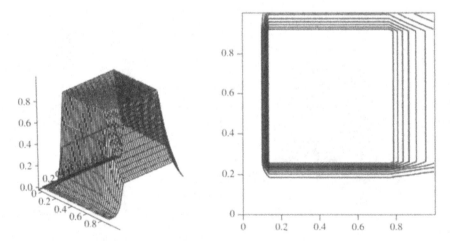

Figure 12.16. A fuzzy box representing the geometry of the data.

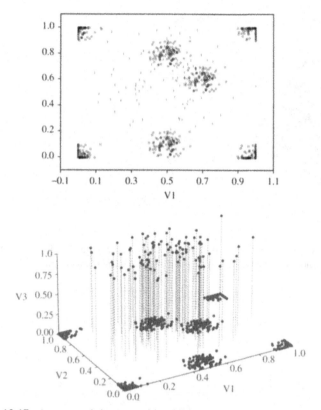

Figure 12.17. Augmented data set with additional clusters of patterns (class-0).

region, as shown in Figure 12.18. Note that the edges of the fuzzy box reflect the distribution of the patterns; if any possibility exists, the edges become less steep; this helps the box partially occupy some regions of the data space.

Example 4. Now we consider data in which we can identify two regions (boxes) positioned along the sides of the unit hyperbox (Figure 12.19).

In this case, we proceed with genetic optimization that involves two fuzzy boxes. The final fitness function is equal to 0.993191 (which is very close to an ideal case of the fitness function equal to 1), meaning that the network is able to capture the geometry of the data. The components of the network—that is, two boxes and their OR aggregation—are shown in Figure 12.20.

When carrying out genetic optimization using only one fuzzy box, we end up with the lower fitness function (equal to 0.8879) and the box that attempts to capture most of the data (but is not fully successful) (Figure 12.21). This attempt is visualized by the box whose core is located at the intersection of the two original boxes (Figure 12.20) and the steepness of the walls, which tend to stretch to the regions where some patterns in class 1 are still not covered by the box.

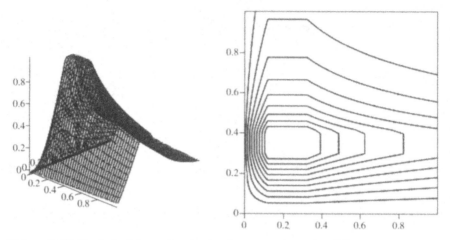

Figure 12.18. Optimized fuzzy box in the data space; note the difference between the steepness of the walls of the box along the two coordinates (features).

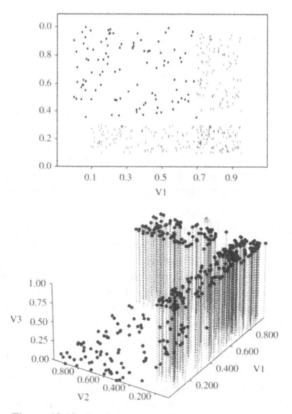

Figure 12.19. Synthetic data with two-box geometry.

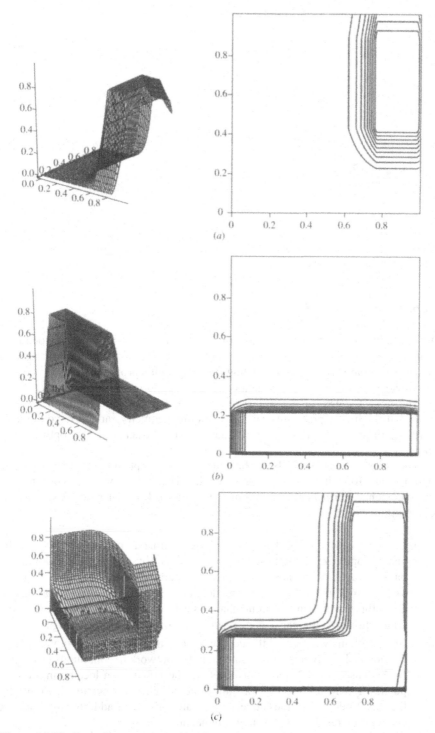

Figure 12.20. Fuzzy boxes generated by the neural network: two boxes shown separately (*a, b*) and their OR aggregation (*c*).

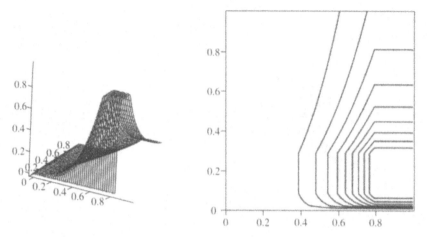

Figure 12.21. Capturing the geometry of the patterns by making use of a single genetically optimized fuzzy box.

12.6. CONCLUSIONS

We introduced and studied a class of logic networks composed of tolerance neurons. These networks are examples of the synergy of neurocomputing and fuzzy sets seamlessly accommodating learning, with profound transparency of the underlying fuzzy hyperbox data constructs supported by the network.

The main features of this approach are (a) the logic-based fabric of the network, (b) high interpretability of the resulting geometry, and (c) evolutionary learning of the network supporting the global character of the optimization processes.

The chapter focused mainly on the underlying concept and developed a genetic optimization for this class of logic networks. There are several expansions and enhancements worth pursuing within the framework of tolerance-based neurocomputing:

- Learning can be viewed as a two-phase optimization process in which the genetic optimization is followed by a gradient-based refinement (a type of backpropagation learning) in which the parameters of the network (W_1, $W_2, \ldots W_c$) are refined in more detail. This learning phase is attractive and efficient, bearing in mind that it starts with optimal collection of the connections, so the chances of further improvement are quite high.

- The problems with which the network deals involve two-class binary classification tasks. Obviously, one can apply the network to continuous problems. In this case, it is advantageous to equip the OR neuron located in the output layer with a collection of adjustable weights (connections). Similarly, the AND neurons in the hidden layer can come with additional calibration capabilities residing within their connections.

REFERENCES

B. Gabrys, A. Bargiela, General fuzzy Min-Max neural network for clustering and classification, *IEEE Trans. on Neural Networks*, 11, 2000, 769–783.

H. Ishibuchi, K. Nozaki, N. Yamamoto, H. Tanaka, Selecting fuzzy if-then rules for classification problems using genetic algorithms, *IEEE Trans. on Fuzzy Systems*, 3, 1995, 260–270.

W. Pedrycz, M.H. Smith, A. Bargiela, A granular signature of data, *Proc. 19th Int. Conf. NAFIPS'2000*, Atlanta, July 2000, 69–73.

W. Pedrycz, A.V. Vasilakos, Linguistic models and linguistic modeling, *IEEE Trans. on Systems, Man, and Cybernetics*, 29, 1999, 745–757.

P.K. Simpson, Fuzzy Min-Max neural networks—Part 1: classification, *IEEE Trans. on Neural Networks*, 3, 5 1992, 776–786.

P.K. Simpson, Fuzzy Min-Max neural networks—Part 2: clustering, *IEEE Trans. on Neural Networks*, 4, 1 1993, 32–45.

T.A. Sudkamp, R.J. Hammel II, Granularity and specificity in fuzzy function approximation, *Proc. NAFIPS-98*, 1998, 105–109.

13 Granular Prototyping

In this chapter, we introduce a logic-driven clustering in which prototypes are formed and evaluated in a sequential manner. A structure in data is revealed by maximizing a certain performance index (objective function) that takes into consideration an overall level of matching (to be maximized) and a similarity level between the prototypes (the component to be minimized). We show how the relevance of the prototypes translates into their granularity. The clustering method helps identify and quantify the anisotropy of the feature space. We also show how each prototype is equipped with its own weight vector describing the anisotropy property and thus implying some ranking of the features in the data space.

13.1. INTRODUCTION

Granular computing is an important method that focuses substantially on fuzzy clustering, especially its niche-addressing aspects of granular prototypes and granular constructs in general. There have been several studies in this area (Bargiela, 2001; Gabrys and Bargiela, 2000; Ishibuschi et al., 1995; Pedrycz and Bargiela, 2002; Simpson, 1992, 1993; Sudkamp, 1993; Sudkamp and Hammel, 1998; Zadeh, 1997), but it is still in its early development stage. This chapter proposes a comprehensive design for logic-driven clustering culminating in granular prototypes. There are several objectives. First, we wish to build prototypes in a sequential manner so that they can be ranked with respect to their relevance. Second, we want the clustering algorithm to exhibit significant explorative capabilities. This will be done by defining a suitable performance index (objective function). Third, the way in which the prototypes are formed should lead to their granular extension.

From the methodological standpoint, we proceed with a top-down presentation by first discussing the essence of the method and then elaborating on all pertinent details. The examples that follow consist of low-dimensional (mainly two-dimensional) patterns, as our intent is to illustrate the efficacies of the proposed clustering and granulation mechanisms. We contrast the algorithm with the FCM, which is treated as a de facto standard in fuzzy clustering.

Knowledge-Based Clustering, by Witold Pedrycz
ISBN 0-471-46966-1   Copyright © 2005 John Wiley & Sons, Inc.

The material is organized into four sections. First, in Section 13.2 we formulate the problem and elaborate on the underlying terminology and notation (which are consistent with those in fuzzy sets). The two concepts fundamental to the general clustering approach are matching (comparison) of fuzzy sets and construction of an objective function (performance index) guiding development of the structure of data. Section 13.3 is devoted to prototype optimization, where we show detailed derivations of formulas for the prototypes. These derivations and resulting formulas show the overall flow of computations: the essence of our approach consists of iterative construction of clusters guided by the performance index (so that they can be added if appropriate) without any up-front commitment to a certain number of clusters. This is in contrast to other methods such as FCM. The development of a granular version of the prototypes that builds on the numeric prototypes designed earlier is discussed in Section 13.4. It is shown that this design splits into two phases in which the performance index associated with each prototype is transformed into its granular (interval) envelope by solving an inverse matching problem. Conclusions are presented in Section 13.5.

13.2. PROBLEM FORMULATION

From now on, we are concerned with data (patterns) distributed in an n-dimensional $[0,1]$ hypercube. In what follows, we will be treating the data as points in $[0,1]^n$, say, $\mathbf{x} \in [0,1]^n$. In general, we are concerned with N patterns (data points) $\mathbf{x}_1, \mathbf{x}_2, \ldots, \mathbf{x}_N$. The standard objective of the clustering method (regardless of its realization) is to reveal a structure in the data set and to present it in a readable, easily comprehensible format. In general, we consider a collection of prototypes to be a tangible and compact reflection of the overall structure. In the approach used here, we adhere to the same principle. The prototypes representing each cluster are selected as elements of the data set. They are selected in such a way that they (a) match (represent) the data to the highest extent while (b) being distinct from each other. These two requirements are represented in the objective function guiding the clustering process. We then define the detailed components of the optimization. Since the elements in the unit hypercube can be viewed as fuzzy sets, we can take advantage of well-known logic operations developed in this domain. The notion of similarity (equality) between membership grades plays a pivotal role, and this concept is crucial to the development of the clustering mechanisms.

13.2.1. Expressing Similarity Between Two Fuzzy Sets

The measure of similarity between two fuzzy sets (in this case, a datum and a prototype) $\mathbf{x} = [x_1 \ x_2 \ldots x_n]^T$ and $\mathbf{v}(= [v_1 \ v_2 \ldots v_n]^T)$ is defined by incorporating the operation of matching ($\equiv$) encountered in fuzzy sets. The following definition will be used (Pedrycz, 1997; Pedrycz and Rocha, 1993):

$$\operatorname{sim}(\mathbf{x}; \mathbf{v}, \mathbf{w}) = \mathop{\mathrm{T}}_{i=1}^{n} (w_i^2 s(x_i \equiv v_i)) \tag{13.1}$$

In (13.1), $T(.)$ and $s(.)$ denote a t-norm and an s-norm, respectively. The weights (w_i) quantify the impact of each coordinate of the feature space $[0,1]^n$ on the final value of the similarity index sim(.). When convenient, we will use the notation sim($\mathbf{x}$, $\mathbf{v}$; $\mathbf{w}$) to emphasize the role played by the weight vector. A careful look at (13.1) reveals that it is nothing but a referential logic neuron with the similarity operation; the computational difference lies in the way in which the connections (w_i) enter the formula of the neuron. To achieve full compatibility, one should consider that $w_i' = w_i^2$.

It is worth stressing that the similarity between two membership grades is rooted in the concept of similarity (or equivalence) of two fuzzy sets (or sets). Given two membership grades a and b (the values of a and b are confined to the unit interval), a similarity level $a \equiv b$ is computed in the form (Pedrycz, 1990)

$$a \equiv b = (a \rightarrow b)t(b \rightarrow a) \tag{13.2}$$

where the implication operation ($\rightarrow$) is defined as a residuation (ϕ-operator) (Pedrycz, 1997):

$$a \rightarrow b = \sup\{c \in [0, 1] \mid atc \le b\} \tag{13.3}$$

The above expression of the residuation is induced by a certain t-norm. The implication models a property of inclusion; referring to (13.3), we note that it just quantifies the degree to which a is *included* in b. The *and* connective used in (13.2) translates it into a verbal expression

$$(a \text{ is included in } b) \text{ and } (b \text{ is included in } a) \tag{13.4}$$

that in essence quantifies the extent to which two membership grades are equal. In fact, this definition traces back to what we know well in set theory: we say that two sets A and B are equal if A is included in B and B is included in A. The reader can refer to other measures used to compare two fuzzy sets; see, for example, Bouchon-Meunier et al. (1996) and Hoppner et al. (1999).

The similarity index is substantially affected by the residuation operation (to be more precise, a specific t-norm is used to induce it). For example, the Lukasiewicz implication (induced by the Lukasiewicz t-norm) produces a series of piecewise linear characteristics.

The similarity index in the case of two variables ($n = 2$) is illustrated in Figure 13.1. The intent is to visualize the impact of the weights on the performance of the index. It is apparent that high values of the weight reduce the impact of the corresponding variable.

13.2.2. Performance Index (Objective Function)

The performance index reflects the character of the underlying clustering philosophy. In this work, we use a performance index that can be concisely described

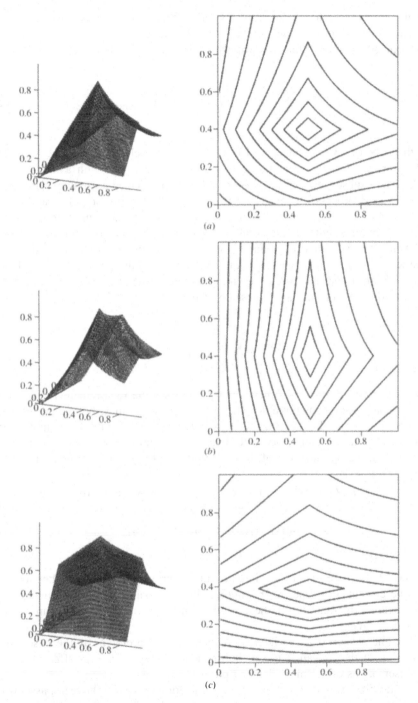

Figure 13.1. Similarity index (three-dimensional plot and two-dimensional contours) for selected values of weight factors: (a) $w_1 = 0.5$, $w_2 = 0.5$; (b) $w_1 = 0.2$, $w_2 = 0.8$; (c) $w_1 = 0.8$, $w_2 = 0.2$. In all cases, $\mathbf{v} = [0.5, 0.4]$.

in the following manner. A prototype of the first cluster $\mathbf{v}_1$ is selected as one of the elements of the data set ($\mathbf{v}_1 = \mathbf{x}_j$ for some $j = 1, 2, \ldots, N$) so that it maximizes the sum of the similarity measures of the form

$$\sum_{k=1}^{N} \text{sim}(\mathbf{x}_k, \mathbf{v}_1; \mathbf{w}_1) \Rightarrow \text{Max}_{v_1, w_1} \qquad (13.5)$$

with $\text{sim}(\mathbf{x}_k, \mathbf{v}_1, \mathbf{w}_1)$ defined by (13.1). Once the first cluster (prototype) has been determined (through a direct search across the data space with a fixed weight vector and subsequent optimization of the weights treated as another part of the optimization process), we move on to the next cluster (prototype) $\mathbf{v}_2$ and repeat the cycle. The form of the objective function remains the same throughout the iterative process, but we now combine the maximization of the sum of similarity measures (13.5) with a constraint on the relative positioning of the new prototype. We want this new prototype, say $\mathbf{v}_2$, not to "duplicate" the first prototype by being too close to it and thus not representing any new part of the data. To avoid this effect, we now consider an expression of the form

$$(1 - \text{sim}(\mathbf{v}_2, \mathbf{v}_1; \mathbf{0})) \sum_{k=1}^{N} \text{sim}(\mathbf{x}_k, \mathbf{v}_2; \mathbf{w}_2) \qquad (13.6)$$

where the first factor $1 - \text{sim}(\mathbf{v}_1, \mathbf{v}_2; \mathbf{0})$ expresses the requirement for $\mathbf{v}_2$ to be as far apart from $\mathbf{v}_1$ as possible. The above expression must be maximized with respect to $\mathbf{v}_2$, and this optimization has to be carried out with the weight vector ($\mathbf{w}_2$) involved. We then proceed to determine the third prototype $\mathbf{v}_3$, and so on. In general, the optimization of the Lth prototype follows the expression

$$Q(L) = (1 - \text{sim}(\mathbf{v}_L, \mathbf{v}_{L-1}; \mathbf{0})) (1 - \text{sim}(\mathbf{v}_L, \mathbf{v}_{L-1}; \mathbf{0})) \ldots$$
$$(1 - \text{sim}(\mathbf{v}_L, \mathbf{v}_1; \mathbf{0})) \sum_{k=1}^{N} \text{sim}(\mathbf{x}_k, \mathbf{v}_L; \mathbf{w}_L) \qquad (13.7)$$

As noted, this expression takes into account all previous prototypes when looking for the current prototype. Interestingly, the performance index to be maximized is a decreasing function of the prototype index, that is, $L_1 < L_2$ implies that $Q(L_1) \leq Q(L_2)$.

Another observation of interest is that the first prototype is the best representative of the overall data set. Subsequent prototypes are, in effect, the best representatives of the more detailed partitions of data.

So far, we have not considered the optimization of the weight vector associated with the prototype that is an integral part of the overall clustering. The next section provides a solution to this problem.

13.3. PROTOTYPE OPTIMIZATION

Let us concentrate on the optimization of the performance index in its general form given by (13.7). Apparently, the optimization consists of two phases: (a) determination of the prototype ($\mathbf{v}_L$) and (b) optimization of the weight vector ($\mathbf{w}_L$). These two phases are intertwined, yet they exhibit different characters. The prototype concerns enumeration out of a finite number of options (patterns in the data set). The weight optimization has not been formulated in detail and now requires a prudent formulation as a constraint type of optimization (with no constraint, the task may return a trivial solution). Referring to (13.7), we observe that it can be written in the form

$$Q(L) = G \sum_{k=1}^{N} \text{sim}(\mathbf{x}_k, \mathbf{v}_L; \mathbf{w}_L) \tag{13.8}$$

Note that the first part of the original expression does not depend on $\mathbf{w}_L$ and can be treated as a constant in this regard:

$$G = (1 - \text{sim}(\mathbf{v}_L, \mathbf{v}_{L-1}; \mathbf{0})(1 - \text{sim}(\mathbf{v}_L, \mathbf{v}_{L-1}; \mathbf{0}) \dots (1 - \text{sim}(\mathbf{v}_L, \mathbf{v}_1; \mathbf{0})) \tag{13.9}$$

We impose the following constraint on $\mathbf{w}_L$, requesting that its components be located in the unit interval and sum to 1:

$$\sum_{j=1}^{n} w_{Lj} = 1 \tag{13.10}$$

The optimization of (13.8) with respect to $\mathbf{w}_L$ for a fixed prototype $\mathbf{v}_L$ takes the following format:

$$\max Q(L) = G \sum_{k=1}^{N} \text{sim}(\mathbf{x}_k, \mathbf{v}_L; \mathbf{w}_L) \tag{13.11}$$

subject to

$$\sum_{j=1}^{n} w_{Lj} = 1$$

The detailed derivations of the weight vector are completed with the use of Lagrange multipliers. First, we produce an augmented form of the performance index:

$$V = G \sum_{k=1}^{N} \left\{ \prod_{j=1}^{n} (w_j^2 s(x_{kj} \equiv v_{Lj})) \right\} - \lambda \left(\sum_{j=1}^{n} w_{Lj} - 1 \right) \tag{13.12}$$

To shorten the expression, we introduce the notation $u_{ks} = x_{ks} \equiv v_{Ls}$. The derivative of V taken with respect to w_{Ls} (the sth coordinate of the weight vector) is set to 0 and the solution of the resulting equations gives rise to the optimal weight vector:

$$\frac{dV}{dw_{Ls}} = 0 \qquad \frac{dV}{d\lambda} = 0 \tag{13.13}$$

The derivatives can be computed once we specify t- and s-norms. For the sake of further derivations (and ensuing experiments), we consider a product and a probabilistic sum as the corresponding models of these operations. Furthermore, we introduce the abbreviated notation $u_{ks} = x_{ks} \equiv v_s$ for $k = 1, 2, \ldots, N$ and $s = 1, 2, \ldots, n$. Taking all of these into account, we have

$$\frac{dV}{dw_s} = G \sum_{k=1}^{N} \frac{d}{dw_s} \{A_{ks} w_s^2 s u_{ks}\} - \lambda = 0 \tag{13.14}$$

where

$$A_{ks} = \mathop{\mathrm{T}}_{\substack{j=1 \\ j \neq s}}^{n} (w_j^2 s u_{ks})$$

The use of the probabilistic sum (s-norm) in (13.14) leads to the expression

$$\frac{d}{dw_s} \{A_{ks} w_s^2 s u_{ks}\} = A_{ks} \frac{d}{dw_s} (w_s^2 + u_{sk} - w_s^2 u_{sk}) = 2A_{ks} w_s (1 - u_{ks}) \tag{13.15}$$

and, in the sequel

$$\frac{dV}{dw_s} = 2G w_s \sum_{k=1}^{N} A_{ks} (1 - u_{ks}) - \lambda = 0 \tag{13.16}$$

From (13.16) we have

$$w_s = \frac{\lambda}{2G \sum_{k=1}^{N} A_{ks} (1 - u_{ks})} \tag{13.17}$$

The form of the constraint, $\sum_{j=1}^{c} w_j = 1$, produces the following expression:

$$\frac{\lambda}{2} \sum_{j=1}^{c} \frac{1}{G \sum_{k=1}^{N} A_{kj} (1 - u_{kj})} = 1 \tag{13.18}$$

or

$$\frac{\lambda}{2} = \frac{1}{\displaystyle\sum_{j=1}^{c} \frac{1}{G \displaystyle\sum_{k=1}^{N} A_{kj}(1 - u_{kj})}}$$
(13.19)

Finally, when (13.19) is inserted in (13.17), the sth coordinate of the weight vector is

$$w_s = \frac{1}{\displaystyle\sum_{j=1}^{c} \frac{\displaystyle\sum_{k=1}^{N} A_{ks}(1 - u_{ks})}{\displaystyle\sum_{k=1}^{N} A_{kj}(1 - u_{kj})}}$$
(13.20)

Summarizing the algorithm, it consists essentially of two steps. We try all patterns as a potential prototype, for each choice optimize the weights, and find a maximal value of $Q(L)$ out of N available options. The one that maximizes this performance index is treated as a prototype. It comes with an optimal weight vector $\mathbf{w}_L$. Each prototype comes with its own weight vector, which may vary from prototype to prototype. Bearing in mind the interpretation of these vectors, we can say that they articulate the "local" characteristics of the feature space of the patterns. As seen in Figure 13.1, the lower the value of the weight for a certain feature (variable), the more essential the corresponding feature is. Note that the importance of the features is not the same across the entire space. The space becomes highly anisotropic where prototypes have different ranking of the features (see Figure 13.2).

We now discuss a number of low-dimensional synthetic data sets that will help us grasp the meaning of the resulting prototypes and interpret their weights.

Figure 13.2. Anisotropy of the feature space of patterns represented by weight vectors associated with prototypes.

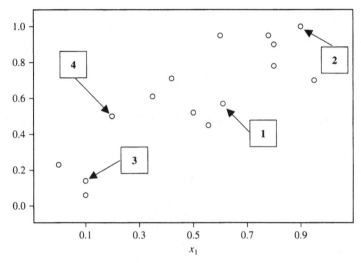

Figure 13.3. Synthetic data; successive detected prototypes are identified by arrows and corresponding numbers.

TABLE 13.1. Prototypes and Their Characterization

Cluster No.	Prototype	Performance Index	w
1	[0.61 0.57]	8.490796	[0.43 0.57]
2	[0.90 1.00]	4.755118	[0.37 0.63]
3	[0.10 0.14]	3.523316	[0.25 0.75]
4	[0.20 0.50]	2.951507	[0.20 0.80]
5	[0.00 0.00]	2.193254	[0.30 0.70]
6	[0.00 0.23]	2.024915	[0.11 0.89]
7	[0.10 0.06]	1.845740	[0.26 0.74]

Note: The starting point where the values of the performance index stabilize has been highlighted.

Example 1. The two-dimensional data set shown in Figure 13.3 exhibits several not very strongly delineated clusters.

The clustering is completed by forming additional clusters one at a time. The values of the performance index associated with the clusters, the position of the prototypes, and their respective weights are summarized in Table 13.1. As expected, the performance of successive clusters gets lower and the prototypes start to move close to each other. This feature of the clustering approach helps us investigate the relevance of the clusters on the fly and stop the search for more structure once the respective performance indexes start assuming low values. In this example, this happens for $c = 5$, at which point the values of the performance index stabilize.

Table 13.1 also includes the weight vectors associated with the prototypes. They reflect upon the "local" properties of the feature space. From their analysis (recall that a lower value of the weight means higher relevance of the feature in the neighborhood of the given prototype), we learn that the first feature (x_1) is more relevant than the second one. This is a quantification of the visual

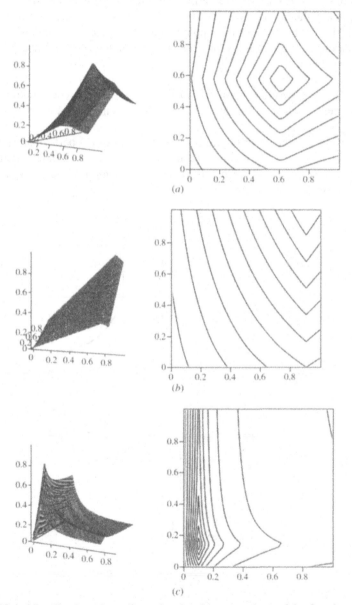

Figure 13.4. Visualization (three-dimensional and contour plots) of the first three clusters in the feature space: cluster no. 1 (a), no. 2 (b), and no. 3 (c).

inspection: as seen in Figure 13.4, when are projected the data on x_2, they tend to be more "crowded" (start overlapping) in comparison with their projection on x_1.

The prototypes produce nonlinear classification boundaries, as shown in Figure 13.5.

For comparative reasons, we carried out clustering using FCM; the resulting prototypes and the boundaries between the clusters are presented in Figure 13.6. It can be seen that the nonlinear boundaries between the clusters identified through maximization of the similarity measure provide much more refined partition of the pattern space.

Example 2. The two-dimensional data in Figure 13.7 show a structure that has three condensed clusters but also includes two points that are somewhat apart from the clusters.

The results are shown in Figure 13.8. The values of the performance index are visualized in Figure 13.9. It can be seen that the performance index "flattens out" for five clusters, which corresponds to the identification of significantly distinct data groupings.

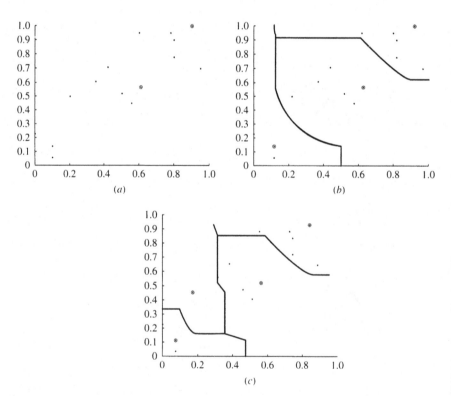

Figure 13.5. Classification regions for (*a*) two clusters, (*b*) three clusters, and (*c*) four clusters, identified through maximization of the similarity measure.

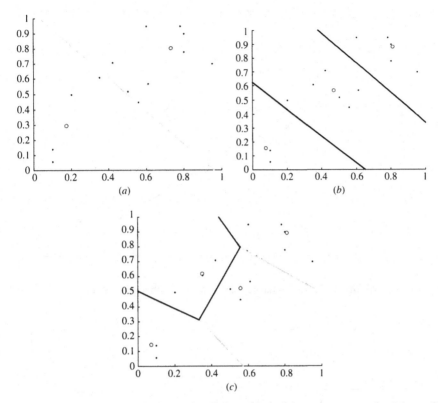

Figure 13.6. FCM clustering and the implied partition of the pattern space for (a) $c = 2$, (b) $c = 3$, and (c) $c = 4$ clusters.

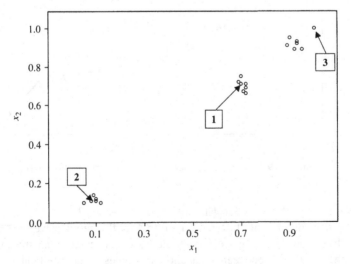

Figure 13.7. Two-dimensional synthetic data with the first three prototypes identified by the clustering algorithm.

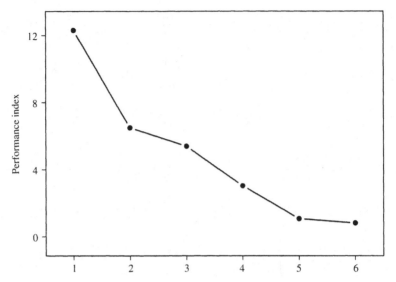

Figure 13.8. Performance index versus number of clusters (c).

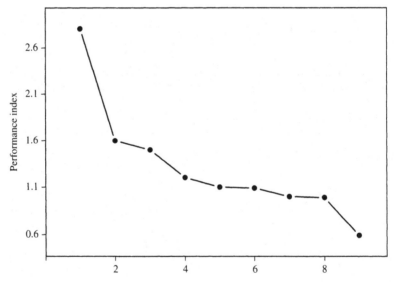

Figure 13.9. Performance index versus number of clusters (c).

Example 3. Four-dimensional data are given in Table 13.2.

The optimal number of clusters is equal to 4 (at this number we see "flattening out" of the values of the performance index, which means that the maximization of the similarity between data and prototypes is counterbalanced by the increase in similarity between the prototypes; Figure 13.11). The weight vectors

TABLE 13.2. Four-Dimensional Synthetic Patterns

Pattern No.	Coordinates	Prototype
1	0.80 0.10 0.60 0.30	
2	0.50 0.20 0.40 0.31	
3	0.60 0.30 0.10 0.35	
4	0.40 0.18 0.87 0.40	←2
5	0.90 0.15 0.50 0.32	←4
6	0.20 0.95 0.65 0.30	
7	0.20 0.40 0.30 0.31	←3
8	0.70 0.20 0.63 0.28	←1
9	1.00 0.00 1.00 0.31	
10	0.05 0.15 0.42 0.33	

TABLE 13.3. Weight Vectors of the First Four Prototypes

Prototype No.	Weight Vector
1	[0.12 0.12 0.16 **0.60**]
2	[0.16 0.18 0.14 **0.52**]
3	[0.04 0.04 0.06 **0.86**]
4	[0.06 0.09 0.15 **0.70**]

of the prototypes in Table 13.3 tell an interesting story: the feature space is quite isotropic, and in all cases the first feature (x_1) carries a higher level of relevance (the first coordinate of the weight vector of each prototype is constantly lower than the other). This is highly intuitive, as the patterns are more "distributed" along the first axis (x_1), which makes it more relevant (discriminatory) in this problem.

Example 4. This two-dimensional data set reveals two very unbalanced clusters. The first group is evidently dominant (100 patterns) over the second cluster (which consists of 5 data points) (Figure 13.10).

As we start building the prototypes, they start representing both clusters in more detail. The second prototype in the sequence has been assigned to the small cluster, meaning that the method is searching for still unrepresented parts of the data structure. We may say that the form of the performance index promotes a vigorous exploration of the data space and acts against "crowding" of the clusters in close proximity to each other. The consecutive clusters are constructed in such a way that they start unveiling some more specific substructures. Notably, the sixth prototype is assigned to the small cluster (Table 13.4).

It is instructive to compare these results with the structure revealed by the FCM. As anticipated (and this point was raised in the literature), FCM ignores the smaller cluster and focuses on the larger cluster. With an increase in the

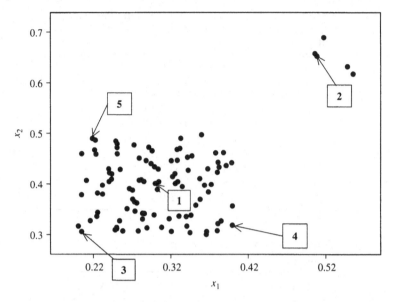

Figure 13.10. Two-dimensional data set with two unequal clusters; the consecutive prototypes produced by the method are identified by numbers.

number of clusters we start capturing the smaller clusters, but this happens later than with the previous method.

Example 5. The glass data set comes from the repository of Machine Learning (http://www.ics.uci.edu/~mlearn/MLRepository.html) and concerns classification of several categories of glass. The study was motivated by criminology investigations. Nine attributes (features) are used in the classification, including the refractive index and the content of iron, magnesium, aluminum, and other elements in the samples. Seven classes (categories) are identified in the problem.

In the experiment, we use the first 100 patterns. The performance index for the individual prototypes is shown in Figure 13.12. The plausible number of clusters is five since the performance index again "flattens out" for larger number of clusters.

The weight vectors of the individual prototypes (we confine ourselves to the five most dominant prototypes) show some level of anisotropy, with the features being ranked quite consistently in the context of the individual prototypes. The mean values and standard deviations of the weights of the first five prototypes are as follows:

Mean values

0.059757   0.025592   0.055397   0.063844   0.053884   0.070637   0.042023
0.610937   0.018282

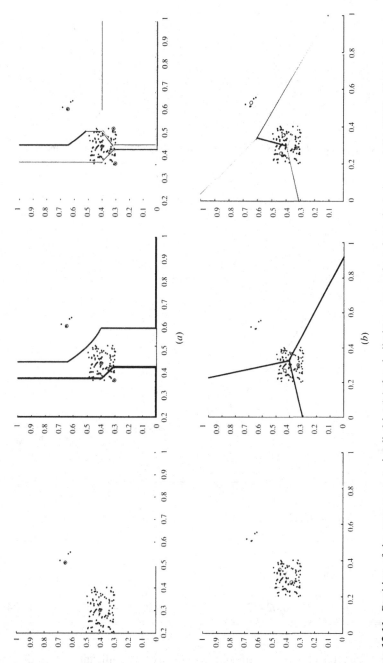

Figure 13.11. Partition of the pattern space implied by (*a*) the similarity measure based on clustering and (*b*) FCM clustering for two, three, and four clusters.

TABLE 13.4. Prototypes of the Clusters, Their Performance Index, and Their Weight Vectors

Prototype No.	Location	Performance Index	Weight Vector
1	(0.300100 0.400800)	83.694626	[0.44 0.56]
2	(0.508700 0.652100)	33.989677	[0.48 0.52]
3	(0.205200 0.305700)	26.135773	[0.39 0.61]
4	(0.399500 0.318400)	9.132071	[0.42 0.58]
5	(0.219200 0.489600)	5.200340	[0.42 0.58]
6	(0.555300 0.617200)	1.585717	[0.41 0.59]
7	(0.359900 0.496900)	0.598020	[0.51 0.48]

Note: The shadowed row highlights a sharp drop in the values of the performance index.

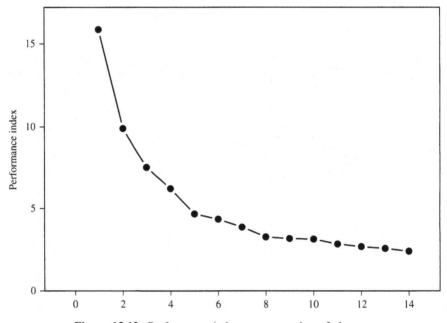

Figure 13.12. Performance index versus number of clusters.

Standard deviations

0.027003   0.013059   0.024531   0.030855   0.020684   0.039291   0.024701
0.162367   0.007705

Feature 8 can be clearly identified as relatively insignificant, while the most essential features are {2, 7, 9}. Their standard deviation is also quite low.

13.4. DEVELOPMENT OF GRANULAR PROTOTYPES

The inherently logical nature of the clustering technique helps us handle another interesting issue arising in the context of data summarization (and clustering per se is aimed at this important target). As is obvious from the previous sections, the prototypes, like the original data, are elements in the unit hypercube. One may question whether this is the only valid way of representing them. Naturally, one would expect the prototypes, as a form of summarization of the data, to reflect the fact that the patterns being represented by them occupy a certain region in the feature space. This lends itself to the notion of a granular prototype, that is, a prototype that spreads in the feature space, where its spread is related to the spatial characteristics of the original data. In a nutshell, we would like to develop prototypes that are represented as Cartesian products of intervals in the feature space. We believe that the granularity of the prototypes gives us greater insight into the nature of the data, as well as the relevance of the prototype itself. The formal framework for building granular prototypes is as follows. Consider $\mathbf{v}$ to be the prototype $\mathbf{v} \in [0, 1]^n$ already determined in the way discussed in Section 13.3. It has a weight vector $\mathbf{w}$. We can compute an average of similarity (q) between this prototype and all patterns by taking the following sum:

$$q = \frac{1}{N} \sum_{k=1}^{N} \text{sim}(\mathbf{x}_k, \mathbf{v}; \mathbf{w}) \tag{13.21}$$

(Note that (13.21) is analogous to (13.8), except that here we do not consider an interaction of $\mathbf{v}$ with other prototypes and we normalize the result.) This average similarity serves as a useful indicator of the relevance of the prototype. Now let us determine such values of u_i for which (13.21) holds. As u_i is effectively a similarity level between x_i and v_i, in essence it implies the interval built around v_i. To see this, note that $u_i = x_i \equiv v_i$, so if v_i and u_i are given, one can determine the range into which x_i should fall in order to satisfy this equality. This range is just an interval (along the ith coordinate) that contains the prototype.

To form the granular prototype, we repeat the process for all features, $i = 1$, $2, \ldots, n$, and we formulate and handle explicitly two optimization tasks arising here. The first one is to determine the values of $u_i, i = 1, \ldots, n$, so that they satisfy (13.21). The second task is an inverse problem emerging in the setting of the similarity index.

13.4.1. Optimization of the Similarity Levels

As part of the construction of the granular prototypes, we encounter the problem of determining the matching levels along individual features given the weight vector $\mathbf{w}$ and the overall matching level q. In other words, we are looking for $\mathbf{u} = [u_1 u_2 \ldots u_n]$ such that

$$\overset{n}{\underset{j=1}{\text{T}}} (w_j^2 s u_j) = \gamma \tag{13.22}$$

where **u** collects the matching levels between a given prototype **v** and some other pattern **x**. The above problem is not trivial, and no closed-form solution can be derived. Some iterative optimization should be used here. Bearing this in mind, we reformulate (13.22) as a standard mean squared error approximation problem

$$P = \left[\mathop{\mathrm{T}}_{j=1}^{n} (w_j^2 s u_j) - \gamma \right]^2 \rightarrow \mathrm{Min}(\mathbf{u}) \qquad (13.23)$$

whose solution is obtained by a series of modifications of **u** through the gradient-based scheme:

$$\mathbf{u}(\text{new}) = \mathbf{u} - \alpha \, \nabla_{\mathbf{u}} P \qquad (13.24)$$

where α denotes a positive learning rate. The detailed expression for the update can be derived for some predefined form of the triangular norm. Again, using the product and the probabilistic sum, we produce a detailed expression for the gradient:

$$u_k(\text{new}) = u_k - \alpha \frac{\partial P}{\partial u_k} \qquad (13.25)$$

$k = 1, 2, \ldots, n$. The detailed expression for the derivative is

$$\frac{\partial P}{\partial u_k} = 2 \left[\mathop{\mathrm{T}}_{j=1}^{n} (w_j^2 s u_j) - \gamma \right] \frac{\partial}{\partial u_k} \left(\mathop{\mathrm{T}}_{j=1}^{n} (w_j^2 s u_j) \right) \qquad (13.26)$$

The inner derivative can be handled for specific t- and s-norms. For a certain pair of them (t-norm: product, s-norm: probabilistic sum), we have

$$\frac{\partial}{\partial u_k} (\mathop{\mathrm{T}}_{j=1}^{n} (w_j^2 s u_j)) = \frac{\partial}{\partial u_k} (B_k(w_k^2 + u_k - w_k^2 u_k)) = B_k(1 - w_k^2)$$

where B_k is computed using the t-norm when excluding the index of interest (k):

$$B_k = \mathop{\mathrm{T}}_{\substack{j=1 \\ j \neq k}}^{n} (w_j^2 s u_j)$$

13.4.2. An Inverse Similarity Problem

The inverse problem with the similarity index can be formulated as follows: given b and γ (both in the unit interval), determine all possible values of 'x' such that $x \equiv b = \gamma$. The character of the solution can be easily envisioned by augmenting this equality by its graphical interpretation, Figure 13.13.

This figure shows that the problem being formulated above requires some refinement in order to enhance the interpretability of the solution and ensure that it always exists. This can be done by moving from the equality to the inequality format of the relationship (Pedrycz, 1990):

$$x \equiv b \leq \gamma \qquad (13.27)$$

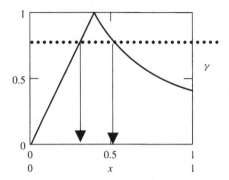

Figure 13.13. Inverse matching problem: computing an interval of solutions to $x \equiv b \leq \gamma$.

The solution to it arises in the form of a confidence interval (or simply interval) implied by a certain value of γ. This solution (interval) is a manifestation of the granularity of the prototype for a given feature. The solution to (13.27) can be obtained analytically for a specific type of the t-norm (or implication). As shown in Figure 13.15, the solution always exists (that is, there is always a nonempty interval for any given value of γ). The granularity of the prototype is a monotonic function of γ: higher values of γ imply higher values of granularity, that is, narrow intervals of the granular prototype. For some critical (low enough) value of γ, the interval expands to the entire unit interval, so we have a granular prototype of the lowest possible level of granularity.

Moving on to the detailed calculations, the interval of the granular prototype $[x_-, x_+]$ is equal to

for $x \rightarrow b = \min(1, b/x)$

$$x_- = \gamma b, \quad x_+ = \min(1, b/\gamma) \tag{13.28}$$

for $x \rightarrow b = \min(1, 1 - x + b)$

$$x_- = \max(0, \gamma - 1 + b), \quad x_+ = \min(1, 1 - \gamma + b) \tag{13.29}$$

the above expressions are determined by considering the increasing and decreasing portions of the matching index as illustrated in Figure 13.15.

Continuing the previous examples, the resulting granular prototypes are shown in Figures 13.14 and 13.15 for Examples 1 and 2, respectively. The granular prototypes summarized as triples of the form {lower_bound, mode, upper_bound} are included in Table 13.5. Note that, by the mode, we mean the original numeric value around which the granular prototype is constructed. The optimization of the degrees of matching (**u**) was completed by running the gradient-based learning with $\alpha = 0.05$ for 100 iterations. The initial values of **u**'s are set up as small (near-zero) random numbers.

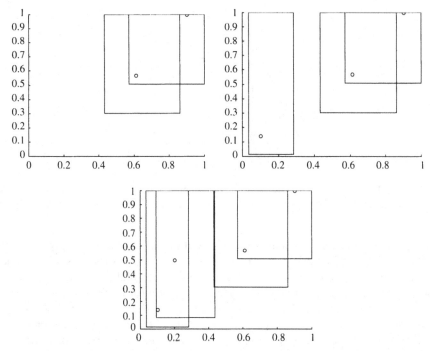

Figure 13.14. Two-, three-, and four-granular prototypes calculated for data from Example 1.

These granular prototypes reinforce and quantify our perception of structural dependencies in data. In the first case, Figure 13.14, we note that the first component of the structure resides in the right upper quadrant of the coordinates, and this is shown very clearly in the distribution of the granules. In fact, prototypes 1 and 2 overlap (meaning that there is some redundancy). The next granule (implied by the third cluster) is essential to the quantification of the structure; it occupies the area close to the origin. The fourth cluster overlaps the third one. Notably, all granules are elongated along the second variable, and this quantifies our observation about the limited relevance of this variable (note that all corresponding weights for the second variable are quite high). The conclusion is that the granules tend to "expand" and occupy the space wherever possible; this expansion is visible for x_2. The granular character of the prototypes shown in Figure 13.15 is again a meaningful manifestation of the structure. The two first granules are far apart (and represent the two evidently distinct groups of data). The boxes do not discriminate between the variables, viewing them as equally essential. The third granule overlaps with the first one, as these two clusters are relatively close. The fourth cluster has a strong resemblance (and overlap) to the second granule.

As this analysis reveals, we can envision the structure of the data by inspecting the resulting granular prototypes. First, these granules help us position clusters in the data space (it is worth stressing that the numeric representation does not

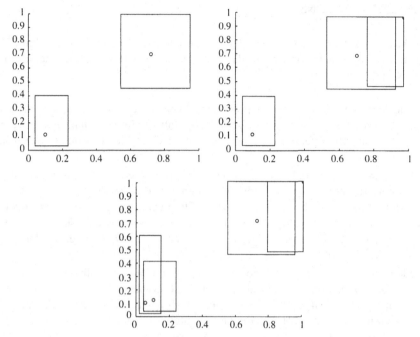

Figure 13.15. Two-, three-, and four-granular prototypes calculated for data from Example 2.

TABLE 13.5. Granular Prototypes Represented as Triples of Lower Bounds, Modes (Numeric Values of the Prototypes), and Upper Bounds: (a) Example 1 and (b) Example 2

(a)
Prototype 1: {0.431784 0.610000 0.861773} {0.302657 0.570000 1.000000}
Prototype 2: {0.570009 0.900000 1.000000} {0.507569 1.000000 1.000000}
Prototype 3: {0.035332 0.100000 0.283029} {0.015797 0.140000 1.000000}
Prototype 4: {0.091757 0.200000 0.435936} {0.081955 0.500000 1.000000}
(b)
Prototype 1: {0.545043 0.720000 0.951118} {0.457757 0.710000 1.000000}
Prototype 2: {0.042248 0.100000 0.236699} {0.035519 0.120000 0.405423}
Prototype 3: {0.784826 1.000000 1.000000} {0.477658 1.000000 1.000000}
Prototype 4: {0.017296 0.050000 0.144543} {0.016742 0.100000 0.597287}

support this form of analysis). Second, we can envision a general geometry of data that could be helpful in designing more detailed classifiers or other models. The granules may exhibit some level of overlap (no matter how such overlap is expressed in a formal fashion). This may help us reason about the possible relevance and redundancy of some of these clusters.

13.5. CONCLUSIONS

We have introduced a new logic-based approach to data analysis by building a certain clustering environment. Their main (and unique) features include the following:

The logic-based character of processing. The search for structure in data is accomplished by exploiting fuzzy set operations. In particular, this concerns the matching operation, which is easily interpretable and comes with well-defined semantics.

Successive (sequential) construction of the prototypes and an assessment of their representation capabilities. The number of clusters is not fixed in advance but can be adjusted dynamically, depending upon the performance of the already constructed prototypes. The prototypes themselves are constructed starting from the most "significant" (relevant) one, so that they come ranked.

Identification and quantification of possible anisotropy of the feature space. The weight vectors with the individual prototypes help quantify the importance of the features. The importance of the features can be local, and their ranking can vary from prototype to prototype.

Development of granular prototypes based on the results of clustering. We showed how the relevance of the prototype can be translated into its granular extension.

These features of the clustering method could be of interest to data analysis. We should stress, however, that the organization of the search for the structure as arranged here could be computationally intensive, especially for large data, sets so this method could be considered a complement to other clustering techniques.

REFERENCES

A. Bargiela, Interval and ellipsoidal uncertainty models, in: W. Pedrycz (ed.), *Granular Computing*, Physica-Verlag, Heidelberg, 2001, 23–57.

B. Bouchon-Meunier, M. Rifqi, S. Bothorel, Towards general measures of comparison of objects, *Fuzzy Sets and Systems*, 84, 2, 1996, 143–153.

B. Gabrys, A. Bargiela, General fuzzy Min-Max neural network for clustering and classification, *IEEE Trans. on Neural Networks*, 11, 3, 2000, 769–783.

F. Hoppner et al., *Fuzzy Cluster Analysis*, John Wiley, Chichester, England, 1999.

H. Ishibuchi, K. Nozaki, N. Yamamoto, H. Tanaka, Selecting fuzzy if-then rules for classification problems using genetic algorithms, *IEEE Trans. on Fuzzy Systems*, 3, 3, 1995, 260–270.

W. Pedrycz, Direct and inverse problems in comparison of fuzzy data, *Fuzzy Sets and Systems*, 34, 1990, 223–235.

W. Pedrycz, *Computational Intelligence: An Introduction*, CRC Press, Boca Raton, FL, 1997.

W. Pedrycz, A. Bargiela, Granular clustering: a granular signature of data, *IEEE Trans. on Systems, Man, and Cybernetics, Part B*, 32, 2, 2002, 212–224.

W. Pedrycz, A. Rocha, Knowledge-based neural networks, *IEEE Trans. on Fuzzy Systems*, 1, 1993, 254–266.

P.K. Simpson, Fuzzy Min-Max neural networks—Part 1: classification, *IEEE Trans. on Neural Networks*, 3, 5, 1992, 776–786.

P.K. Simpson, Fuzzy Min-Max neural networks—Part 2: clustering, *IEEE Trans. on Neural Networks*, 4, 1, 1993, 32–45.

T. Sudkamp, Similarity, interpolation, and fuzzy rule construction, *Fuzzy Sets and Systems*, 58, 1, 1993, 73–86.

T.A. Sudkamp, R.J. Hammell II, Granularity and specificity in fuzzy function approximation, *Proc. NAFIPS-98*, 1998, 105–109.

L.A. Zadeh, Toward a theory of fuzzy information granulation and its centrality in human reasoning and fuzzy logic, *Fuzzy Sets and Systems*, 90, 1997, 111–117.

14 Granular Mappings

In this chapter, we are concerned with the granular representation of mappings (or experimental data) in the form $R: \mathbf{R} \rightarrow [0, 1]$ (for one-dimensional cases) and $R: \mathbf{R}^n \rightarrow [0, 1]$ (for multivariable cases), with $\mathbf{R}$ being a set of real numbers. As the name implies, granular mapping is defined over information granules and maps them into a collection of granules expressed in some output space. The design of the granular mapping is discussed in terms of set- and fuzzy set–based granulation. It is regarded as a two-phase process: (a) the definition of an interaction between information granules and experimental evidence or existing numeric mapping and (b) the use of these measures of interaction in building an explicit expression for the granular mapping. We show how to develop information granules in the case of multidimensional numeric data using fuzzy clustering (FCM). Experimental results serve as an illustration of the proposed approach.

14.1. INTRODUCTION AND PROBLEM STATEMENT

The notion of granularity has been formalized in several languages of information granulation including sets (quite often alluded to in interval analysis), fuzzy sets, rough sets, and shadowed sets. While information granules are the building entities, the fundamental construct is mapping. When dealing with information granules, the mapping can be referred to as a granular mapping, defined as a transformation from an input space to an output space characterized at the granular level; that is, the mapping operates on information granules defined in the corresponding spaces. Granular mappings are frequently encountered in rule-based systems (e.g., fuzzy rule-based systems), where the mapping is given in the form of "if-then" statements. There are two fundamental types of problems in such settings: (a) analysis of granular mappings (say, rule-based systems), which is associated with various interpretation aspects, and (b) design of these mappings, which requires the development of the (experimentally meaningful and transparent) associations between the information granules.

More formally, the problem we are interested in is posed as follows:

Given (a) a function R or a collection of experimental data (D) defined in some input space $\mathbf{X}$ ($\mathbf{X} \subset \mathbf{R}$) and (b) a finite collection of information granules $A = \{A_1, A_2, \ldots, A_c\}$ defined in the same input space $\mathbf{X}$, represent $R(D)$ as a granular

Knowledge-Based Clustering, by Witold Pedrycz
ISBN 0-471-46966-1   Copyright © 2005 John Wiley & Sons, Inc.

mapping. That is, construct a transformation $\mathbf{A} \rightarrow \mathcal{G}(\mathbf{Y})$, with $\mathcal{G}(\mathbf{Y})$ denoting a family of information granules (sets, fuzzy sets, rough sets, etc.) defined in the output space $\mathbf{Y}$.

To ensure clarity of presentation, the material is organized in a bottom-up format. First, we elaborate on the concept and realization of interaction of granular probes with the experimental environment and demonstrate that possibility and necessity measures can be viewed as vehicles that help quantify such a concept. We then move to the simplest one-dimensional scenario in which such granular mappings arise and consider a family of sets (intervals) treated as granular probes in this granular environment. Then in Section 14.4 we discuss a multidimensional case in which fuzzy clustering plays an important role.

14.2. POSSIBILITY AND NECESSITY MEASURES AS THE COMPUTATIONAL VEHICLES OF GRANULAR REPRESENTATION

In light of the overall scenario outlined in Section 14.1, we can see that a sound starting point pertains to a way in which information granules interact with the mapping (R) or experimental data over which such a granular mapping has to be realized. Intuitively, one of the first realizations involves the attempt to quantify the interaction between A_i's and the mapping. As the granules adhere to the fundamental concepts of granular mechanisms and computing such as inclusion, intersection, complement, etc. (no matter how these concepts become implemented, which, in turn, depends upon the nature of the granular environment), it is appealing to revisit two of them. First, the notion of intersection concerns the interaction between the probe (A_i) and the given mapping. Second, a way in which inclusion of the probes in the mapping has been realized leads to another point of view at this interaction. Interestingly, these two ideas already exist in the literature in the form of possibility and necessity measures. More specifically, given two information granules A and R (here A stands for one of the granules from the family of the granular probes **A**, while R is a mapping whose granular format we are about to develop), the possibility measure, $\mathrm{Poss}(A, X)$, describes a level of overlap between these two. The necessity measure, $\mathrm{Nec}(A, X)$, captures a level of inclusion of A in X. While these descriptors are quite generic, their realization needs to be described in more detail, depending upon the character of the granular environment. In the case of fuzzy sets (and sets), we have (Dubois and Prade, 1980)

$$\mathrm{Poss}(A, R) = \sup_{x \in X} [A\ (x)\, t\, R(x)] \qquad (14.1)$$

The plot visualizing the computations of the possibility and necessity measures is shown in Figure 14.1. Computationally, we note that the possibility measure looks at the intersection between A and R and then takes an optimistic aggregation of the intersection by picking up the highest values among the intersection grades of A and R that are taken over all elements of the universe of discourse

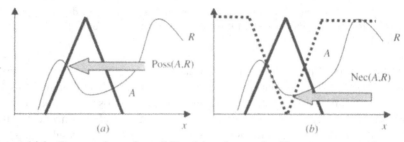

Figure 14.1. Computations of possibility (a) and necessity (b) measures treated as a vehicle of interaction between a granular probe A and the numeric mapping R; t-norm: minimum, s-norm: maximum. The dotted line in (b) shows a complement of A, $1 - A(x)$.

X. The necessity measure expresses a pessimistic degree of inclusion of A in R and is computed as follows:

$$\mathrm{Nec}(A, R) = \inf_{x \in X}[(1 - A(x))sR(x)] \qquad (14.2)$$

The computational details are presented in Figure 14.1. In contrast to the possibility measure, the necessity measure is asymmetric (which is obvious, as we are concerned with the inclusion predicate).

These two definitions are applied in the case of R given in an analytic (explicit) fashion. If we are given the experimental data (namely, input-output pairs $D = \{(x_k, y_k)\}, k = 1, 2, \ldots, N$), then the above calculations are modified, with the supremum and infimum operations being replaced by the maximum and minimum operations taken over all data D:

$$\mathrm{Poss}(A, D) = \max_{(x_k, y_k)}[A(x_k)ty_k] \qquad (14.3)$$

$$\mathrm{Nec}(A, D) = \min_{(x_k, y_k)}[(1 - A(x_k))sy_k] \qquad (14.4)$$

The possibility and necessity measures articulate a way in which A interacts with R or experimental data.

Considering the family of the information granules A, we compute the possibility and necessity measures with respect to R or D and end up with the $2c$-tuple representation

$$\lambda_i = \mathrm{Poss}(A_i, R), \mu_i = \mathrm{Nec}(A_i, R) \qquad (14.5)$$

which is a manifestation of R (or D) expressed in the granular language of A. Notably, by changing elements of A, we end up with a different representation of the same mapping. Let us reiterate that different types of A's provide us with different points of view (perspectives) on the same mapping R or experimental evidence D.

14.3. BUILDING THE GRANULAR MAPPING

So far, we have arrived at the representation (manifestation) of R expressed in the languages of A_i's. This is a prerequisite to the construction of a granular mapping, as the elements of A are the basis for the granules in the output space. They are related to A_i's and result from the reconstruction process guided by (λ_i, μ_i). From the computational standpoint, we can view this as a solution to a certain inverse problem. Let us start with a single information granule (A) for which the values of λ and μ are known. There is no unique solution to this problem. There is, however, a maximal information granule denoted here by $\hat{R}$ (14.6), whose construction is supported by the theory of fuzzy relational equations (in fact, (14.5) is a sup-t composition of R and A). The membership (characteristic) function of this maximal fuzzy set (mapping) induced by A is (Bargiela and Pedrycz, 2002; Bortolan and Pedrycz, 1997; Di Nola et al., 1989)

$$\hat{R}(x) = A(x) \rightarrow \lambda = \begin{cases} 1 & \text{if } A(x) \leq \lambda \\ \lambda & \text{otherwise} \end{cases} \qquad (14.6)$$

The above formula applies for a t-norm realized as a minimum operator. In general, (14.6) is in the form

$$\hat{R}(x) = A(x) \rightarrow \lambda = \sup[a \in [0, 1] | at A(x) \leq \lambda] \qquad (14.7)$$

When using the entire family of A_i's (which leads to the intersection of $\hat{R}_i$'s) we obtain

$$\hat{R} = \bigcap_{i=1}^{c} \hat{R}_i \qquad (14.8)$$

From the theoretical point of view that arises in the setting of fuzzy relational equations, we are dealing here with a system of equations $\lambda_i = \text{Poss}(A_i, R), i = 1, 2, \ldots, c$ to be solved with respect to R for a given λ_i and A_i.

The theory of fuzzy relational equations plays the same dominant role in the case of necessity computations. It is worth noting that we are faced with so-called dual fuzzy relational equations. Here the minimal solution to (14.5) for A and μ_i is

$$\tilde{R}(x) = (1 - A(x))\varepsilon\mu = \begin{cases} \mu & \text{if } 1 - A(x) < \mu \\ 0 & \text{otherwise} \end{cases} \qquad (14.9)$$

Again, the above formula applies to the maximum realization of the s-norm. The general formula takes the form

$$\tilde{R}(x) = (1 - A(x))\varepsilon\mu = \inf[a \in [0, 1] | as(1 - A(x)) \geq \mu\} \qquad (14.10)$$

Because of the minimal solution, the collection of granular probes leads us to the partial results that are afterward combined through a union operation:

$$\tilde{R} = \bigcup_{i=1}^{c} \tilde{R}_i \qquad (14.11)$$

In conclusion, (14.7) and (14.10) become the granular representations of the mapping (R) arising in the collection of the information granules A given in advance. The obvious containment relationship holds:

$$\tilde{R} \subseteq R \subseteq \hat{R} \qquad (14.12)$$

where the granularity of the mapping manifests through the two different bounds (lower and upper approximations of R).

As a simple yet highly illustrative example, consider a collection of sets (intervals) regarded as granular probes of some nonlinear numeric mapping (see Figure 14.2). A single information granule produces the result shown in Figure 14.3a, which in fact produces membership values equal to 1 over any

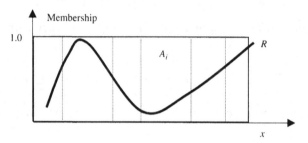

Figure 14.2. Mapping (R) with a collection of superimposed sets (intervals) A_i.

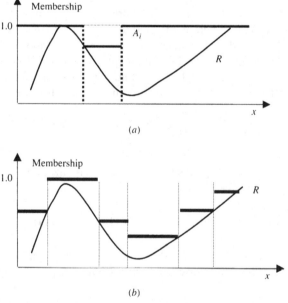

(a)

(b)

Figure 14.3. Computing the upper bound of R (solid staircase line) with the use of a single set (A) (a) and a family $\{A_i\}$ (b).

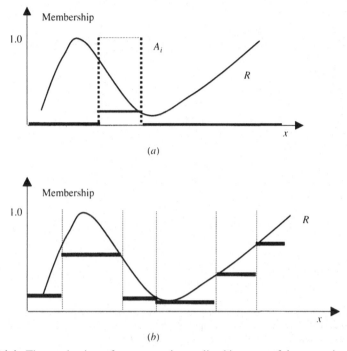

(a)

(b)

Figure 14.4. The mechanism of reconstruction realized in terms of the necessity measure.

argument not belonging to A. The aggregation of all partial results gives more specific results (see again Figure 14.3).

Interestingly, the reconstructed fuzzy set exhibits a stairwise type of membership function where the height of the individual jumps and their distribution across the space depend on the distribution of A_i's. The same effect that concerns the lower bound of R is present in Figure 14.4. When combined, the result is a granular mapping, Figure 14.5. It is worth noting that by changing the position of the cutoff points (intervals), we end up with different granular mappings. Eventually the mapping can be subject to optimization in which we develop the collection of A_i's in such a way that the granular mapping is as specific as possible (so that the bounds are made tight).

14.4. DESIGNING MULTIVARIABLE GRANULAR MAPPINGS THROUGH FUZZY CLUSTERING

The one-dimensional case can be generalized with the same design objective as before. The primary step is to find a collection of information granules in $\mathbf{R}^n$ so that they are meaningful constructs (in light of the available data). The dimensionality of the input space suggests treating all inputs at the same time rather than discussing each variable separately (which is impractical and leads to a significant number pf combinations of such granules). In other words, we

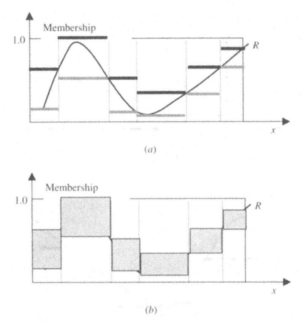

Figure 14.5. Building a granular mapping: upper and lower bounds of the mapping (*a*) and interval-valued granules in the output space (*b*).

focus on forming granular relations (e.g., relations, fuzzy relations), and this immediately raises the idea of fuzzy clustering. We recall that a fuzzy partition matrix is in essence a collection of discrete fuzzy relations. Once the clusters have been determined, the next part of the construction is the same as in the one-dimensional case. To draw the linkages, note that the partition matrix can be represented as c fuzzy relations $A_1, A_2, \ldots, A_c$ defined in the finite data set:

$$U = \begin{bmatrix} A_1 \\ A_2 \\ \vdots \\ A_c \end{bmatrix}$$

By doing this, we explicitly form the collection of the information granules A, as was done in the one-dimensional case. More specifically, the possibility and necessity computed for the granules are governed by the expressions $\mathrm{Poss}(A_i, D) = \max_{x_k}[A_i(\mathbf{x}_k)ty_k]$ and $\mathrm{Nec}(A_i, D) = \min_{x_k}[(1 - A_i(\mathbf{x}_k))sy_k]$. Alluding to the way in which information granules interact with the mapping or data, we note that most of the interaction occurs in the region where the granules are normal (viz., they assume a membership grade equal to 1). As the FCM imposes the unity constraint (viz., the membership grades in all clusters assumed for the same point sum to 1), we normalize these grades by raising the highest membership grade to 1.

Two essential parameters of the granular mapping result from the use of fuzzy clustering in the development of the underlying construct: the number of clusters (c) and the shape of the probes and possible patterns of behavior (interaction) between them.

We have already seen that the number of information granules dictates the granularity of the mapping. This has been clearly demonstrated in the case of sets (intervals) in the one-dimensional problem, in which a certain portion of the data (or function) falls under the realm of the granule (become identified by it) and then implies the granularity of the mapping itself. We have learned that in limit (which may not be of practical interest) the granularity of the mapping is very high. This effect could be quantified by computing the ratio of the average value of the σ-count (cardinality) of the information granules in the output space to the granularity of the probes (fuzzy relations in the input space).

As the clusters in the design of the granular mapping play a primary role (as our intuition might have already suggested), it is instructive to understand more clearly the possibility of control over the shape of the clusters. An illustration in a one-dimensional case is the best option. Figure 14.6 shows the membership

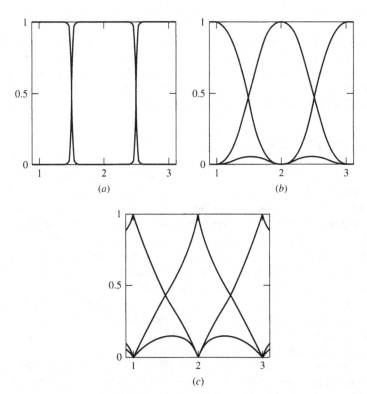

Figure 14.6. Plots of membership functions of information granules generated by FCM clustering for selected values of m: (a) $m = 1.1$, (b) $m = 2.0$, (c) $m = 3$.

functions for selected values of the fuzzification coefficient (the values of the prototypes are kept fixed and equal to 1, 2, and 3, respectively). Depending upon the value of this coefficient, the resulting fuzzy sets tend to resemble sets (when m approaches 1) or exhibit a significant overlap and interact among themselves (which is quite obvious for the values of m around 2) or get "spiky" and lose most of the interaction as the value of the fuzzification coefficient increases (a good example is $m = 3$).

14.5. QUANTIFICATION OF GRANULAR MAPPINGS

As stated before, each segment of the granular mapping is formed within the individual information granule occurring in the input space. The granularity of the mapping (quantified at the end of the output space) is directly related to the upper and lower bounds of the interval formed by the possibility and necessity values. To quantify this granularity and relate it to the experimental data (which is the entry point of the overall design), it is beneficial to introduce the following index:

$$\bar{g} = \frac{1}{N} \sum_{k=1}^{N} (u_k - l_k) \tag{14.13}$$

which tells us the average spread of the interval, with l_k and u_k denoting the lower and upper bound, respectively. The average over 'N' data points helps us quantify the granulation effect and abstract it from the individual variations. Higher values of 'g' imply lower granularity of the resulting mapping (that is, larger differences between the bounds of the mapping).

14.6. EXPERIMENTAL STUDIES

The experiments presented here are designed to visualize the performance of the granular mappings in numeric data. The data set comes from the Machine Learning repository and deals with the fuel economy (in miles per gallon) of various vehicles described by a series of seven parameters including displacement, weight, number of cylinders, and so on. The output has been normalized to the unit interval. Furthermore, the data set was split randomly into 50–50% training and testing sets. The training set was used to carry out clustering, and then we computed the possibility-necessity characteristics of the granular mapping. The FCM was run for 20 iterations (at which point there were no practically visible changes in the values of the objective function). The distance function was assumed to be the normalized Euclidean distance (the weights were computed as the standard deviations of the individual inputs).

There are two essential parameters of the mapping associated with the clustering mechanism: the number of clusters (information granules to achieve the mapping) and the fuzzification coefficient (m). We explore various combinations of these parameters to obtain a sense of their role and some general tendencies.

The main results are displayed in Figure 14.7. The figure helps us draw several conclusions. First, as we could have intuitively expected, the increased number of clusters yields better performance of the mapping (the bounds start to become tighter). This tendency becomes quite apparent; the changes in performance become more evident for the smaller number of clusters (hence, there is a significant difference when moving from two to six clusters and far less pronounced differences for higher values of 'c'). The number of clusters also affects the optimal values of the fuzzification coefficient. In general, optimal values of 'm' tend to become lower as the number of clusters increases. Relating this to the shape of the membership functions (see Figure 14.6), we could envision that with more clusters their boundaries need to be shapes, as more set-oriented, and with less spread and overlap with the neighboring information granules. More specifically, the optimal fuzzification factors are given as: $c = 2$, $m = 2.6$; $c = 4$, $m = 1.6$; $c = 6$, $m = 1.6$; $c = 10$, $m = 1.2$.

The testing set was used to assess the generalization abilities of the granular mapping. The results shown in Table 14.1 indicate that the differences in performance on the training and testing sets are not substantial, which leads us to conclude that the mapping exhibits sound generalization abilities. In essence, with $c = 10$, the performance on the training and testing sets is practically the same.

Figures 14.8 and 14.9 visualize the behavior of the granular mapping for selected numbers of clusters (with the optimal values of the fuzzification coefficients). In

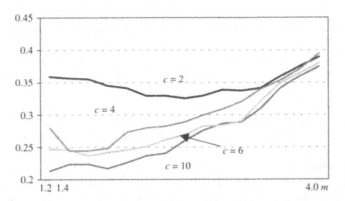

Figure 14.7. Performance (granulation) index of the mapping as a function of the fuzzification coefficient of the FCM for selected numbers of clusters (c).

TABLE 14.1. Performance of the Granular Mapping on a Training and Testing Set

c	2	4	6	10
Training set	0.326	0.245	0.237	0.213
Testing set	0.354	0.256	0.256	0.212

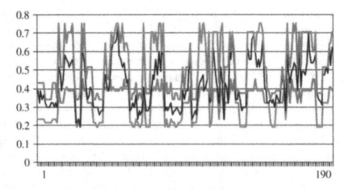

Figure 14.8. Training data and bounds of the granular mapping ($c = 2$).

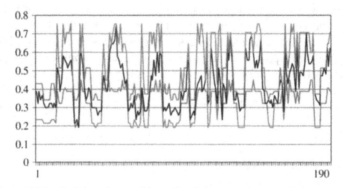

Figure 14.9. Training data and bounds of the granular mapping ($c = 10$).

these two cases, where $c = 2$ and $c = 10$, the essential components of the granular mapping—possibility and necessity measures—are included in Table 14.2.

We can present the distribution of bounds by plotting individual differences between the bounds (lower and upper) and the numeric experimental data (see Figure 14.10).

Notably, the bounds are asymmetric; the two plots above help us in two ways. First, they are useful in identifying data points, with the broadest intervals that could be revisited as potential outliers. Second, they help us learn about the distribution of granularity of the realized mapping and quantify which bound (lower or upper) is "tighter" with respect to the experimental data (Table 14.2).

14.7. CONCLUSIONS

We have introduced and studied the concept, analyzed the properties, and discussed the design of granular mappings viewed as one of the fundamental constructs of granular computing. The developed granular mapping help establish

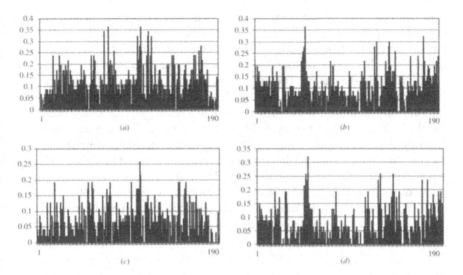

Figure 14.10. Differences between the bounds and numeric experimental data for $c = 4$, $m = 1.4$: (a) upper bound and data, (b) lower bound and data; $c = 18$, $m = 1.2$: (c) upper bound and data, (d) lower bound and data.

TABLE 14.2. Possibility and Necessity Measures of Granular Mapping: (a) $c = 2$ and (b) $c = 10^*$

Cluster No.	Possibility	Necessity
	(a)	
1	0.483	0.193
2	0.751	0.321
	(b)	
1	0.601	0.451
2	0.644	0.408
3	0.708	0.386
4	0.386	0.322
5	0.429	0.236
6	0.451	0.386
7	0.376	0.193
8	0.751	0.386
9	0.515	0.322
10	0.343	0.215

*In both cases, the results are reported for the optimal values of the fuzzification factor.

more general view of the data or detailed numeric functions, which for this purpose are perceived as a collection of information granules. In a nutshell, the introduced construct is formed by probing the function or data by a series of granular "probes" (sets or fuzzy sets), recording the results of this interaction, and aggregating the results. The underlying logical framework is based upon the calculus of fuzzy relational equations. The probing results consist of possibility and necessity measures. These then give rise to a set of fuzzy relational equations, and the granular mapping is formed by solving a system of such equations. The resulting lower and upper bounds fully describe the mapping. We elaborated on the role and optimization of the granular probes and showed how their selection is achieved through fuzzy clustering (and FCM in particular). The choice of the shape of the fuzzy relations (which is controlled by the values of the fuzzification coefficient) and the number of clusters (granules) have been studied in detail, with a quantitative demonstration of their impact on the specificity of the granular mapping (Bargiela and Pedrycz, 2003; Zadeh and Kacprzyk, 1999).

REFERENCES

A. Bargiela, W. Pedrycz, *Granular Computing: An Introduction*, Kluwer Academic Publishers, Boston 2002.

A. Bargiela, W. Pedrycz, Recursive information granulation: aggregation and interpretation issues, *IEEE Trans. on Systems, Man, and Cybernetics SMC-B*, 33, 1, 2003, 96–112.

G. Bortolan, W. Pedrycz, Reconstruction problem and information granularity, *IEEE Trans. on Fuzzy Systems*, 2, 1997, 234–248.

A. Di Nola, W. Pedrycz, S. Sessa, E. Sanchez, *Fuzzy Relational Equations and Their Applications in Knowledge Engineering*, Kluwer Academic Publishers, Dordrecht, 1989.

D. Dubois, H. Prade, *Fuzzy Sets and Systems: Theory and Applications*, Academic Press, New York, 1980.

L.A. Zadeh, J. Kacprzyk (eds.), *Computing with Words in Information/Intelligent Systems*, vol. 1–2, Physica-Verlag, Heidelberg, 1999.

15 Linguistic Modeling

In this chapter, we are concerned with the important paradigm of linguistic modeling. As the name indicates, linguistic (granular) modeling is concerned with models of complex systems consisting of information granules, particularly fuzzy sets. The purpose of the models is to reveal associations (links) between fuzzy sets defined in input and output spaces. Rule-based models occupy a central position in this broad spectrum of linguistic models. Our objective is to demonstrate the central role of fuzzy clustering—after all, fuzzy clusters contribute to a "blueprint" of linguistic modeling.

15.1. INTRODUCTION

In granular computing, one of the most important applications is system modeling. The quest for models that are accurate, transparent, and user-friendly has characterized intelligent systems for decades. Granularity of information arose as a fundamental concept that plays a pivotal role in any modeling endeavor. To achieve transparency, models must be designed with components that are meaningful to the user. More emphatically, well-defined semantics of the information granules used in system modeling is essential when designing a user-centric model. As the domain knowledge conveyed by information granules and/or the character of relationships captured in the language of logic dependencies is an indispensable component of such models, we refer to such human-centric models as knowledge-based models and to the entire domain as knowledge-based modeling.

Interestingly, in this capacity, rule-based systems have assumed a central role in system modeling. Not surprisingly, these systems focus on information granules. Regardless of the specific format of the rules, they all involve granules in their conditions and conclusion:

$$\text{if condition}_1 \text{ is } A \text{ and condition}_2 \text{ is } B \dots \text{ then conclusion is } W$$

with $A, B, \dots W$ denote information granules defined in their respective spaces. Depending on their specific realization, we refer to such systems as fuzzy, rough, or rule-based systems. Interestingly, the change in the granularity level of

Knowledge-Based Clustering, by Witold Pedrycz
ISBN 0-471-46966-1   Copyright © 2005 John Wiley & Sons, Inc.

the information granules in the rules implies a certain level of accuracy and transparency or user friendliness. Intuitively, we note that higher granularity (specificity) of the condition end of the rules implies more specific rules. If the granularity at the condition end decreases (so that we start using more general information granules), this implies rules of higher generality (that become applicable or "fire" in more cases). Having more general rules, we require fewer of them in describing the system or experimental data. However, by the same token, accuracy may be reduced by the readability or transparency of the resulting construct. Again, these two criteria, their quantification, and possible trade-offs have been a subject of many investigations. See Kim et al. (1997), Pork et al. (2002) Sugeno and Yasukawa (1993), Takagi and Sugeno (1985), and Farag et al. (1998), to name only a few of the studies in this realm.

In this chapter, we discuss the methodological and algorithmic issues of knowledge-based clustering in the design of granular models and elaborate on a variety of architectures of models emerging therein. It is convenient to view the development of granular model as a multiphase development process (Figure 15.1) in which we position clustering as a basic vehicle supporting the design of granular models. Notably, by choosing a different clustering platform or even changing the level of specificity at which we intend to construct information granules, we may end up with different models. The refinement phase focuses on the use of more specialized and numerically inclined optimization vehicles such as regression models (Box and Jenkins, 1970), neural networks (Hecht Nielsen, 1990), splines, local regression models, polynomial networks (Park et al., 2002), wavelets, and the like.

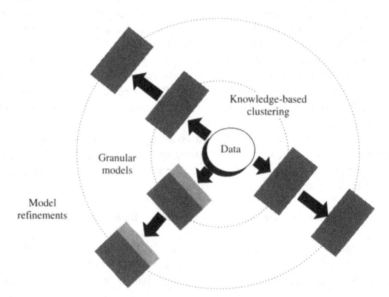

Figure 15.1. From data to granular models: the main design phases. Note the role of clustering in the development of a design blueprint of granular models.

We start with the commonly encountered fuzzy modeling scenario, in which we construct a nonlinear mapping based on fuzzy clusters (or, more specifically, their prototypes).

15.2. CLUSTER-BASED REPRESENTATION OF INPUT-OUTPUT MAPPING

Fuzzy clusters (Gonzalez et al., 2002; Pedrycz and Vasilakos, 1999) establish a solid basis for constructing fuzzy models. This framework is formed by distributing clusters (more specifically, their prototypes) throughout the feature space. There is also a corresponding set of prototypes defined in the output space (whatever this space may be; we may have continuous membership grades of classes or a continuous output variable). These prototypes are regarded as a structural "skeleton" or a blueprint of the model. There are several ways of presenting the detailed formula of the resulting model. The one commonly used in the literature takes the weighted sum of the prototypes in the output space and combines them linearly by using the membership grades of some given input x, that is, $u_1(\mathbf{x}), u_2(\mathbf{x}), \ldots, u_c(\mathbf{x})$. Let $z_1, z_2, \ldots, z_c$ denote the prototypes in the output space; the input space contains a series of corresponding prototypes $\mathbf{v}_1, \mathbf{v}_2, \ldots \mathbf{v}_c$. The output of the model (y) is determined as a weighted sum

$$y = \sum_{i=1}^{c} z_i u_i(\mathbf{v}) \tag{15.1}$$

with $u_i(\mathbf{x})$ being the "activation" level of the ith cluster in the input space resulting from the use of $\mathbf{x}$ computed in the standard form

$$u_i(\mathbf{x}) = \frac{1}{\sum\limits_{j=1}^{c} \left(\dfrac{\|\mathbf{x} - \mathbf{v}_i\|}{\|\mathbf{x} - \mathbf{v}_j\|} \right)^{\frac{2}{m-1}}} \tag{15.2}$$

The reader familiar with radial basis function (RBF) neural networks (Joo et al., 2002; Karyannis and Mi, 1997; Ridella et al., 1998; Rovetta and Zunino, 2000) will easily recognize that the above expression plays a role similar to that of any RBF in this category of neural networks. The striking difference is that here these receptive fields are automatically constructed, without any need for their further adjustment. It is educational to visualize the characteristics of the model, which realizes some nonlinear mapping from the input to the output space. The character of nonlinearity depends upon the values of the prototypes. This nonlinearity is easily affected by the values of the fuzzification factor (m). Figure 15.2 provides several plots of the input-output characteristics of the fixed values of the prototypes. The values of the fuzzification factor are varied. We have included its typical value of 2. Undoubtedly, this design parameter has a significant impact on the character of the nonlinearity produced, demonstrating

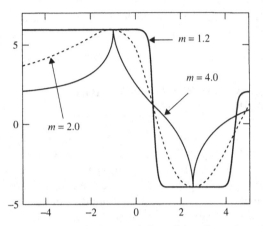

Figure 15.2. Nonlinear input-output characteristics of the cluster-based model. The prototypes are fixed ($v_1 = -1$, $v_2 = 2.5$, $v_3 = 6.1$; $z_1 = 6$, $z_2 = -4$, $z_3 = 2$), while the fuzzification factor assumes several selected values.

the flexibility of this category of models. The values of m close to 1 produce the stepwise nature of the mapping; we note significant jumps at the points where the individual rules switch. In this fashion, the impact of each rule is clearly delineated. The typical value of the fuzzification factor, 2, produces a very smooth transition between the rules, manifesting in smooth nonlinearities of the input-output relationships of the model. The increase in the value of m, as shown in Figure 15.2, yields characteristics that are quite spiky. We reach some modal values when moving close to the prototypes; in the remaining cases, the characteristics switch between them in a relatively abrupt manner, appearing close to the averages of the modes.

Figure 15.3 illustrates the characteristics of the model with different values of the prototypes. Again, it is apparent that in moving the prototypes, we can adjust the nonlinear mapping of the model to the existing experimental data.

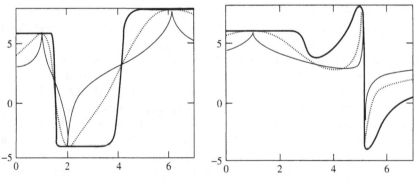

Figure 15.3. Nonlinear input-output characteristics of the cluster-based model. The prototypes in the input space vary; $z_1 = 6$, $z_2 = -4$, $z_3 = 8$; $m = 2$.

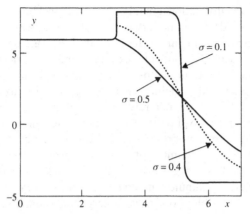

Figure 15.4. Nonlinear input-output characteristics of an RBF network; the spread of the receptive fields has an effect somewhat similar to that of the fuzzification factor used in the previous construct. The other parameters are $z_1 = 6$, $z_2 = -4$, $z_3 = 8$; $v_1 = 1$; $v_2 = 5.2$, $v_3 = 5.1$.

It is worth noting that the fuzzy clusters defined in the input space play a role similar to that of receptive fields in RBF neural networks. However, there is a substantial design difference. Fizzy clustering produces a set of fully developed membership functions. In RBF networks, we usually need to optimize the parameters of the receptive fields. In Gaussian receptive fields, we must adjust their centers and spreads (radii). A typical example of a model in this category is governed by the expression

$$y = \frac{\sum_{i=1}^{c} z_i G(x; v_i, \sigma)}{\sum_{i=1}^{c} G(x; v_i, \sigma)} \tag{15.3}$$

where $G(x; m, \sigma)$ denotes a certain Gaussian function characterized by its modal value (m) and spread (σ). (Note that we usually require some normalization, as the sum of these receptive fields may not always generate a value equal to 1.) As shown in Figure 15.4, there is no guarantee that the model coincides with the prototypes defined in the output space. This can be attributed to the smoothing effect of the Gaussian receptive fields. This is in contrast to the nonlinear relationship formed by the fuzzy partition; due to their design, these receptive fields coincide with the values of the prototypes.

15.3. CONDITIONAL CLUSTERING IN THE DEVELOPMENT OF A BLUEPRINT OF GRANULAR MODELS

Clustering plays a crucial role in granular modeling. First, it helps convert numeric data into information granules (with or without some hints of domain knowledge). These information granules form the backbone or blueprint of the model. While the model could be further refined, it is aimed at capturing the most essential, numerically dominant features of data by summarizing them. Obviously, the more clusters we intend to capture, the more detailed the resulting blueprint is. Second, clustering helps manage dimensionality, which is usually a critical issue in rule-based modeling. As they are based on fuzzy relations (rather than fuzzy sets), clusters do not rely on the number of variables, and those variables are not shown explicitly in the information granules as they emerge at the level of the Cartesian product of the variables. Let us remember that in any rule-based system, the number of input is critical in determining the dimensionality of the rule base. A complete rule base consists of p^n, where p is the number of information granules in each input variable and n is the number of variables. Even in a problem of fairly modest dimensionality, (say $n = 10$) and very few information granules defined for each variable (say, $p = 4$), we end up with a significant number of rules: $4^{10} = 1,049*10^6$. Keeping the same number of variables but using eight linguistic terms (information granules) substantially increases the size of the rule base; in this case, we end up with $2.825*10^8$ rules. The effect of combinatorial explosion is clear (see Figure 15.5).

There are several ways of handling the dimensionality problem. The obvious one is to recognize that we do not need a complete rule base because there are various combinations of conditions that never occur in practice and are not supported by any experimental evidence. While this seems pretty straightforward, it is not easy to be confident about the nature of such unnecessary rules. The second, more feasible approach is to treat all variables at the same time and apply fuzzy clustering. The number of clusters is far smaller than the number of rules involving individual variables.

In what follows, we discuss various modes of incorporating clustering results into the skeleton of the model. It is important to understand the implications of using the clustering technique in forming information granules, especially

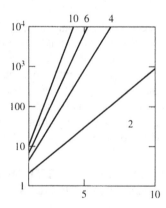

Figure 15.5. Number of rules treated as a function of input variables; the dependency is illustrated for several values of p.

within the models. The most critical observation concerns the relational aspects of clustering and the directional features of models. By its nature (unless properly endowed), clustering treats multivariable data as relational constructs, so the final products of cluster analysis are clusters as descriptors of data where each variable is treated in the same manner, regardless of where it is positioned as a modeling entity. In a nutshell, in typical clustering algorithms we do not distinguish between input and output variables. This stands in sharp contrast to system modeling. There the role of the variable is critical, as most practical models are directional constructs, viz., they model a mapping from independent to dependent variables. The distinction between these two categories of variables requires modifications to the clustering algorithm to accommodate this requirement. As an example, let us consider a many input–many output (MIMO) model involving input and output variables $\mathbf{x}$ and $\mathbf{y}$, respectively, that is, $\mathbf{y} = f(\mathbf{x})$. The experimental data come in the usual format $\{(\mathbf{x}_k, \mathbf{y}_k)\}$, $k = 1, 2, \ldots, N$. If we are to ignore directionality, the immediate approach to clustering the data would be to concatenate the vectors so that $\mathbf{z}_k = [\mathbf{x}_k \mathbf{y}_k]$ and carry out clustering in the augmented feature space. Obviously, we concentrate on the relational aspects of data, ignoring the possible mapping component that is of interest. To alleviate the problem, we may attenuate the output variables by assigning higher weight to them. In essence, the clustering needs to pay more attention to the differences (and similarities) occurring in the output spaces (that is, vectors $\mathbf{y}_1, \mathbf{y}_2, \ldots, \mathbf{y}_n$). This issue was raised and has resulted in so-called D-fuzzy clustering (Hirota and Pedrycz, 1995). An easy implementation would be to admit the distance function computed differently, depending upon the coordinates of z, by using a weight factor g. For instance, the Euclidean distance between $\mathbf{z}$ and $\mathbf{z}'$ would be

$$\|\mathbf{z} - \mathbf{z}'\| = \|\mathbf{x} - \mathbf{x}'\| + \gamma\|\mathbf{y} - \mathbf{y}'\| \qquad (15.4)$$

where $\gamma > 0$. The higher the value of γ, the more attention is given to the output variables. As usual, the dimensionality of the input space is higher than that of the output space; the value of g needs to be properly selected to ensure a suitable balance. Even though we become aware of the main direction, the choice of the weight factor is a matter of intensive experimentation. Furthermore, we must understand the implications of the possible normalization effect of the input variables.

Conditional clustering (Pedrycz, 1998) is naturally geared to direction-aware clustering. The context variable(s) are the output variables. Defining contexts over them is a separate task. Once the contexts are given, the ensuing clustering is induced (or directed) by the fuzzy set (relation) provided. Recall that the rules achieve the granular model "if condition then conclusion." In context-based clustering, the role of the conclusion is played by the context fuzzy set. Given the context, we use fuzzy clustering to reflect on the pertinent portion of data in the input space and find a conditional structure there. By changing the context, we continue to search by focusing on other parts of the data. The result produced in this manner is a web of information granules developed conditionally upon

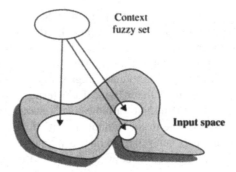

Figure 15.6. A blueprint of a linguistic model induced by predefined fuzzy sets (or relations) of the context defined in the input space.

the assumed collection of contexts. Hence the directional aspects of the construct we want to develop on the basis of the information granules become evident. We do not need to worry about any weight factor (as presented above). The design of contexts is quite intuitive. First, these are fuzzy sets whose semantics is well defined. We can use the terms *low, medium,* and *large* output. We can move achieve further refinements of these terms and introduce more linguistic categories. They could be assessed with regard to their experimental evidence by expressing their sigma count with respect to numeric data (Pedrycz, 1998). To ensure full coverage of the output space, the fuzzy sets of contexts could form a fuzzy partition. Obviously, we can carry out clustering of data in the output space and arrive at membership functions generated in automatically. This option is particularly attractive when there are many output variables and the manual definition of the context fuzzy relations could be tedious or impractical.

The blueprint of the model (Figure 15.6) must be further formalized to capture the mapping between the information granules. This leads to a detailed architecture of an inherently linguistic network whose outputs are information granules. A linguistic or granular neuron is an interesting construct worth exploring in this setting.

15.4. THE GRANULAR NEURON AS A GENERIC PROCESSING ELEMENT IN GRANULAR NETWORKS

As its name indicates, a granular neuron is a neuron with granular connections. More precisely, we consider the transformation of many numeric inputs $u_1, u_2, \ldots, u_c$ (confined to the unit interval) of the form

$$Y = N(u_1, u_2, \ldots, u_c, W_1, W_2, \ldots, W_c) = \sum_{\oplus} W_i \otimes u_i \qquad (15.5)$$

with $W_1, W_2, \ldots, W_c$ denoting granular weights (connections; see Figure 15.7). The symbols of generalized addition and multiplication ($\oplus, \otimes$) are used here

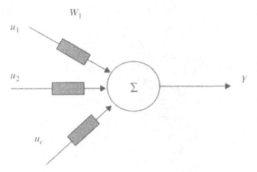

Figure 15.7. Computational model of a granular neuron; note the granular character of the connections.

to emphasize the granular character of the arguments used in this aggregation. Depending on the specific realization, these connections can realized as intervals, fuzzy sets, shadowed sets, rough sets, and so on. As a result of processing, the output Y is also a granular neuron whose granular character is associated with the nature of the connections of the neuron.

The computations performed by the neuron depend on the underlying formalism of the granular computing achieved there. Having the general form of the transformation (15.5), we still need to come up with a realization of the summation and multiplication operations to produce a final expression for the output of the neuron. For instance, if the connections have an interval character, the resulting granular input Y is completely characterized by its bounds:

$$\text{Lower bound:} \quad y_- = \sum_{i=1}^{c} w_{i-} u_i$$

$$\text{Upper bound:} \quad y_+ = \sum_{i=1}^{c} w_{i+} u_i$$

Here we have assumed that the connections are described in the interval form $W_i = [w_{i-}, w_{i+}], i = 1, 2, \ldots, c.$

The plot of the characteristics of the neuron for two inputs $u_1 = \alpha$ and $u_2 = 1 - \alpha$ is shown in Figure 15.8. Here $W_1 = [0.3, 3]$ and $W_2 = [1.4, 7]$. As anticipated, the result (output of the neuron) is an interval that moves along the y-axis as α changes from 0 to 1.

In fuzzy sets used to implement the connections, we end up with more complicated formulas. They can be simplified profoundly if we confine ourselves to triangular fuzzy sets (fuzzy numbers) of the connections. Following the calculus of fuzzy numbers, we note that the multiplication of W_i by a positive constant scales the fuzzy number yet retains the piecewise character of the membership function. Furthermore, the summation operation does not affect the shape of the

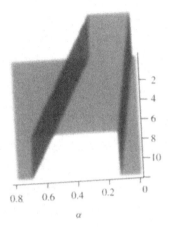

Figure 15.8. The interval output of the linguistic neuron with interval connections.

membership function, so the final result can be again described in the following format:

$$Y = \left\langle \sum_{i=1}^{c} a_i u_i, \sum_{i=1}^{c} m_i u_i, \sum_{i=1}^{c} b_i u_i \right\rangle \tag{15.6}$$

where each connection W_i is fully characterized by the triple of real numbers $W_i = \langle a_i, m_i, b_i \rangle$. Here m_i denotes a modal value of the connection, while a_i and b_i stand for the lower and upper bound, respectively, of the triangular number describing this fuzzy set of the connection. The plot of the output of the neuron for u_1 and u_2 defined as above is presented in Figure 15.9.

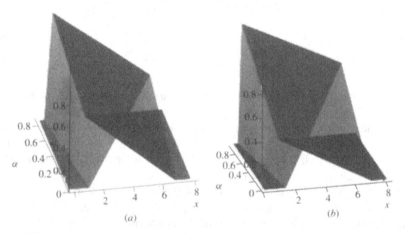

Figure 15.9. The output of the linguistic neuron with connections realized as triangular fuzzy numbers: $W_1 = \langle 0.3, 0.5, 3.0 \rangle$, $W_2 = \langle 1.4, 1.5, 7.0 \rangle$ (a) and $W_1 = \langle 0.3, 2.0, 3.0 \rangle$, $W_2 = \langle 1.4, 5.0, 7.0 \rangle$ (b).

The granular neuron exhibits several interesting properties that generalize the characteristics of (numeric) neurons. Adding a nonlinearity component (g) to the linear aggregation does not change the essence of computing; in a monotonically increasing relationship ($g(Y)$) we end up with a transformation of the original output interval or fuzzy set (in this case, we have to follow the calculations using the well-known extension principle).

15.5. THE ARCHITECTURE OF LINGUISTIC MODELS BASED ON CONDITIONAL FUZZY CLUSTERING

Conditional fuzzy clustering has provided us with a backbone of the linguistic model (Pedrycz and Vasilakos, 1999). Following the principle of conditional clustering, we end up with a general topology of the model shown in Figure 15.10.

The development of the linguistic model consists of two main phases: (a) forming fuzzy sets or relations of the contexts and (b) conditional clustering for the already available collection of contexts. The two phases are linked: once contexts are given, the clustering uses this information in the directed search for the structure.

Much research has focused on RBF neural networks, fuzzy neural networks, and counterpropagation networks, which seem to be similar to the linguistic model discussed here. These models such as RBF neural networks fuzzy neural networks etc. in spite of some differences, exhibit strong similarities. In one way or another, they dwell on the concept of hybrid unsupervised-supervised learning. Consider, for example, RBF neural networks. Receptive fields forming the input layer of the network are initially constructed using clustering, which yields prototypes (centers) of the RBFs and provides an initial estimate of the spread of

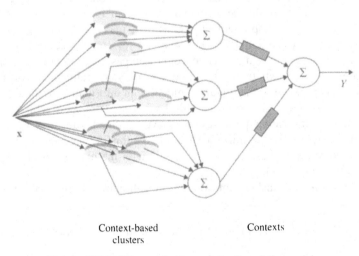

Context-based   Contexts
clusters

Figure 15.10. The architecture of the linguistic models.

the functions. This is done in unsupervised mode with clustering, SOMs, or vector quantization algorithms. The ensuing linear part of the network is optimized using the standard least square error method; the optimal values of connections are easily computed in an analytical fashion. Further refinements are possible based on the values of the connections; the position and spread of the receptive fields are updated. From the architectural perspective, fuzzy models tend to be similar to the RBF neural networks. They involve the design of receptive fields (which in this case tend to carry some semantics relevant to fuzzy sets) and optimization of the connections. The receptive fields can be more sophisticated, with well-defined meanings, but the hybrid mechanism of unsupervised-supervised learning is present there. The linguistic models, by contrast, have two clear differences. First, we link the output with the clustering process (as the contexts provide all necessary guidance); second, the output of the model is granular. This is a result of the use of granular constructs of the contexts. Notably, by changing the granularity of the contexts, we can control the size of the fuzzy set of the output and adjust the specificity of the overall modeling process. The computations of the output fuzzy sets are completed in two steps: (a) aggregation of activation levels (membership grades) of all clusters associated with a given context and (b) linear combination of the activation levels with the parameters of the context fuzzy sets. In triangular membership functions of the context fuzzy sets, the calculations follow the scheme described by (15.6).

15.6. REFINEMENTS OF LINGUISTIC MODELS

The conditional FCM has produced the prototypes (or, equivalently, clusters) in the input space, and using these we generate inputs to the granular neuron. The connections of the neuron are just the fuzzy sets of the context. In essence, the parameters of the network are downloaded from the phase of fuzzy clustering. Further optimization of the network may not seem to be required. However, there is still room for improvement. Refinement may be necessary because of the fact that each conditional FCM is realized for some specific context, and these tasks are separate; as a consequence, prototypes may require some shifting. Furthermore, the contexts themselves may require some refinement and refocus. Note also that the result of the linguistic model is an information granule (interval, fuzzy set, fuzzy relation, etc.), and this has to be compared with a numeric datum y_k. For illustrative purposes, we have confined ourselves to a single-output linguistic model. Optimization has to take this into account. For instance, one can require the highest matching possible with the highest specificity of the result (see Figure 15.11).

This requirement translates into one of the following optimization problems:

$$\frac{1}{N} \sum_{k=1}^{N} T_k(y_k) \to \text{Max (maximization of average}$$

agreement with numeric data)

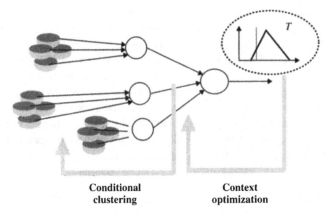

Conditional   Context
clustering   optimization

Figure 15.11. Optimization of the linguistic model; see the detailed description in text.

or

$$\frac{1}{N} \sum_{k=1}^{N} (b_k - a_k) \rightarrow \text{Min (minimization of the average spread}$$

of the granular output of the network)

whose solutions have to be searched for over the parameters of the context fuzzy sets. In this sense, optimization can be confined to the portion of the network requiring the refinement of the contexts. Furthermore, we can make the optimization more manageable since the successive contexts have a one-half overlap. Given this condition, optimization concentrates on the modal values of the triangular fuzzy sets of the context. Once these have been modified, the conditional FCM repeats and the iterative optimization moves forward.

15.7. CONCLUSIONS

Linguistic (granular) models are based on the concept of information granules regarded as semantically meaningful conceptual entities that are crucial to the overall framework of user-centric modeling. The user is can cast all modeling activities in a way that directly reflects the main purpose of the modeling problem. For instance, in data mining, the user is ultimately interested in revealing relationships that could be of potential interest to the given problem.

The algorithm of fuzzy clustering is particularly well suited to the key objectives of linguistic modeling. Fuzzy clusters fully reflect the character of the data. The search for the structure is ultimately affected by the well-articulated modeling needs of the user. We have demonstrated that fuzzy sets of context can play an important role in shaping modeling activities and help combat dimensionality problems (by decomposing the original problem into a series of subproblems guided by specific contexts). By linking context fuzzy sets and the induced clusters, we form the blueprint of the model. Further refinements of the connections (links) of this web of connections become a straightforward task given the reduced, limited dimensionality of the problem and the transparency of the model. It is relevant to note that the results of linguistic models are expressed at a certain level of granularity. Instead of single numeric results (which are typical of numeric models), the user is provided with a fuzzy set of results that can be easily communicated in a linguistic format (compatible with the vocabulary of linguistic terms originally used in the design of the model) and presented visually in the format of respective membership functions.

REFERENCES

G.E.P. Box, G.M. Jenkins, *Time Series Analysis, Forecasting and Control*, Holden Day, San Francisco, 1970.

W.A. Farag, V.H. Quintana, G. Lambert-Torres, A genetic-based neuro-fuzzy approach for modeling and control of dynamical systems, *IEEE Trans. on Neural Networks*, 9, 1998, 756–767.

J. Gonzalez, I. Rojas, H. Pomares, J. Ortega, A. Prieto, A new clustering technique for function approximation, *IEEE Trans. on Neural Networks*, 13, 2002, 132–142.

R. Hecht Nielsen, *Neurocomputing*, Addison-Wesley, Menlo Park, CA, 1990.

K. Hirota, W. Pedrycz, D-fuzzy clustering, *Pattern Recognition Letters*, 16, 1995, 193–200.

M. Joo, S. Wu, J. Lu, H.L. Toh, Face recognition with radial basis function (RBF) neural networks, *IEEE Trans. on Neural Networks*, 13, 2002, 697–710.

N.B. Karyannis, G.W. Mi, Growing radial basis neural networks: merging supervised and unsupervised learning with network growth techniques, *IEEE Trans. on Neural Networks*, 8, 1997, 1492–1506.

E. Kim, M. Park, S. Ji, M. Park, A new approach to fuzzy modeling, *IEEE Trans. on Fuzzy Systems*, 5, 1997, 328–337.

B.-J. Park, W. Pedrycz, S.-K. Oh, Fuzzy polynomial neural networks: hybrid architectures of fuzzy modeling, *IEEE Trans. on Fuzzy Systems*, 10, 2002, 607–621.

W. Pedrycz, Conditional fuzzy C-Means, *Pattern Recognition Letters*, 17, 1996, 625–632.

W. Pedrycz, Conditional fuzzy clustering in the design of radial basis function neural networks, *IEEE Trans. on Neural Networks*, 9, 1998, 601–612.

W. Pedrycz, A.V. Vasilakos, Linguistic models and linguistic modeling, *IEEE Trans. on Systems, Man, and Cybernetics*, 29, 1999, 745–757.

S. Ridella, S. Rovetta, R. Zunino, Plastic algorithm for adaptive vector quantization, *Neural Computing and Applications*, 7, 1998, 37–51.

S. Rovetta, R. Zunino, Vector quantization for license-plate location and image coding, *IEEE Trans. on Industrial Electronics*, 47, 2000, 159–167.

M. Sugeno, T. Yasukawa, A fuzzy-logic-based approach to qualitative modeling, *IEEE Trans. on Fuzzy Systems*, 1, 1993, 7–31.

T. Takagi, M. Sugeno, Fuzzy identification of systems and its applications to modeling and control, *IEEE Trans. on Systems, Man, and Cybernetics*, 15, 1985, 116–132.

Bibliography

J. Abonyi, R. Babuska, F. Szeifert, Modified Gath-Geva fuzzy clustering for identification of Takagi-Sugeno fuzzy models, *IEEE Transactions on Systems, Man, and Cybernetics, Part B*, Vol. 32, No. 5, October 2002, 612–621.

J. Abonyi, F. Szeifert, Supervised fuzzy clustering for the identification of fuzzy classifiers, *Pattern Recognition Letters*, Vol. 24, No. 14, October 2003, 2195–2207.

P. Alam, D. Booth, K. Lee, T. Thordarson, The use of fuzzy clustering algorithms and self-organizing neural networks for identifying potentially failing banks: an experimental study, *Expert Systems with Applications*, Vol. 18, No. 3, April 2000, 185–199.

D. Altman, Efficient fuzzy clustering of multi-spectral images, *Proc. IEEE 1999 Int. Geoscience and Remote Sensing Symposium, IGARSS '99*, 28 June–2 July 1999, Vol. 3, 1594–1596.

M.J. Anderson, E. Micheli-Tzanakou, Auditory stimulus optimization with feedback from fuzzy clustering of neuronal responses, *IEEE Transactions on Information Technology in Biomedicine*, Vol. 6, No. 2, June 2002, 159–170.

K. Arakawa, Y. Arakawa, A nonlinear digital filter using fuzzy clustering, *Proc. IEEE Int. Conference on Acoustics, Speech, and Signal Processing*, 23–26 March 1992, vol. 4, 309–312.

S. Asharaf, M.N. Murty, An adaptive rough fuzzy single pass algorithm for clustering large data sets, *Pattern Recognition*, Vol. 36, No. 12, December 2003, 3015–3018.

G.P. Babu, M.N. Murty, Clustering with evolution strategies, *Pattern Recognition*, Vol. 27, No. 2, February 1994, 321–329.

A. Baraldi, P. Blonda, A survey of fuzzy clustering algorithms for pattern recognition, *IEEE Transactions on Systems, Man, and Cybernetics, Part B*, Vol. 29, No. 6, December 1999, 778–785.

A. Bargiela, W. Pedrycz, Recursive information granulation: aggregation and interpretation issues, *IEEE Transactions on Systems, Man, and Cybernetics SMC-B*, Vol. 33, No. 1, January 2003, 96–112.

G. Barkó, J. Abonyi, J. Hlavay, Application of fuzzy clustering and piezoelectric chemical sensor array for investigation on organic compounds, *Analytica Chimica Acta*, Vol. 398, No. 2–3, October 1999, 219–226.

M. Barni, V. Cappellini, A. Mecocci, Comments on "A possibilistic approach to clustering," *IEEE Transactions on Fuzzy Systems*, Vol. 4, No. 3, August 1996, 393–396.

M. Barni, R. Gualtieri, A new possibilistic clustering algorithm for line detection in real world imagery, *Pattern Recognition*, Vol. 32, No. 11, November 1999, 1897–1909.

V. Barra, J.-Y. Boire, Segmentation of fat and muscle from MR images of the thigh by a possibilistic clustering algorithm, *Computer Methods and Programs in Biomedicine*, Vol. 68, No. 3, June 2002, 185–193.

B. Barshan, B. Ayrulu, Fuzzy clustering and enumeration of target type based on sonar returns, *Pattern Recognition*, Vol. 37, No. 2, February 2004, 189–199.

N. Belacel, P. Hansen, N. Mladenovic, Fuzzy J-Means: a new heuristic for fuzzy clustering, *Pattern Recognition*, Vol. 35, No. 10, October 2002, 2193–2200.

G. Beni, X. Liu, A least biased fuzzy clustering method, *IEEE Transactions on Pattern Analysis and Machine Intelligence*, Vol. 16, No. 9, September 1994, 954–960.

A.M. Bensaid, L.O. Hall, J.C. Bezdek, L.P. Clarke, Partially supervised clustering for image segmentation, *Pattern Recognition*, Vol. 29, No. 5, May 1996, 859–871.

A.M. Bensaid, L.O. Hall, J.C. Bezdek, L.P. Clarke, M.L. Silbiger, J.A. Arrington, R.F. Murtagh, Validity-guided (re)clustering with applications to image segmentation, *IEEE Transactions on Fuzzy Systems*, Vol. 4, No. 2, May 1996, 112–123.

L. Bentabet, Y.M. Zhu, O. Dupuis, V. Kaftandjian, D. Babot, M. Rombaut, Use of fuzzy clustering for determining mass functions Dempster-Shafer theory, *Proc. 5th Int. Conference on Signal Processing, WCCC-ICSP 2000*, 21–25 Aug. 2000, vol. 3, 1462–1470.

J.C. Bezdek, R. Ehrlich, W. Full, FCM: the fuzzy C-Means clustering algorithm, *Computers and Geosciences*, Vol. 10, No. 2–3, 1984, 191–203.

J.C. Bezdek, J.D. Harris, Fuzzy partitions and relations: an axiomatic basis for clustering, *Fuzzy Sets and Systems*, Vol. 1, No. 2, April 1978, 111–127.

J.C. Bezdek, R.J. Hathaway, Clustering with relational C-Means partitions from pairwise distance data, *Mathematical Modelling*, Vol. 9, No. 6, June 1987, 435–439.

J.C. Bezdek, R.J. Hathaway, Optimization of fuzzy clustering criteria using genetic algorithms, *Proc. 1st IEEE Conference on Evolutionary Computation*, 27–29 June 1994, Vol. 2, 589–594.

B. Bhanu, A. Dong, Concepts learning with fuzzy clustering and relevance feedback, *Engineering Applications of Artificial Intelligence*, Vol. 15, No. 2, April 2002, 123–138.

Z. Bide, S. Caixin, H. Xuesong, Z. Yinchun, A new method of diagnosing turbogenerator's vibration fault based on fuzzy clustering analysis and wavelet packets transform, *Proc. 5th Int. Conference on Electrical Machines and Systems, ICEMS 2001*, 18–20 August 2001, vol. 1, 339–342.

C. Borgelt, H. Timm, R. Kruse, Using fuzzy clustering to improve naive Bayes classifiers and probabilistic networks, *Proc. 9th IEEE Int. Conference on Fuzzy Systems*, 7–10 May 2000, vol. 1, 53–58.

S. de Bruin, A. Stein, Soil-landscape modelling using fuzzy C-Means clustering of attribute data derived from a Digital Elevation Model (DEM), *Geoderma*, Vol. 83, No. 1–2, April 1998, 17–33.

B.P. Buckles, F.E. Petry, D. Prabhu, R. George, R. Srikanth, Fuzzy clustering with genetic search, *Proc. 3rd IEEE Conference on Fuzzy Systems*, 26–29 June 1994, vol. 1, 46–50.

D. Cabello, S. Barro, R. Ruiz, E.L. Zapata, J. Mira, Fuzzy clustering: application to the diagnosis of ventricular arrhythmias, *Proc. Annual Int. Conference of the IEEE Engineering in Medicine and Biology Society*, 4–7 November 1988, vol. 1, 5–6.

Y.Y. Cai, H.T. Loh, A.Y.C. Nee, Qualitative primitive identification using fuzzy clustering and invariant approach, *Image and Vision Computing*, Vol. 14, No. 7, July 1996, 451–464.

G. Castellano, A.M. Fanelli, C. Mencar, A fuzzy clustering approach for mining diagnostic rules, *Proc. IEEE Int. Conference on Systems, Man, and Cybernetics, SMC 2003*, October 5–8, 2003, Vol. 2, 2007–2012.

A.L. Cechin, D.R.P. Simon, K. Stertz, State automata extraction from recurrent neural nets using k-means and fuzzy clustering, *Proc. 23rd Int. Conference of the Chilean Computer Science Society SCCC 2003*, 6–7 November 2003, 73–78.

B.B. Chaudhuri, P.R. Bhowmik, An approach of clustering data with noisy or imprecise feature measurement, *Pattern Recognition Letters*, Vol. 19, No. 14, December 1998, 1307–1317.

G. Chen, Q. Wei, H. Zhang, Discovering similar time-series patterns with fuzzy clustering and DTW methods, *Proc. Joint 9th IFSA World Congress and 20th NAFIPS Int. Conference*, 25–28 July 2001, vol. 4, 2160–2164.

M.-S. Chen, S.-W. Wang, Fuzzy clustering analysis for optimizing fuzzy membership functions, *Fuzzy Sets and Systems*, Vol. 103, No. 2, 16 April 1999, 239–254.

T.W. Cheng, D.B. Goldgof, L.O. Hall, Fast fuzzy clustering, *Fuzzy Sets and Systems*, Vol. 93, No. 1, January 1998, 49–56.

V. Chepoi, D. Dumitrescu, Fuzzy clustering with structural constraints, *Fuzzy Sets and Systems*, Vol. 105, No. 1, July 1999, 91–97.

J.H. Chiang, P.Y. Hao, A new kernel-based fuzzy clustering approach: support vector clustering with cell growing, *IEEE Transactions on Fuzzy Systems*, Vol. 1, No. 4, August 2003, 518–527.

K.K. Chintalapudi, M. Kam, A noise-resistant fuzzy C-Means algorithm for clustering, *Proc. IEEE International Conference on Fuzzy Systems*, 4–9 May 1998, vol. 2, 1458–1463.

Y. Chtioui, D. Bertrand, D. Barba, Y. Dattee, Application of fuzzy C-Means clustering for seed discrimination by artificial vision, *Chemometrics and Intelligent Laboratory Systems*, Vol. 38, No. 1, August 1997, 75–87.

M.C. Clark, L.O. Hall, D.B. Goldgof, L.P. Clarke, R.P. Velthuizen, M.S. Silbiger, MRI segmentation using fuzzy clustering techniques, *IEEE Engineering in Medicine and Biology Magazine*, Vol. 13, No. 5, November–December 1994, 730–742.

R. Coppi, P. D'Urso, Three-way fuzzy clustering models for LR fuzzy time trajectories, *Computational Statistics and Data Analysis*, Vol. 43, No. 2, June 2003, 149–177.

R.N. Dave, Boundary detection through fuzzy clustering, *Proc. IEEE Int. Conference on Fuzzy Systems*, 8–12 March 1992a, 127–134.

R.N. Dave, Generalized fuzzy c-shells clustering and detection of circular and elliptical boundaries, *Pattern Recognition*, Vol. 25, No. 7, July 1992b, 713–721.

R.N. Dave, Robust fuzzy clustering algorithms, *Proc. 2nd IEEE Int. Conference on Fuzzy Systems*, 28 March–1 April 1993, vol. 2, 1281–1286.

R.N. Dave, T. Fu, Robust shape detection using fuzzy clustering: practical applications, *Fuzzy Sets and Systems*, Vol. 65, No. 2–3, August 1994, 161–185.

R.N. Dave, S. Sen, Robust fuzzy clustering of relational data, *IEEE Transactions on Fuzzy Systems*, Vol. 10, No. 6, December 2002, 713–727.

M. Delgado, A. Gomez-Skarmeta, H.M. Barbera, A tabu search approach to the fuzzy clustering problem, *Proc. 6th IEEE Int. Conference on Fuzzy Systems*, 1–5 July 1997, vol. 1, 125–130.

M. Delgado, A.F. Gómez-Skarmeta, F. Martín, A methodology to model fuzzy systems using fuzzy clustering in a rapid-prototyping approach, *Fuzzy Sets and Systems*, Vol. 97, No. 3, August 1998, 287–301.

M. Delgado, A. Gomez-Skarmeta, M.A. Vila, Hierarchical clustering to validate fuzzy clustering, *Proc. Int. Joint Conference of the 4th IEEE Int. Conference on Fuzzy Systems and 2nd Int. Fuzzy Engineering Symposium on Fuzzy Systems*, 20–24 March 1995, vol. 4, 1807–1812.

B.B. Devi, V.V.S. Sarma, A fuzzy multistage evolutionary (FUME) clustering technique, *Pattern Recognition Letters*, Vol. 2, No. 3, March 1984, 139–145.

A. Devillez, P. Billaudel, G.V. Lecolier, A fuzzy hybrid hierarchical clustering method with a new criterion able to find the optimal partition, *Fuzzy Sets and Systems*, Vol. 128, No. 3, June 2002, 323–338.

J.A. Dickerson, Y. Daaboul, T.H. Jobe, C.M. Helgason, Analysis of concomitant mechanisms in stroke pathogenesis using fuzzy clustering techniques, *Proc. Conference of the Fuzzy Information Processing Society, NAFIPS '97*, 21–24 September 1997, 211–216.

J.A. Dickerson, R. Matalon, C. Helgason, Fuzzy clustering of vitamin and homocysteine levels in patients with history of ischemic stroke, *Proc. Conference of the North American Fuzzy Information Processing Society, NAFIPS*, 20–21 August 1998, 231–236.

M. Doroodchi, A.M. Reza, Implementation of fuzzy cluster filter for nonlinear signal and image processing, *Proc. 5th IEEE Int. Conference on Fuzzy Systems*, 8–11 September 1996a, vol. 3, 2117–2122.

M. Doroodchi, A.M. Reza, Fuzzy cluster filter, *Proc. Int. Conference on Image Processing*, 16–19 September 1996b, vol. 2, 939–942.

B. Due, P. Schroeter, J. Bigun, Motion estimation and segmentation by fuzzy clustering, *Proc. Int. Conference on Image Processing*, 23–26 October 1995, vol. 3, 472–475.

M.A. Egan, Locating clusters in noisy data: a genetic fuzzy C-Means clustering algorithm, *Proc. Conference of the North American Fuzzy Information Processing Society, NAFIPS*, 20–21 August 1998, 178–182.

M.A. Egan, M. Krishnamoorthy, K. Rajan, Comparative study of a genetic fuzzy C-Means algorithm and a validity guided fuzzy C-Means algorithm for locating clusters in noisy data, *Proc. IEEE Int. Conference on Evolutionary Computation*, 4–9 May 1998, vol. 1, 440–445.

A. El Imrani, A. Bouroumi, H. Zine El Abidine, M. Limouri, A. Essaïd, A fuzzy clustering–based niching approach to multimodal function optimization, *Cognitive Systems Research*, Vol. 1, No. 2, June 2000, 119–133.

S. Eschrich, J. Ke, L.O. Hall, D.B. Goldgof, Fast accurate fuzzy clustering through data reduction, *IEEE Transactions on Fuzzy Systems*, Vol. 11, No. 2, April 2003, 262–270.

M.J. Fadili, S. Ruan, D. Bloyet, B. Mazoyer, Unsupervised fuzzy clustering analysis of fMRI series, *Proc. 20th Annual Int. Conference on Engineering in Medicine and Biology Society*, 1998, 29 October–1 November 1998, vol. 2, 696–699.

J.-L. Fan, W.-Z. Zhen, W.-X. Xie, Suppressed fuzzy C-Means clustering algorithm, *Pattern Recognition Letters*, Vol. 24, No. 9–10, June 2003, 1607–1612.

A. Flores-Sintas, J.M. Cadenas, F. Martin, Membership functions in the fuzzy C-Means algorithm, *Fuzzy Sets and Systems*, Vol. 101, No. 1, January 1999, 49–58.

F.M. Frattale Mascioli, A. Rizzi, M. Panella, G. Martinelli, Clustering with unconstrained hyperboxes, *Proc. IEEE Int. Conference on Fuzzy Systems, FUZZ-IEEE '99*, 22–25 August 1999, vol. 2, 1075–1080.

M. Friederichs, O. Fränzle, A. Salski, Fuzzy clustering of existing chemicals according to their ecotoxicological properties, *Ecological Modelling*, Vol. 85, No. 1, February 1996, 27–40.

H. Frigui, P. Gader, J. Keller, Fuzzy clustering for land mine detection, *Proc. Conference of the North American Fuzzy Information Processing Society, NAFIPS*, 20–21 August 1998, 261–265.

H. Frigui, R. Krishnapuram, Competitive fuzzy clustering, *Proc. 6th IEEE Int. Conference on Fuzzy Systems*, 1–5 July 1997a, vol. 1, 225–228.

H. Frigui, R. Krishnapuram, Clustering by competitive agglomeration, *Pattern Recognition*, Vol. 30, No. 7, July 1997b, 1109–1119.

I. Gath, A.B. Geva, Unsupervised optimal fuzzy clustering, *IEEE Transactions on Pattern Analysis and Machine Intelligence*, Vol. 11, No. 7, July 1989, 773–780.

I. Gath, D. Hoory, Detection of elliptic shells using fuzzy clustering: application to MRI images, *Proc. 12th IAPR Int. Conference on Pattern Recognition, Conference B: Computer Vision & Image Processing*, 9–13 October 1994, vol. 2, 251–255.

I. Gath, D. Hoory, Fuzzy clustering of elliptic ring-shaped clusters, *Pattern Recognition Letters*, Vol. 16, No. 7, July 1995, 727–741.

I. Gath, A. Smolyak Iskoz, B. Van Cutsem, Data induced metric and fuzzy clustering of non-convex patterns of arbitrary shape, *Pattern Recognition Letters*, Vol. 18, No. 6, June 1997, 541–553.

H. Genther, A. Konig, M. Glesner, Rule weight generation for a fuzzy classification system based on fuzzy clustering methods, *Proc. 3rd IEEE Conference on Fuzzy Systems*, 26–29 June 1994, vol. 1, 614–617.

A.B Geva, Non-stationary time-series prediction using fuzzy clustering, *Proc. 18th Int. Conference of the North American Fuzzy Information Processing Society, NAFIPS 99*, 10–12 June 1999a, 413–417.

A.B. Geva, Hierarchical unsupervised fuzzy clustering, *IEEE Transactions on Fuzzy Systems*, Vol. 7, No. 6, December 1999b, 723–733.

A.B. Geva, Hierarchical-fuzzy clustering of temporal-patterns and its application for time-series prediction, *Pattern Recognition Letters*, Vol. 20, No. 14, December 1999c, 1519–1532.

A.B. Geva, D.H Kerem, Forecasting generalized epileptic seizures from the EEG signal by wavelet analysis and dynamic unsupervised fuzzy clustering, *IEEE Transactions on Biomedical Engineering*, Vol. 10, No. 10, October 1998, 1205–1216.

M. Gil, E.G. Sarabia, J.R. Llata, J.P. Oria, Fuzzy C-Means clustering for noise reduction, enhancement and reconstruction of 3D ultrasonic images, *Proc. 7th IEEE Int. Conference on Emerging Technologies and Factory Automation, ETFA '99*, 18–21 October 1999, vol. 1, 465–472.

P. Golding, A.E. Jones, Cochannel interference in wireless LANs and fuzzy clustering, *Proc. 8th IEEE Int. Symposium on Personal, Indoor and Mobile Radio Communications, PIMRC '97*, 1–4 September 1997, vol. 3, 1171–1175.

A.F. Gómez-Skarmeta, M. Delgado, M.A. Vila, About the use of fuzzy clustering techniques for fuzzy model identification, *Fuzzy Sets and Systems*, Vol. 106, No. 2, September 1999, 179–188.

P.J.F. Groenen, K. Jajuga, Fuzzy clustering with squared Minkowski distances, *Fuzzy Sets and Systems*, Vol. 120, No. 2, June 2001, 227–237.

T. Gu, B. Dubuisson, Similarity of classes and fuzzy clustering, *Fuzzy Sets and Systems*, Vol. 34, No. 2, January 1990, 213–221.

F. Guoyao, Optimization methods for fuzzy clustering, *Fuzzy Sets and Systems*, Vol. 93, No. 3, February 1998, 301–309.

Y. Hata, S. Kobashi, S. Hirano, Medical image segmentation by fuzzy logic techniques, *Proc. IEEE Int. Conference on Systems, Man, and Cybernetics*, 11–14 October 1998, vol. 4, 4098–4103.

R.J. Hathaway, J.C. Bezdek, Local convergence of the fuzzy C-Means algorithms, *Pattern Recognition*, Vol. 19, No. 6, 1986, 477–480.

R.J. Hathaway, J.C. Bezdek, Switching regression models and fuzzy clustering, *IEEE Transactions on Fuzzy Systems*, Vol. 1, No. 3, August 1993, 195–204.

R.J. Hathaway, J.C. Bezdek, NERF C-Means: non-Euclidean relational fuzzy clustering, *Pattern Recognition*, Vol. 27, No. 3, March 1994, 429–437.

R.J. Hathaway, J.C. Bezdek, Clustering incomplete relational data using the non-Euclidean relational fuzzy C-Means algorithm, *Pattern Recognition Letters*, Vol. 23, No. 1–3, January 2002, 151–160.

R.J. Hathaway, J.W. Davenport, J.C. Bezdek, Relational duals of the C-Means clustering algorithms, *Pattern Recognition*, Vol. 22, No. 2, 1989, 205–212.

J. Henriques, A. Cardoso, A. Dourado, Supervision and C-Means clustering of PID controllers for a solar power plant, *International Journal of Approximate Reasoning*, Vol. 22, Issues 1–2, September–October 1999, 73–91.

K. Hirota, K. Iwama, Application of modified FCM with additional data to area division of images, *Information Sciences*, Vol. 45, No. 2, July 1988, 213–230.

K. Hirota, W. Pedrycz, Directional fuzzy clustering and its application to fuzzy modelling, *Fuzzy Sets and Systems*, Vol. 80, No. 3, June 1996, 315–326.

K. Honda, N. Sugiura, H. Ichihashi, Simultaneous approach to principal component analysis and fuzzy clustering with missing values, *Proc. Joint 9th IFSA World Congress and 20th NAFIPS Int. Conference*, 25–28 July 2001, vol. 3, 1810–1815.

F. Höppner, Speeding up fuzzy C-Means: using a hierarchical data organisation to control the precision of membership calculation, *Fuzzy Sets and Systems*, Vol. 128, No. 3, June 2002, 365–376.

F. Höppner, F. Klawonn, Obtaining interpretable fuzzy models from fuzzy clustering and fuzzy regression, *Proc. 4th Int. Conference on Knowledge-Based Intelligent Engineering Systems and Allied Technologies*, 30 August–1 September 2000, vol. 1, 162–165.

F. Höppner, F. Klawonn, A contribution to convergence theory of fuzzy C-Means and derivatives, *IEEE Transactions on Fuzzy Systems*, Vol. 11, No. 5, October 2003, 682–694.

X. Hua, Z. Yannan, Y. Zehong, S. Xiaomin, J. Peifa, An adaptive method of signal selecting based on the strategy of fuzzy clustering and expert reasoning, *Proc. 3rd World Congress on Intelligent Control and Automation*, 28 June–2 July 2000, vol. 3, 1726–1729.

E. Ikeda, T. Imaoka, H. Ichihashi, T. Miyoshi, Nonlinear component analysis by fuzzy clustering and multidimensional scaling methods, *Proc. Int. Joint Conference on Neural Networks, IJCNN '99*, 10–16 July 1999, vol. 4, 2539–2543.

P.T. Im, B. Qiu, M. Wingate, L. Herron, Linear boundary detection by cluster prototype centring based on fuzzy memberships, *Proc. 3rd Australian and New Zealand Conference on Intelligent Information Systems, ANZIIS-95*, 27 November 1995, 210–213.

I. Ishimaru, H. Saito, T. Asano, T. Furuhashi, A new reformation scheme of steel bars and a learning method for various plastic characteristic using fuzzy clustering, *Proc. IEEE Int. Conference on Systems, Man, and Cybernetics, IEEE SMC '99*, 12–15 October 1999, vol. 5, 185–190.

K. Jajuga, L_1-norm based fuzzy clustering, *Fuzzy Sets and Systems*, Vol. 39, No. 1, January 1991, 43–50.

C.V. Jawahar, P.K. Biswas, A.K. Ray, Investigations on fuzzy thresholding based on fuzzy clustering, *Pattern Recognition*, Vol. 30, No. 10, October 1997, 1605–1613.

G.Y. Jiang, H.F. Wang, X.D. Zhang, L.H. Gang, The study on the application of fuzzy clustering analysis in the dynamic identification of road traffic state, *Proc. IEEE Intelligent Transportation Systems*, 12–15, October 2003, vol. 2, 1449–1452.

R.I. John, P.R. Innocent, M.R. Barnes, Neuro-fuzzy clustering of radiographic tibia image data using type 2 fuzzy sets, *Information Sciences*, Vol. 125, No. 1–4, June 2000, 65–82.

K. Jongwoo, R. Krishnapuram, R. Davé, Application of the least trimmed squares technique to prototype-based clustering, *Pattern Recognition Letters*, Vol. 17, Issue 6, May 1996, 633–641.

A. Joshi, S. Auephanwiriyakul, R. Krishnapuram, On fuzzy clustering and content based access to networked video databases, *Proc. 8th Int. Workshop on Research Issues in Data Engineering, Continuous-Media Databases and Applications*, 23–24 February 1998, 42–49.

K. Kamei, D.M. Auslander, K. Inoue, A fuzzy clustering method for multidimensional parameter selection in system with uncertain parameters, *Proc. IEEE Int. Conference on Fuzzy Systems*, 8–12 March 1992, 355–362.

M. Kamel, B. Hadfield, M. Ismail, Fuzzy query processing using clustering techniques, *Information Processing and Management*, Vol. 26, No. 2, 1990, 279–293.

M.S. Kamel, S.Z. Selim, A thresholded fuzzy C-Means algorithm for semi-fuzzy clustering, *Pattern Recognition*, Vol. 24, No. 9, 1991, 825–833.

H. Kamimura, M. Kurano, Clustering by a fuzzy metric, *Fuzzy Sets and Systems*, Vol. 120, No. 2, June 2001, 249–254.

N.B. Karayiannis, Maximum entropy clustering algorithms and their application in image compression, *Proc. IEEE Int. Conference on Systems, Man, and Cybernetics, "Humans, Information and Technology,"* 2–5 October 1994, vol. 1, 337–342.

N.B. Karayiannis, Fuzzy and possibilistic clustering algorithms based on generalized reformulation, *Proc. 5th IEEE Int. Conference on Fuzzy Systems*, 8–11 September 1996, vol. 2, 1393–1399.

S.N. Kavuri, V. Venkatasubramanian, Using fuzzy clustering with ellipsoidal units in neural networks for robust fault classification, *Computers and Chemical Engineering*, Vol. 17, No. 8, August 1993, 765–784.

U. Kaymak, M. Setnes, Fuzzy clustering with volume prototypes and adaptive cluster merging, *IEEE Transactions on Fuzzy Systems*, Vol. 10, No. 6, December 2002, 705–712.

P.R. Kersten, Including auxiliary information in fuzzy clustering, *Proc. 1996 Biennial Conference of the North American Fuzzy Information Processing Society, NAFIPS*, 19–22 June 1996, 221–224.

G. Keswani, L.O. Hall, Text classification with enhanced semi-supervised fuzzy clustering, *Proc. IEEE Int. Conference on Fuzzy Systems, 2002, FUZZ-IEEE'02*, 12–17 May 2002, 621–626.

D.-W. Kim, K.H. Lee, D. Lee, Fuzzy cluster validation index based on inter-cluster proximity, *Pattern Recognition Letters*, Vol. 24, No. 15, November 2003, 2561–2574.

E. Kim, H. Lee, M. Park, A simply identified Sugeno-type fuzzy model via double clustering, *Information Sciences*, Vol. 110, No. 1–2, September 1998, 25–39.

M.H. Kim, H.S. Choi, W.D. Lee, Fuzzy clustering using extended MFA for continuous-valued state space, *Proc. Int. Joint Conference on Neural Networks*, 7–11 June 1992, vol. 2, 733–738.

T. Kim, J.C. Bezdek, R.J. Hathaway, Optimality tests for fixed points of the fuzzy C-Means algorithm, *Pattern Recognition*, Vol. 21, No. 6, 1988, 651–663.

Y.S. Kim, S. Mitra, Integrated adaptive fuzzy clustering (IAFC) algorithm, *Proc. 2nd IEEE Int. Conference on Fuzzy Systems*, 28 March–1 April 1993, vol. 2, 1264–1268.

Y.S. Kim, S. Mitra, An adaptive integrated fuzzy clustering model for pattern recognition, *Fuzzy Sets and Systems*, Vol. 65, No. 2–3, August 1994, 297–310.

F. Klawonn, R. Kruse, Automatic generation of fuzzy controllers by fuzzy clustering *Proc. IEEE Int. Conference on Systems, Man, and Cybernetics, 1995, "Intelligent Systems for the 21st Century,"* 22–25 October 1995, vol. 3, 2040–2045.

X. Kong, R. Wang, G. Li, Fuzzy clustering algorithms based on resolution and their application in image compression, *Pattern Recognition*, Vol. 35, No. 11, November 2002, 2439–2444.

D.H. Kraft, J. Chen, A. Mikulcic, Combining fuzzy clustering and fuzzy inferencing in information retrieval, *Proc. 9th IEEE Int. Conference on Fuzzy Systems, FUZZ IEEE 2000*, 7–10 May 2000, vol. 1, 375–380.

V. Kreinovich, H.T. Nguyen, S.A. Starks, Y. Yeung, Decision making based on satellite images: optimal fuzzy clustering approach, *Proc. 37th IEEE Conference on Decision and Control*, 16–18 December 1998, vol. 4, 4246–4251.

R. Krishnapuram, H. Frigui, O. Nasraoui, The Fuzzy C Quadric Shell clustering algorithm and the detection of second-degree curves, *Pattern Recognition Letters*, Vol. 14, No. 7, July 1993, 545–552.

R. Krishnapuram, H. Frigui, O. Nasraoui, Fuzzy and possibilistic shell clustering algorithms and their application to boundary detection and surface approximation, *IEEE Transactions on Fuzzy Systems*, Vol. 3, No. 1, February 1995, 29–43.

R. Krishnapuram, A. Joshi, L. Yi, A fuzzy relative of the k-medoids algorithm with application to web document and snippet clustering, *Proc. 1999 IEEE Int. Conference on Fuzzy Systems, FUZZ-IEEE '99*, 22–25 August 1999, vol. 3, 1281–1286.

R. Krishnapuram, J. Kim, A note on the Gustafson-Kessel and adaptive fuzzy clustering algorithms, *IEEE Transactions on Fuzzy Systems*, Vol. 7, No. 4, August 1999, 453–461.

H.J. Kweon, J.W. Suk, J.S. Song, M.H. Lee, Intelligent QRS typification using fuzzy clustering, *Proc. 17th Annual Conference on Engineering in Medicine and Biology Society*, 20–23 September 1995, vol. 1, 199–200.

S.H. Kwon, Cluster validity index for fuzzy clustering, *Electronics Letters*, Vol. 34, No. 22, October 1998, 2176–2177.

P. Lagacherie, D.R. Cazemier, P. van Gaans, P.A. Burrough, Fuzzy k-means clustering of fields in an elementary catchment and extrapolation to a larger area, *Geoderma*, Vol. 77, No. 2–4, June 1997, 197–216.

H.S. Lee, On system identification via fuzzy clustering for fuzzy modeling, *Proc. IEEE Int. Conference on Systems, Man, and Cybernetics, IEEE SMC'98*, 11–14 October 1998, vol. 2, 1956–1961.

R.P. Li, M. Mukaidono, A maximum-entropy approach to fuzzy clustering, *Proc. Int. Joint Conference of the 4th IEEE Int. Conference on Fuzzy Systems and 2nd Int. Fuzzy Engineering Symposium*, 20–24 March 1995, vol. 4, 2227–2232.

T.W. Liao, A.K. Celmins, R.J. Hammell II, A fuzzy C-Means variant for the generation of fuzzy term sets, *Fuzzy Sets and Systems*, Vol. 135, No. 2, 16 April 2003, 241–257.

A.W.C. Liew, S.H. Leung, W.H. Lau, Fuzzy image clustering incorporating spatial continuity, *IEE Proceedings—Vision, Image and Signal Processing*, Vol. 147, No. 2, April 2000, 185–192.

A.W.C. Liew, H. Yan, Adaptive spatially constrained fuzzy clustering for image segmentation, *Proc. 10th IEEE Int. Conference on Fuzzy Systems*, 2–5 December 2001, vol. 3, 801–804.

A.W.C. Liew, H. Yan, An adaptive spatial fuzzy clustering algorithm for 3-D MR image segmentation, *IEEE Transactions on Medical Imaging*, Vol. 22, No. 9, September 2003, 1063–1075.

G. Lim, J.C. Bezdek, Small targets in LADAR images using fuzzy clustering, *Proc. IEEE Int. Conference on Fuzzy Systems*, 4–9 May 1998, vol. 1, 61–66.

J.S. Lin, K.S. Cheng, C.W. Mao, A modified Hopfield neural network with fuzzy C-Means technique for multispectral MR image segmentation, *Proc. Int. Conference on Image Processing*, 16–19 September 1996, vol. 1, 327–330.

D.A Linkens, M.Y. Chen, Hierarchical fuzzy clustering based on self-organising networks, *Proc. IEEE Int. Conference on Fuzzy Systems*, 4–9 May 1998, vol. 2, 1406–1410.

A. Linusson, S. Wold, B. Nordén, Fuzzy clustering of 627 alcohols, guided by a strategy for cluster analysis of chemical compounds for combinatorial chemistry, *Chemometrics and Intelligent Laboratory Systems*, Vol. 44, No. 1–2, December 1998, 213–227.

H. Liu, S.T. Huang, Evolutionary semi-supervised fuzzy clustering, *Pattern Recognition Letters*, Vol. 24, No. 16, December 2003, 3105–3113.

Y. Liu, Y. Shen, Z. Liu, An approach to fault diagnosis for non-linear system based on fuzzy cluster analysis, *Proc. 17th IEEE Instrumentation and Measurement Technology, IMTC 2000*, 1–4 May 2000, vol. 3, 1469–1473.

C.-C. Lo, S.-J. Wang, Video segmentation using a histogram-based fuzzy C-Means clustering algorithm, *Computer Standards and Interfaces*, Vol. 23, No. 5, November 2001, 429–438.

V. Loia, W. Pedrycz, S. Senatore, P-FCM: a proximity-based fuzzy clustering for user-centered web applications, *International Journal of Approximate Reasoning*, Vol. 34, No. 2–3, November 2003, 121–144.

C.G. Looney, Interactive clustering and merging with a new fuzzy expected value, *Pattern Recognition*, Vol. 35, No. 11, November 2002, 2413–2423.

Y.H. Man, I. Gath, Detection and separation of ring-shaped clusters using fuzzy clustering, *IEEE Transactions on Pattern Analysis and Machine Intelligence*, Vol. 16, No. 8, August 1994, 855–861.

R.L. de Mantaras, L. Valverde, New results in fuzzy clustering based on the concept of indistinguishability relation, *IEEE Transactions on Pattern Analysis and Machine Intelligence*, Vol. 10, No. 5, September 1988, 754–757.

J.F. Martínez-Trinidad, J. Ruiz-Shulcloper, Fuzzy clustering of semantic spaces, *Pattern Recognition*, Vol. 34, No. 4, April 2001, 783–793.

F. Masulli, A. Schenone, A fuzzy clustering based segmentation system as support to diagnosis in medical imaging, *Artificial Intelligence in Medicine*, Vol. 16, No. 2, June 1999, 129–147.

M.A. Matos, N.D. Hatziargriou, J.A.P. Lopes, Multicontingency steady state security evaluation using fuzzy clustering techniques, *IEEE Transactions on Power Systems*, Vol. 15, No. 1, February 2000, 177–183.

U. Maulik, S. Bandyopadhyay, Fuzzy partitioning using a real-coded variable-length genetic algorithm for pixel classification, *IEEE Transactions on Geoscience and Remote Sensing*, Vol. 41, No. 5, May 2003, 1075–1081.

D. Melcher, M. Matthies, Application of fuzzy clustering to data dealing with phytotoxicity, *Ecological Modelling*, Vol. 85, No. 1, February 1996, 41–49.

W.M. Melek, M.R. Emami, A.A. Goldenberg, An improved robust fuzzy clustering algorithm, *Proc. IEEE Int. Conference on Fuzzy Systems, FUZZ-IEEE '99*, 22–25 August 1999, vol. 3, 1261–1265.

M. Ménard, Fuzzy clustering and switching regression models using ambiguity and distance rejects, *Fuzzy Sets and Systems*, Vol. 122, No. 3, September 2001, 363–399.

M. Ménard, Extension of the objective functions in fuzzy clustering, *Proc. IEEE Int. Conference on Fuzzy Systems, FUZZ-IEEE '02*, 12–17 May 2002, vol. 2, 1450–1455.

M. Ménard, V. Courboulay, P.-A. Dardignac, Possibilistic and probabilistic fuzzy clustering: unification within the framework of the non-extensive thermostatistics, *Pattern Recognition*, Vol. 36, No. 6, June 2003, 1325–1342.

M. Ménard, C. Demko, P. Loonis, The fuzzy c+2-means: solving the ambiguity rejection in clustering, *Pattern Recognition*, Vol. 33, No. 7, July 2000, 1219–1237.

M. Ménard, M. Eboueya, Extreme physical information and objective function in fuzzy clustering, *Fuzzy Sets and Systems*, Vol. 128, No. 3, June 2002, 285–303.

S. Miyamoto, An overview and new methods in fuzzy clustering, *Proc. 2nd Int. Conference on Knowledge-Based Intelligent Electronic Systems, KES '98*, 21–23 April 1998, vol. 1, 33–40.

S. Miyamoto, Fuzzy multisets and fuzzy clustering of documents, *Proc. 10th IEEE Int. Conference on Fuzzy Systems*, 2–5 December 2001, 1539–1542.

S. Miyamoto, Information clustering based on fuzzy multisets, *Information Processing and Management*, Vol. 39, No. 2, March 2003, 195–213.

A. Mikulcic, J. Chen, Experiments on using fuzzy clustering for fuzzy control system design, *Proc. 5th IEEE Int. Conference on Fuzzy Systems*, 8–11 September 1996, vol. 3, 2168–2174.

L. Min, J.C. Liu, J. Ming, Incoherent radar target recognition using semi-fuzzy clustering, *Proc. IEEE 1992 National Aerospace and Electronics Conference, NAECON 1992*, 18–22 May 1992, vol. 1, 265–269.

J. Moghaddas, N.R. Prasad, Fuzzy clustering approach to evaluating power system security, *Proc. 42nd Midwest Symposium on Circuits and Systems*, 8–11 August 1999, vol. 2, 925–928.

T.K. Moon, Temporal pattern recognition using fuzzy clustering, *Proc. 3rd IEEE Conference on Fuzzy Systems*, 26–29 June 1994, vol. 1, 432–435.

A. Murgu, Fuzzy clustering of aggregated flows for traffic control in ATM networks, *Proc. 6th IEEE Int. Conference on Fuzzy Systems*, 1–5 July 1997, vol. 1, 235–240.

S. Nascimento, B. Mirkin, F. Moura-Pires, A fuzzy clustering model of data with proportional membership, *Proc. 19th Int. Conference of the North American Fuzzy Information Processing Society, NAFIPS 2000*, 13–15 July 2000, 261–266.

H.H. Nguyen, P. Cohen, A fuzzy clustering approach to texture segmentation, *Proc. IEEE Int. Conference on Systems, Man, and Cybernetics, IEEE SMC 90*, 4–7 November 1990, 200–205.

S. Nitsuwat, J.S. Jin, H.M. Hudson, Motion-based video segmentation using fuzzy clustering and classical mixture model, *Proc. 2000 Int. Conference on Image Processing*, 10–13 September 2000, vol. 1, 300–303.

J.C. Noordam, W.H.A.M. van den Broek, L.M.C. Buydens, Multivariate image segmentation with cluster size insensitive fuzzy C-Means, *Chemometrics and Intelligent Laboratory Systems*, Vol. 64, No. 1, October 2002, 65–78.

B.J. Oh, K.C. Kwak, S.S. Kim, J.W. Ryu, A method of designing nonlinear channel equalizer using conditional fuzzy C-Means clustering, *Proc. 6th Int. Conference on Signal Processing*, 26–30 August 2002, vol. 2, 1355–1358.

C.H. Oh, K. Honda, H. Ichihashi, Fuzzy clustering for categorical multivariate data *Proc. Joint 9th IFSA World Congress and 20th NAFIPS Int. Conference*, 25–28 July 2001, vol. 4, 2154–2159.

S.-K. Oh, W. Pedrycz, H.-S. Park, Rule-based multi-FNN identification with the aid of evolutionary fuzzy granulation, *Knowledge-Based Systems*, Vol. 17, No. 1, January 2004, 1–13.

M.K. Pakhira, S. Bandyopadhyay, U.L. Maulik, Validity index for crisp and fuzzy clusters, *Pattern Recognition*, Vol. 37, No. 3, March 2004, 487–501.

S.K. Pal, S. Mitra, Fuzzy dynamic clustering algorithm, *Pattern Recognition Letters*, Vol. 11, No. 8, August 1990, 525–535.

R.D. Pascual-Marqui, A.D. Pascual-Montano, K. Kochi, J.M. Carazo, Smoothly distributed fuzzy C-Means: a new self-organizing map, *Pattern Recognition*, Vol. 34, No. 12, December 2001, 2395–2402.

W. Pedrycz, Fuzzy sets in pattern recognition: methodology and methods, *Pattern Recognition*, Vol. 23, No. 1–2, 1990, 121–146.

W. Pedrycz, Conditional fuzzy C-Means, *Pattern Recognition Letters*, Vol. 17, No. 6, May 1996, 625–631.

W. Pedrycz, Fuzzy set technology in knowledge discovery, *Fuzzy Sets and Systems*, Vol. 98, No. 3, September 1998a, 279–290.

W. Pedrycz, Conditional fuzzy clustering in the design of radial basis function neural networks, *IEEE Transactions on Neural Networks*, Vol. 9, No. 4, 1998, 601–612.

W. Pedrycz, G. Vukovich, Feature analysis through information granulation and fuzzy sets, *Pattern Recognition*, Vol. 35, No. 4, April 2002a, 825–834.

W. Pedrycz, Collaborative fuzzy clustering, *Pattern Recognition Letters*, Vol. 23, No. 14, December 2002b, 1675–1686.

W. Pedrycz, Fuzzy clustering with a knowledge-based guidance, *Pattern Recognition Letters*, Vol. 25, No. 4, March 2004, 469–480.

W. Pedrycz, J.C. Bezdek, R.J. Hathaway, G.W. Rogers, Two nonparametric models for fusing heterogeneous fuzzy data, *IEEE Transactions on Fuzzy Systems*, Vol. 6, No. 3, August 1998, 411–425.

W. Pedrycz, A.V. Vasilakos, Linguistic models and linguistic modeling, *IEEE Transactions on Systems, Man, and Cybernetics*, Vol. 29, No. 6, 1999, 745–757.

W. Pedrycz, J. Waletzky, Fuzzy clustering with partial supervision, *IEEE Transactions on Systems, Man, and Cybernetics-B*, Vol. 27, No. 5, October 1997, 787–795.

S. Pemmaraju, S. Mitra, Y.Y. Shieh, G.H. Roberson, Multiresolution wavelet decomposition and neuro-fuzzy clustering for segmentation of radiographic images, *Proc. 8th IEEE Symposium on Computer-Based Medical Systems*, 9–10 June 1995, 142–149.

D.L. Pham, Spatial models for fuzzy clustering, *Computer Vision and Image Understanding*, Vol. 84, No. 2, November 2001, 285–297.

D.L. Pham, Fuzzy clustering with spatial constraints, *Proc. Int. Conference on Image Processing*, 22–25 September 2002, vol. 2, II-65–II-68.

T.D. Pham, H. Yan, Color image segmentation using fuzzy integral and mountain clustering, *Fuzzy Sets and Systems*, Vol. 107, No. 2, October 1999, 121–130.

N.J. Pizzi, R.A. Vivanco, R.L. Somorjai, EvIdent™: a functional magnetic resonance image analysis system, *Artificial Intelligence in Medicine*, Vol. 21, No. 1–3, January–March 2001, 263–269.

S. Policker, A.B. Geva, Prediction of time varying composite sources by temporal fuzzy clustering, *Proc. 11th IEEE Signal Processing Workshop on Statistical Signal Processing*, 6–8 August 2001, 329–332.

M. Ramze Rezaee, B.P.F. Lelieveldt, J.H.C. Reiber, A new cluster validity index for the fuzzy C-Mean, *Pattern Recognition Letters*, Vol. 19, No. 3–4, March 1998, 237–246.

G. Rantitsch, Application of fuzzy clusters to quantify lithological background concentrations in stream-sediment geochemistry, *Journal of Geochemical Exploration*, Vol. 71, No. 1, October 2000, 73–82.

L. Ren, G.W. Irwin, D. Flynn, Nonlinear identification of turbogenerator AVR loop dynamics using fuzzy clustering, *Proc. Int. Conference on Power System Technology, PowerCon*, 13–17 October 2002, vol. 3, 1503–1508.

M.R. Rezaee, P.M.J. van der Zwet, B.P.E. Lelieveldt, R.J. van der Geest, A. Reiber, Multiresolution image segmentation technique based on pyramidal segmentation and fuzzy clustering, *IEEE Transactions on Image Processing*, Vol. 9, No. 7, July 2000, 1238–1248.

H.S. Rhee, K.W. Oh, A performance measure for the fuzzy cluster validity, *Proc. 1996 Asian Fuzzy Systems Symposium "Soft Computing in Intelligent Systems and Information Processing"*, 11–14 December 1996, 364–369.

M. Ronen, Y. Shabtai, H. Guterman, Rapid process modeling—model building methodology combining unsupervised fuzzy-clustering and supervised neural networks, *Computers and Chemical Engineering*, Vol. 22, Supplement 1, March 1998, S1005–S1008.

P.J. Rousseeuw, L. Kaufman, E. Trauwaert, Fuzzy clustering using scatter matrices, *Computational Statistics and Data Analysis*, Vol. 23, No. 1, November 1996, 135–151.

X. Ruan, A pattern recognition machine with fuzzy clustering analysis, *Proc. 3rd World Congress on Intelligent Control and Automation*, 28 June–2 July 2000, vol. 4, 2530–2534.

M. Ruben, G. Mireille, J. Diego, C. Carlos, T. Javier, Segmentation of ventricular angiographic images using fuzzy clustering, *Proc. IEEE 17th Annual Conference on Engineering in Medicine and Biology Society*, 20–23 September 1995, vol. 1, 405–406.

T.A. Runkler, J.C. Bezdek, Web mining with relational clustering, *International Journal of Approximate Reasoning*, Vol. 32, No. 2–3, February 2003, 217–236.

T.A. Runkler, R.H. Palm, Identification of nonlinear systems using regular fuzzy c-elliptotype clustering, *Proc. 5th IEEE Int. Conference on Fuzzy Systems*, 2, 8–11 September 1996, vol. 2, 1026–1030.

E.H. Ruspini, New experimental results in fuzzy clustering, *Information Sciences*, Vol. 6, 1973, 273–284.

S. Russell, W. Lodwick, Fuzzy clustering in data mining for telco database marketing campaigns, *Proc. 18th Int. Conference of the North American Fuzzy Information Processing Society, NAFIPS 1999*, 10–12 June 1999, 720–726.

J. Sabharwal, J. Chen, Intelligent pH control using fuzzy linear invariant clustering, *Proc. 28th Southeastern Symposium on System Theory*, 31 March–2 April 1996, 514–518.

G.I. Sanchez-Ortiz, A. Noble, Fuzzy clustering driven anisotropic diffusion: enhancement and segmentation of cardiac MR images, *Proc. IEEE Nuclear Science Symposium*, 8–14 November 1998, vol. 3, 1873–1874.

C. Sârbu, H.F. Pop, Fuzzy clustering analysis of the first 10 MEIC chemicals, *Chemosphere*, Vol. 40, No. 5, March 2000, 513–520.

M. Sato-Ilic, On clustering based on homogeneity, *Proc. Joint 9th IFSA World Congress and 20th NAFIPS Int. Conference*, 25–28 July 2001, vol. 5, 2505–2510.

M. Sato-Ilic, Fuzzy regression analysis using fuzzy clustering, *Proc. 2002 Annual Meeting of the North American Fuzzy Information Processing Society, NAFIPS 2002*, 27–29 June 2002, 57–62.

M. Sato-Ilic, Weighted principal component analysis for interval-valued data based on fuzzy clustering, *Proc. IEEE Int. Conference on Systems, Man, and Cybernetics, IEEE SMC 03*, October 5–8, 2003, vol. 5, 4476–4482.

S. Scandizzo, A fuzzy clustering approach for the measurement of operational risk, *Proc. 3rd Int. Conference on Knowledge-Based Intelligent Information Engineering Systems*, 31 August–1 September 1999, 324–328.

S. Schupp, A. Elmoataz, J. Fadili, P. Herlin, D. Bloyet, Image segmentation via multiple active contour models and fuzzy clustering with biomedical applications, *Proc. 15th Int. Conference on Pattern Recognition*, 3–7 September 2000, vol. 1, 622–625.

Y.M. Sebzalli, X.Z. Wang, Knowledge discovery from process operational data using PCA and fuzzy clustering, *Engineering Applications of Artificial Intelligence*, Vol. 14, No. 5, October 2001, 607–616.

M. Setnes, Supervised fuzzy clustering for rule extraction, *IEEE Transactions on Fuzzy Systems*, Vol. 8, No. 4, August 2000, 416–424.

M. Setnes, U. Kaymak, Fuzzy modeling of client preference from large data sets: an application to target selection in direct marketing, *IEEE Transactions on Fuzzy Systems*, Vol. 9, No. 1, February 2001, 153–163.

M. Setnes, H. Roubos, Transparent fuzzy modeling using fuzzy clustering and GAs, *Proc. 18th Int. Conference of the North American Fuzzy Information Processing Society, NAFIPS 1999*, 10–12 June 1999, 198–202.

J.-B. Sheu, A fuzzy clustering-based approach to automatic freeway incident detection and characterization, *Fuzzy Sets and Systems*, Vol. 128, No. 3, June 2002, 377–388.

Y. Shi, M. Mizumoto, An improvement of neuro-fuzzy learning algorithm for tuning fuzzy rules based on fuzzy clustering method, *Proc. IEEE Int. Conference on Fuzzy Systems*, 4–9 May 1998, vol. 2, 991–996.

R. Srikanth, R. George, N. Warsi, D. Prabhu, F.E. Petry, B.P. Buckles, A variable-length genetic algorithm for clustering and classification, *Pattern Recognition Letters*, Vol. 16, No. 8, August 1995, 789–800.

C. Stutz, T.A. Runkler, Classification and prediction of road traffic using application-specific fuzzy clustering, *IEEE Transactions on Fuzzy Systems*, Vol. 10, No. 3, June 2002, 297–308.

F.-C. Su, W.-L. Wu, Y.-M. Cheng, Y.-L. Chou, Fuzzy clustering of gait patterns of patients after ankle arthrodesis based on kinematic parameters, *Medical Engineering and Physics*, Vol. 23, No. 2, March 2001, 83–90.

J. Suckling, T. Sigmundsson, K. Greenwood, E.T. Bullmore, A modified fuzzy clustering algorithm for operator independent brain tumor classification of dual echo MR images, *Magnetic Resonance Imaging*, Vol. 17, No. 7, September 1999, 1065–1076.

I.H. Suh, J.H. Kim, F. Rhee, Fuzzy clustering involving convex polytopes, *Proc. 5th IEEE Int. Conference on Fuzzy Systems*, 8–11 September 1996, vol. 2, 1013–1019.

H.J. Sun, S.R. Wang, Z. Mei, A fuzzy clustering based algorithm for feature selection, *Proc. Int. Conference on Machine Learning and Cybernetics*, 4–5 November 2002, vol. 4, 1993–1998.

S. Supot, S. Manas, Codebook design algorithm for classified vector quantization based on fuzzy clustering, *Proc. IEEE Int. Conference on Industrial Technology, IEEE ICIT '02*, 11–14 December 2002, vol. 2, 751–753.

Y. Susu, Evolutionary search based fuzzy self-organizing clustering, *Proc. 1999 Congress on Evolutionary Computation, CEC 99*, 6–9 July 1999, vol. 1, 1–8.

R. Tagliaferri, A. Staiano, D. Scala, A supervised fuzzy clustering for Radial Basis Function Neural Networks training, *Proc. Joint 9th IFSA World Congress and 20th NAFIPS Int. Conference*, 25–28 July 2001, vol. 3, 1804–1809.

O. Takata, S. Miyamoto, K. Umayahara, Fuzzy clustering of data with uncertainties using minimum and maximum distances based on L_1 metric, *Proc. Joint 9th IFSA World Congress and 20th NAFIPS Int. Conference*, 25–28 July 2001, vol. 5, 2511–2516.

F.D. Tamás, J. Abonyi, J. Borszéki, P. Halmos, Trace elements in clinker: II. Qualitative identification by fuzzy clustering, *Cement and Concrete Research*, Vol. 32, No. 8, August 2002, 1325–1330.

Y. Tao, Y. Xinmiao, Fuzzy comprehensive assessment, fuzzy clustering analysis and its application for urban traffic environment quality evaluation, *Transportation Research Part D: Transport and Environment*, Vol. 3, Issue 1, January 1998, 51–57.

P. Teppola, S.-P. Mujunen, P. Minkkinen, Adaptive Fuzzy C-Means clustering in process monitoring, *Chemometrics and Intelligent Laboratory Systems*, Vol. 45, No. 1–2, January 1999, 23–38.

H. Timm, F. Klawonn, R. Kruse, An extension of partially supervised fuzzy cluster analysis, *Proc. Annual Meeting of the North American Fuzzy Information Processing Society, NAFIPS 2002*, 27–29 June 2002, 63–68.

H. Timm, R. Kruse, Fuzzy cluster analysis with missing values, *Proc. Int. Conference of the North American Fuzzy Information Processing Society, NAFIPS*, 20–21 August 1998, 242–246.

H. Timm, R. Kruse, A modification to improve possibilistic fuzzy cluster analysis, *Proc. IEEE International Conference on Fuzzy Systems, FUZZ-IEEE '02*, 12–17 May 2002, vol. 2, 1460–1465.

Y.A Tolias, S.M. Panas, Image segmentation by a fuzzy clustering algorithm using adaptive spatially constrained membership functions, *IEEE Transactions on Systems, Man, and Cybernetics, Part A*, Vol. 28, No. 3, May 1998, 359–369.

R.E. Trauwaert, On the meaning of Dunn's partition coefficient for fuzzy clusters, *Fuzzy Sets and Systems*, Vol. 25, No. 2, February 1988, 217–242.

R.E. Trauwaert, L. Kaufman, P. Rousseeuw, Fuzzy clustering algorithms based on the maximum likelihood principle, *Fuzzy Sets and Systems*, Vol. 42, No. 2, July 1991, 213–227.

R.E. Trauwaert, L. Kaufman, Fuzzy clustering with high contrast, *Journal of Computational and Applied Mathematics*, Vol. 64, No. 1–2, November 1995, 81–90.

B. Treiger, I. Bondarenko, H. Van Malderen, R. Van Grieken, Elucidating the composition of atmospheric aerosols through the combined hierarchical, nonhierarchical and fuzzy clustering of large electron probe microanalysis data sets, *Analytica Chimica Acta*, Vol. 317, No. 1–3, December 1995, 33–51.

K. Tsuda, M. Minoh, K. Ikeda, Extracting straight lines by sequential fuzzy clustering, *Pattern Recognition Letters*, Vol. 17, No. 6, May 1996, 643–649.

I.B. Turksen, S. Jiang, Fuzzy cluster analysis for multi-antecedent rule base restructuring based on S-implication, *Proc. IEEE Int. Conference on Fuzzy Systems*, 8–12 March 1992, 941–948.

A. Tzes, P.Y. Peng, J. Guthy, Genetic-based fuzzy clustering for DC-motor friction identification and compensation, *IEEE Transactions on Control Systems Technology*, Vol. 6, No. 4, July 1998, 462–472.

B. Van Cutsem, I. Gath, Detection of outliers and robust estimation using fuzzy clustering, *Computational Statistics and Data Analysis*, Vol. 15, No. 1, January 1993, 47–61.

T. Van Le, Evolutionary fuzzy clustering, *Proc. IEEE Int. Conference on Evolutionary Computation*, 29 November–1 December 1995, vol. 2, 753–758.

R.P. Velthuizen, L.O. Hall, L.P. Clarke, M.L. Silbiger, An investigation of mountain method clustering for large data sets, *Pattern Recognition*, Vol. 30, No. 7, July 1997, 1121–1135.

A. Vernet, G.A. Kopp, Classification of turbulent flow patterns with fuzzy clustering, *Engineering Applications of Artificial Intelligence*, Vol. 15, No. 3–4, June–August 2002, 315–326.

E.L. Walker, Combining geometric invariants with fuzzy clustering for object recognition, *Proc. 18th Int. Conference of the North American Fuzzy Information Processing Society, NAFIPS 99*, 10–12 June 1999, 571–574.

C.H. Wei, C.S. Fahn, A distributed approach to fuzzy clustering by genetic algorithms, *Proc. 1996 Asian Fuzzy Systems Symposium, "Soft Computing in Intelligent Systems and Information Processing,"* 11–14 December 1996, 350–357.

M.P. Windham, Cluster validity for fuzzy clustering algorithms, *Fuzzy Sets and Systems*, Vol. 5, No. 2, March 1981, 177–185.

C. Windischberger, M. Barth, C. Lamm, L. Schroeder, H. Bauer, R.C. Gur, E. Moser, Fuzzy cluster analysis of high-field functional MRI data, *Artificial Intelligence in Medicine*, Vol. 29, No. 3, November 2003, 203–223.

G.D. Wozniak, P.R. Kletke, K.S. Kmetik, Physician services mix and physician specialty: an application of fuzzy clustering, *Proc. 5th IEEE Int. Conference on Fuzzy Systems*, 8–11 September 1996, vol. 2, 980–984.

J.-W. Wu, T.-R. Tsai, Weighted quasi-likelihood estimation based on fuzzy clustering analysis method and dimension reduction technique, *Fuzzy Sets and Systems*, Vol. 128, No. 3, June 2002, 353–364.

K.-L. Wu, M.-S. Yang, Alternative C-Means clustering algorithms, *Pattern Recognition*, Vol. 35, No. 10, October 2002, 2267–2278.

X.L. Xie, G. Beni, A validity measure for fuzzy clustering, *IEEE Transactions on Pattern Analysis and Machine Intelligence*, Vol. 13, No. 8, August 1991, 841–847.

Y. Xie, V.V. Raghavan, X. Zhao, 3M algorithm: finding an optimal fuzzy cluster scheme for proximity data, *Proc. IEEE Int. Conference on Fuzzy Systems, FUZZ-IEEE '02*, 12–17 May 2002, vol. 1, 627–632.

X. Xiong, K.L. Chan, Towards an unsupervised optimal fuzzy clustering algorithm for image database organization, *Proc. 15th Int. Conference on Pattern Recognition*, 3–7 September 2000, vol. 3, 897–900.

A. Yamakawa, Y. Kanaumi, H. Ichihashi, T. Miyoshi, Simultaneous application of clustering and correspondence analysis, *Proc. Int. Joint Conference on Neural Networks, IJCNN '99*, 10–16 July 1999, vol. 6, 4334–4338.

J.-T. Yan, P.-Y. Hsiao, A fuzzy clustering algorithm for graph bisection, *Information Processing Letters*, Vol. 52, No. 5, December 1994, 259–263.

J.-T. Yan, P.-Y. Hsiao, A new fuzzy-clustering-based approach for two-way circuit partitioning, *Proc. 8th Int. Conference on VLSI Design*, 4–7 January 1995, 359–364.

J.-F. Yang, S.-S. Hao, P.-C. Chung, Color image segmentation using fuzzy C-Means and eigenspace projections, *Signal Processing*, Vol. 82, No. 3, March 2002, 461–472.

M.-S. Yang, A survey of fuzzy clustering, *Mathematical and Computer Modelling*, Vol. 18, No. 11, 1993, 1–16.

M.-S. Yang, Y.-J. Hu, K. Chia-Ren Lin, C. Chia-Lee Lin, Segmentation techniques for tissue differentiation in MRI of ophthalmology using fuzzy clustering algorithms, *Magnetic Resonance Imaging*, Vol. 20, No. 2, February 2002, 173–179.

M.-S. Yang, P.Y. Hwang, D.H. Chen, Fuzzy clustering algorithms for mixed feature variables, *Fuzzy Sets and Systems*, Vol. 141, No. 2, January 2004, 301–317.

M.-S. Yang, C.-H. Ko, On a class of fuzzy c-numbers clustering procedures for fuzzy data, *Fuzzy Sets and Systems*, Vol. 84, No. 1, November 1996, 49–60.

M.-S. Yang, C.-H. Ko, On cluster-wise fuzzy regression analysis, *IEEE Transactions on Systems, Man, and Cybernetics, Part B*, Vol. 27, No. 1, February 1997, 1–13.

M.-S. Yang, H.-H. Liu, Fuzzy clustering procedures for conical fuzzy vector data, *Fuzzy Sets and Systems*, Vol. 106, No. 2, September 1999, 189–200.

M.-S. Yang, J.-A. Pan, On fuzzy clustering of directional data, *Fuzzy Sets and Systems*, Vol. 91, No. 3, November 1997, 319–326.

M.-S. Yang, C.-F. Su, On parameter estimation for normal mixtures based on fuzzy clustering algorithms, *Fuzzy Sets and Systems*, Vol. 68, No. 1, November 1994, 13–28.

M. Yasuda, T. Furuhashi, M. Matsuzaki, S. Okuma, Fuzzy clustering using deterministic annealing method and its statistical mechanical characteristics, *Proc. 10th IEEE Int. Conference on Fuzzy Systems*, 2–5 December 2001, vol. 3, 797–800.

O.K. Yoon, D.M. Kwak, D.W. Kim, K.H. Park, MR brain image segmentation using fuzzy clustering, *Proc. IEEE Int. Conf. on Fuzzy Systems, FUZZ-IEEE '99*, 22–25 August 1999, vol. 2, 853–857.

Y. Yoshinari, W. Pedrycz, K. Hirota, Construction of fuzzy models through clustering techniques, *Fuzzy Sets and Systems*, Vol. 54, No. 2, March 1993, 157–165.

M. Yoshizawa, H. Takeda, T. Yambe, S. Nitta, Assessing cardiovascular dynamics during ventricular assistance. Use of fuzzy clustering techniques, *IEEE Engineering in Medicine and Biology Magazine*, Vol. 13, No. 5, November–December 1994, 687–692.

X. Yuan, T.M. Khoshgoftaar, E.B. Allen, K. Ganesan, An application of fuzzy clustering to software quality prediction, *Proc. 3rd IEEE Symposium on Application-Specific Systems and Software Engineering Technology*, 24–25 March 2000, 85–90.

N. Zahid, O. Abouelala, M. Limouri, A. Essaid, Unsupervised fuzzy clustering, *Pattern Recognition Letters*, Vol. 20, No. 2, February 1999, 123–129.

N. Zahid, M. Limouri, A. Essaid, A new cluster-validity for fuzzy clustering, *Pattern Recognition*, Vol. 32, No. 7, July 1999, 1089–1097.

E.L. Zapata, F.F. Rivera, O.G. Plata, M.A. Ismail, Parallel fuzzy clustering on fixed size hypercube SIMD computers, *Parallel Computing*, Vol. 11, No. 3, August 1989, 291–303.

D. Zhang, M. Kamel, M.I. Elmasry, Mapping fuzzy clustering neural networks onto systolic arrays, *Proc. 3rd IEEE Conference on Fuzzy Systems*, 26–29 June 1994, vol. 1, 218–222.

D. Zhang, S.K. Pal, A fuzzy clustering neural networks (FCNs) system design methodology, *IEEE Transactions on Neural Networks*, Vol. 11, No. 5, September 2000, 1174–1177.

M. Zhang, L.O. Hall, D.B. Goldgof, A generic knowledge-guided image segmentation and labeling system using fuzzy clustering algorithms, *IEEE Transactions on Systems, Man, and Cybernetics, Part B*, Vol. 32, No. 5, October 2002, 571–582.

L. Zhao, Y. Tsujimura, M. Gen, Genetic algorithm for fuzzy clustering, *Proc. IEEE Int. Conference on Evolutionary Computation*, 20–22 May 1996, 716–719.

G. Zouridakis, N.N. Boutros, B.H. Jansen, A fuzzy clustering approach to study the auditory P50 component in schizophrenia, *Psychiatry Research*, Vol. 69, No. 2–3, March 1997, 169–181.

G. Zouridakis, B.H. Jansen, N.N. Boutros, A fuzzy clustering approach to EP estimation, *IEEE Transactions on Biomedical Engineering*, Vol. 44, No. 8, August 1997, 673–680.

G. Zouridakis, D.C. Tam, Identification of reliable spike templates in multi-unit extracellular recordings using fuzzy clustering, *Computer Methods and Programs in Biomedicine*, Vol. 61, No. 2, February 2000, 91–98.

Index

Knowledge-Based Clustering, by Witold Pedrycz
ISBN 0-471-46966-1 Copyright © 2005 John Wiley & Sons, Inc.

Printed in the United States
By Bookmasters